Functional Foods for Chronic Diseases

Advances in the Development of Functional Foods

Volume 3

Related Titles from D&A Inc. / Functional Foods Center:

Functional Foods for Cardiovascular Diseases (by Danik M. Martirosyan PhD)
(ISBN-10: 0976753502: ISBN-13: 978-0976753506)
Product Dimensions: 10.7 x 8.1 x 0.8 inches.
A collection of reviews on modern approaches in the prevention and treatment of cardiovascular diseases by using the newest achievements in herbal remedies, food additives, functional foods, nutraceuticals, non-traditional plants and new computerized methods in this field is introduced. The book also discusses clinical nutrition during ishchemic heart disease, which creates the most favorable conditions for the processes of recovery, restoration of the functional condition of heart attacks and the reduction of the burden on the cardiovascular system.

Functional Foods for Chronic Diseases (by Danik M. Martirosyan PhD)
ISBN-10: 0976753529; ISBN-13: 978-0976753520
Product Dimensions: 10.6 x 8.4 x 0.7 inches.
This book reviews functional foods for the prevention and treatment of treatment of chronic diseases from multidisciplinary perspective. The report stresses the urgency of modern society to deal with chronic diseases and the need for the creation of functional foods on the basis of natural plant-derived resources. This book represents collections of selected reviews on modern approaches in the prevention and treatment of chronic diseases such as cardiovascular, cancer, diabetes and obesity by using the newest achievements in herbal remedies, food additives and non-traditional plants.

If you would like more details about these books, please contact us through email at ffc_usa@sbcglobal.net or by phone: 469-441-8272.

FUNCTIONAL FOODS FOR CHRONIC DISEASES

Advances in the Development of Functional Foods

Edited by

Danik M. Martirosyan, Ph.D.

D&A Inc.

Edited by Danik M. Martirosyan, PhD

FUNCTIONAL FOODS FOR CHRONIC DISEASES

Advances in the Development of Functional Foods

D&A Inc.
580 W. Arapaho Rd., Suite 130
Richardson, TX 75080
http://www.functionalfoodscenter.net

Manufactured in the United State of America

Copyright ©2008 by D&A Inc. / Danik M. Martirosyan

All rights reserved. No parts of this book may be reproduced in any form or by any means, electronic or mechanical, including photocopy, recording, or any information storage and retrieval system, without written consent of publisher.

Library of Congress Control Number: 2007909262

ISBN 10: 0-9767535-4-5, **ISBN 13:** 978-0-9767535-4-4

For information regarding special discounts for bulk purchases, please contact D&A Inc. Special Sales at 469-441-8272 or ffc_usa@sbcglobal.net

Important Notice:

This publication is neither a medical guide nor manual for self-treatment. If you should suspect that you suffer from a medical problem, you should seek competent medical care. The reader should consult his or her health professional before adopting any of the suggestions in this book.

D&A Inc.
Edited by Danik M. Martirosyan, PhD
2008

ACKNOWLEDGMENTS

As many of you already know, the contributions of various scientists from different parts of the world, including partipants in the 3rd International Confererence "Functional Foods and Phytotherapy for Chronic Diseases," compose this book. The conference, which took place in Dallas, Texas, was organized by the Functional Foods Center, UNESCO Chair Life Sciences International Education Center, Russian Academy of Natural Sciences, Center for Obesity Research and Program Evaluation at Texas A&M University, and the American Botanical Council (media sponsor). The organizing of the conference, and the creation of this book which followed, could not have been possible without the support of these dedicated sponsors.

Thus the writing of *Functional Food for Chronic Diseases* was a team effort and I would like to send our warmest thanks to all the contributors of this book for sharing their articles with us. It is the hope of the editor and contributors that those who read this book will become more knowledgeable about the role of functional foods in modern diet in preventing chronic diseases such as cancer, cardiovascular disorders, diabetes, and obesity.

On behalf of the Functional Foods Center, I would like to thank Dr. Arunas Savitskas from Kaunas Medical University in Kaunas, Litvania, and Dr. Undurti Das, MD, President and CEO of UND Life Sciences from Shaker Heights, Ohio, for their enthusiasm in organizing our international conference "Functional Foods for the Prevention and Treatment of Chronic Diseases" and this book.

I extend my appreciation to Megh Raj Bhandari, PhD, from the Laboratory of Food Biochemistry at Hokkaido University in Sapporo, Japan, for his tremendous support in organizing this conference and great aid in composing this book.

I would also like to thank Mark Blumenthal, the founder and executive director of the American Botanical Council, for organizing our annual conference reports in *HerbalGram* magazine.

The editor and publisher acknowledge Hasmik Martirosyan for the book cover design.

ABOUT THE FUNCTIONAL FOODS CENTER

The Functional Foods Center is a research and development center at D&A Inc. (Dallas, TX, USA), providing research expertise for further development of functional food innovations. The Functional Foods Center combines cutting-edge research expertise in the bio-medical sciences with practical business experience in order to develop and commercialize functional foods for human needs. The Functional Foods Center creates a global network of scientists and functional food experts in an environment conducive to innovative research collaboration.

The Functional Foods Center focuses on health maintenance and risk reduction of disease by researching, developing, and creating new products, which are mostly composed of naturally occurring substances. The goal is to accelerate the process of commercialization of innovative foodstuffs both in domestic and international markets. We are confirming that the positive impact of functional foods on health must be scientifically tested and proved by independent research and investigation. We are also considering that usage of functional foods could be provided on the basis of specific diseases, the physiological condition of individuals, and of pre-existing syndromes and symptoms.

ABOUT THE EDITOR:

Danik M. Martirosyan is the founder of the Functional Foods Center at D&A Inc. He is the Chairman of Annual International Conferences in the series entitled "Functional Foods for the Prevention and Treatment of Chronic Diseases", since 2004, as well as the editor of books in the series *Functional Foods for Chronic Diseases.* Furthermore, he is an Academician of the International Academy of Scientific Discoveries and Inventions and Honorary Associate Professor at the Department of Nutrition and Food Sciences at Texas Woman's University.

For more information about our international conferences and/or books, feel free to contact the Functional Foods Center at:

Functional Foods Center at D&A Inc.
580 W. Arapaho Rd., suite 130
Richardson, TX 75080
Phone: 469-441-8272
Fax: 214-646-0022
Email: ffc_usa@sbcglobal.net
Website: www.functionalfoodscenter.net

CONTENTS

PART TWO
BOTANICALS AND FOOD SUPPLEMENTS AS A SOURCE OF FUNCTIONAL FOODS **95**

PART THREE
REVIEWS **162**

CORRESPONDING AUTHORS:

Editor: Danik M. Martirosyan, PhD
Functional Foods Center,
Richardson, TX 75080, USA

Chapter 1

Farzad Shidfar, Ph.D,
Assistant professor of nutrition,
School of Health,
Iran University of Medical Sciences,
Tehran 15168, Iran

Chapter 2

Igor A. Sobenin, MD, PhD,
Institute of General Pathology and Pathophysiology,
Russian Academy of Medical Sciences,
Moscow, 125315, Russia

Chapter 3

Professor Parviz Ghadirian, PhD,
Department of Nutrition, Faculty of Medicine,
University of Montreal,
Montreal, Quebec H2W 117 Canada

Chapter 4

Professor Asinobi, C.O., PhD
Department of Nutrition and Dietetics,
Faculty of Agriculture and Vetinary Medicine
Imo State University, Owerri, Nigeria

Chapter 5

Professor Arūnas Savickas, PhD
Kaunas University of Medicine,
LT – 44307 Kaunas, Lithuania

Chapter 6

Jaime Uribarri, MD, Professor of Medicine
Annenberg B-1 Room AM414

Mount Sinai Medical Center
New York, NY 10029, USA

Chapter 7

Suresh T. Mathews, PhD
Cellular & Molecular Nutrition,
Dept.of Nutrition & Food Science,
260 Lem Morrison Dr., 101 Poultry Sci Bldg,
Auburn University, Auburn, AL 36849, USA

Chapter 8

Valery P. Kisel, PhD
Institute of Solid State Physics,
Department of Crystal Growth,
Chernogolovka, Moscow District, 142 432, Russia

Chapter 9

Shahidul Islam, PhD
Department of Agriculture,
University of Arkansas at Pine Bluff,
1200 North University Drive,
Mail Slot 4913, Pine Bluff, AR 71601

Chapter 10

Mahmoud Mohammad, PhD
Post Doctoral Associate
Baylor College of Medicine
Department of Pediatrics-Nutrition
1100 Bates, CNRC 7055
Houston, Texas 77030, USA

Chapter 11

Megh Raj Bhandari, PhD
Laboratory of Food Biochemistry,
Division of Applied Bioscience,
Graduate School of Agriculture,
Hokkaido University, Sapporo- 8589, Japan

Chapter 12

Jurga Bernatoniene, PhD
Kaunas University of Medicine,
A. Mickevičiaus 9, LT-44307 Kaunas, Lithuania

Chapter 13

M. Abdul Hannan, Lecturer,
Department of Biochemistry,
Bangladesh Agricultural University,
Mymensingh-2202, Bangladseh

Chapter 14

Larissa I. Weisfeld, PhD
Emanuel Institute of Biochemical Physics,
Russian Academy of Sciences,
Moscow, 119991, Russia

Chapter 15

Professor, Elhadi M. Yahia, PhD
Postharvest Technology and Human Nutrition Programs,
Faculty of Natural Sciences,
Autonomous University of Queretaro, Campus Juriquilla,
Avenida de las Ciencias, Queretaro, 76230, Mexico.

Chapter 16

Sonika Banyal, PhD
Department of Food Science and Nutrition,
Department of Chemistry and Biochemistry,
CSK Himachal Pradesh Krishi Vishvavidyalaya,
Palampur, 176062, India

Chapter 17

Undurti N. Das, MD
UND Life Sciences,
Shaker Heights, OH 44120, USA

Chapter 18

Nikolay P. Timofeev, PhD
Research-Production Enterprise “Collective Farm BIO”;
Koryazhma, 165650, Russia

Chapter 19

Sharon Rabb, PhD
Naturopath and Clinical Nutritionist,
Alternative Healing, Dallas, TX, 75206

Chapter 20

Karl Biel, PhD
Institute of Basic Biological Problems,
Russian Academy of Sciences, Pushchino,
Moscow Region 142290, Russia

Chapter 21

Seema Kafle, Research Scientist
Laboratory of Dairy Science,
Faculty of Agriculture,
Hokkaido University,
Sapporo-0608589, Japan

Chapter 22

M. Jahangir Alam Chowdhury, Research Scientist
Wheat Research Center,
Bangladesh Agricultural Research Institute,
Gazipur, Bangladesh.

.

Chapter 23

Jairam Vanamala, PhD
Center for Obesity Research and Program Evaluation,
Department of Nutrition and Food Science,
Texas A&M University,
College Station, 77843, TX, USA

INTRODUCTION

Simply put, this book provides information for researchers, physicians, and health care professionals occupied in the prevention and treatment of chronic diseases. Furthermore, it provides significant information for people interested in maintianing good health and therefore, a longer, happier life.

Chronic diseases such as hypertension, diabetes, and obesity are a global epidemic in various developed countries and there is an unprecedented level of interest in this area of research. Chronic diseases are by far the leading cause of death in the world and their impact is steadily growing. According to the World Health Organization, global action to prevent chronic disease could save the lives of 36 million people who would otherwise be dead by 2015. Approximately 17 million people die prematurely each year as a result of this global epidemic. Perhaps even more unfortunate, one billion people globally are overweight or obese, a remarkable one sixth of the world population.

The publication of this book serves two great purposes. First, it spreads the word about new functional food products for chronic diseases such as hypertension, diabetes, and obesity to the general public. It not only introduces new functional foods, but also shows the investigations and research that led to their creation. Second, the book preserves the numerous ideas and contributions made in the field. This shows the progress and evolution of this thriving field, with the power to change the lives of millions of people.

The forever growing field of functional foods brings together research scientists, food manufacturers and consumers who are committed to this issue through modern achievements of surgical approaches and potential of drug therapy, where particular emphasis is placed on the unresolved problems of pharmaceutical side effects. The urgency of modern society to deal with chronic diseases is stressed in this book as well as the need for the creation of healthy and functional foods on the basis of natural plant-derived resources, with a view to prevent and treat countless chronic diseases.

Danik M. Martirosyan, Ph.D.

Founder and President of Functional Foods Center at D&A Inc.,
Academician of International Academy of Scientific Discoveries and Inventions,
Chairmen of annual conferences in the series "Functional Foods for Chronic Diseases"

PART ONE

FUNCTIONAL FOODS FOR THE PREVENTION AND TREATMENT OF CHRONIC DISEASES

THE EFFECTS OF SOY PROTEIN ON SERUM PARAOXONASE 1 ACTIVITY AND LIPOPROTEINS IN POSTMENOPAUSAL WOMEN

Farzad Shidfar[1], Elham Ehramphosh[1], Iraj Heydari[1], Ladan Haghighi[1], Sharieh Hosseyni[2], Shahrzad Shidfar[3]

[1]School of Health, Iran University of Medical Sciences, Tehran, Iran
[2]Islamic Azad University, Sari, Iran
[3]Memorial hospital, Worcester, Massachauset U.S.A

Corresponding Author: Farzad Shidfar, Ph.D, assistant professor of nutrition, School of health, Iran University of Medical Sciences, Tehran 15168, Iran, E-mail: farzadshidfar@yahoo.com

Keywords: Soy protein, lipoprotein concentrations, LDL-cholesterol, HDL-cholesterol

INTRODUCTION

A high blood concentration of LDL-cholesterol to HDL-cholesterol ratio (LDL-c/ HDL-c) is an established risk factor for cardiovascular disease.[1] After allowance for measurement error, a reduction in cholesterol concentration of 1mmol /L is associated with a > 21 % reduction in coronary artery disease risk[2]. Menopause is associated with an increase in concentration of triacylglycerol (TG), total cholesterol (TC) and LDL-c in serum and a decrease in concentration of HDL-C. Furthermore, menopausal women are at a higher risk for cardiovascular disease (CVD) [2, 3, 4].

The significant reduction in circulating concentrations of estradiol and estrone are believed to influence hepatic lipid and lipoprotein metabolism. Intervention studies have shown no cardiovascular benefits of hormone replacement therapy (HRT) and that HRT generally increases TG concentrations.[3, 5, 6]

Within the past 25 years, numerous studies have reported an inverse association between soy protein intake and plasma cholesterol concentration. [7-10] This association is particulary evident in hypercholesterolemic men and women but is less consistently evident in normocholesterlemic individuals. [10-12]

Intake of isoflavone – rich soy protein in hypercholesterolemic human subjects decreased TC and LDL-c concentrations. However, the reduction was significant only in a subgroup with the highest LDL-C concentration. [10, 14-17]

Nevertheless, it must be emphasized that isoflavones alone (without soy protein) has not been shown to lower serum lipids. [9,18] Paraoxonase 1 (PON1) is an enzyme associated with HDL particles[19] and protect lipoprotein particles from free radical oxidation because it can hydrolyse oxidized cholestryl esters, phosphatidyl choline core aldehyde and degrade hydrogen peroxide [19-21]. Thus it is thought to attenuate the oxidation of LDL, and HDL and preserve the integrity of HDL, linking PON1 to CVD [22], and the scientific interest in PON1 increased immensely. [19-22]

Currently, data is lacking in women during early stages of menopause on the relationship between soy protein and lipoproteins or PON1 activity. The primary aim of this study was to determine whether 10 wk of isoflavone – rich soy protein diet exerts an effect on lipid and lipoprotein concentrations in postmenopausal women. The secondary aim was to determine the contribution of soy proteins to PON1 activity in these postmenopausal women.

SUBJECTS AND METHODS:

Study design

A double – blind randomized clinical trial of parallel design was used to compare the effect of soy protein diet high in isoflavones with a control group on serum lipids and paraxonase 1 activity.

During a 3 week time period, subjects continued their usual diet and then were stratified by BMI, serum TC, and TG and randomly assigned to one of two groups:

The women in treatment group were asked to consume a total of 50 g/d soy protein containing 164 mg isoflavones (Protein technologies International, St. Louis, MO). For this purpose, each subject consumed 130 g/d cooked roasted soy beans for 10 weeks. The women in control group consumed 50g/d whey protein (Protein Technologies International , St, Louis, MO)

Beginning 2 week earlier and continuing throughout the intervention, subjects were required to avoid all supplements, including vitamins, minerals and herbal remedies. All women were asked to maintain their usual diets and physical activity and not to alter their lifestyle during the intervention.

Subjects were also provided with a list of foods that are isoflavone rich (ie, primarily legumes) and were instructed to avoid these foods during the intervention .The women were informed of their rights as volunteers in this study and signed consent forms

Subject screening, selection and characteristics

Forty five nonsmoking, healthy postmenopausal women were recruited from the endocrine research center of Iran University of Medical Sciences, Tehran, Iran, 2005.

The interviews were conducted to screen postmenopausal women to ensure that they met our exclusion and inclusion criteria: Cessation of menses for over a year, elevation of FSH level, a body mass index (kg/m2) between 19 and 30 plasma TG and TC more than 200 mg/dl were willing randomly assigned to treatment and were able to participate for 10 weeks.

Women were excluded if they had chronic diseases (i.e., heart diseases, cancer, diabetes, diabetes, hepatic, kidnay and thyroid diseases), were taking medication chronically, had taken antihypertensive drugs, antibiotics, lipid lowering drugs, or sex hormone treatment during the past 3 months. This also includes women who were smoking, had a hysterectomy, diet regimen rich in soy protein or isoflavones, were a strict vegetarian, or had fluctuating weight of over 4-5 kg during past year.

Data collection and measurement

Anthropometry data included measurement of height, waist circumferences (using stadiometer) and weight (using a balance scale) was obtained at the beginning and end of the study.

Dietary intake was monitored by the same dietitian throughout the study and subjects were asked to complete a 24 hour diet recall questionnaire in the beginnig, third, sixth and tenth week as well as a lifestyle questionnaire (e.g. physical activity, income, etc.) at the begining and end of the 10th week of intervention. The women were followed up by telephone each week; patients who could not be reached by telephone were instructed to return to the clinic every other week. Subjects were required to provide venous blood samples after fasting overnight (for 12-14 h) a at the beginning and end of intervention. Compliance was based on the number of packets returned at the week 10 visit. All sample were collected while the subject rested in a supine position for 10 minutes. Serum lipoproteins, were assayed with a cobas MIRA analyzer (Roche Diagnostics, Basel, Switzerland). TC and TG levels were measured enzymatically with the triacylglycerol GPO- PAP- cholesterol CHOD-PP kit (MAN Co, Iran). Serum HDL- cholesterol was determined enzymatically using the CHOD-PAP kit after precipitation of the chylomicrons, VLDL and LDL with phosphotungstic acid and mg2+. Serum LDL-cholesterol was detemined enzymatically using the CHOD-PAP kit after precipitation of LDL with heparin and sodium citrates, and then by utilizing the following formula: LDL- cholesterol = total cholesterol- cholesterol in the supernatant.

Paraoxonase1 activity was measured by colorimetric method (23). The intra- assay CV for these assays (n.10) were 1.3,1.3,1.4,1.3 and 1.9% for TC, HDL-c , LDL-c , TG and paraoxonase1 respectively and the inter-assay CV (n.10) were 1.1, 0.9,0.95 1.7 and 1.8% respectively .

Statistical analyses

All data was expressed as means ±SD. The level of significance was chosen $P<0.05$. Statistical analyses were performed with PC SPSS 13.0. The normal distribution of the variables was checked by kolmogorov Smirnov test. In order to test whether the differences between the mean values of the items studied in both groups were significant; the student's t test was used.

Differences in the same postmenopausal women, before and after 10 weeks of intervention were evaluated by paired t-test. Diet records were analyzed by using Food processor II software. [24]

For comparison of means in different intervals of 24-hour diet recall questionnaires, ANOVA was used. For qualitative variables (e.g education, occupation, income, etc.), a chi-square test was used.

RESULTS:

Forty-two of forty-five randomly assigned postmenopausal women completed the study. Baseline characteristics of the postmenopausal women confirmed that they were well matched for the inclusion criteria (Table1). All the women consumed 100% of their soy packets.

Table 1. Baseline characteristics in 42 postmenopausal women

Variable	Control (n= 21)	Soy (n= 21)
Age (year)	56.6 ± 5.08	53.4 ± 4.05
Weight (kg)	71.8 ± 5.5	69.7 ± 6.1

BMI (kg/m2)	26.8 ± 3.06	27.1 ± 3.3
Height (cm)	164.04 ± 7.3	160.8 ± 6.02
Waist (cm)	90.8 ± 9.6	87.6 ± 12.7
Hip (cm)	113.5 ± 16.3	108.8 ± 13.6
Waist / hip (cm)	0.8 ± 0.21	0.86 ± 0.19

The expected potential contributors to lipid and lipoprotein concentration included age, body size and composition and dietary intake of selected nutrients. None of these characteristics at baseline were significantly different between the treatment and control group, and no significant changes took place during the intervention (Tables 1 and 2).

There was no significant difference in serum PON1 and lipoproteins at the beginning of the study between the two groups. Relative to the control group, soy consumption lead to a significant increase in PON1 activity at the end of the study (P= 0.029) but also lead to a significant decrease in LDL-c TC, LDL-c /HDL-c, TG/HDL-c, TC/HDL-c (P=0.0001, P=0.008, P=0.012, P=0.041 and P=0.029 respectively) (Table 3).

There was a significant increase in PON-1 activity and HDL-c after intervention compared to baseline values in the soy group (P=0.015 and P= 0.011 respectively) (Table 3). LDL-c , TC, LDL-HDL-c , TC/HDL-c and TG/HDL-c had a significant decrease after intervention compared to initial values in soy group (P= 0.001, P=0.002 , P=0.001 , P= 0.001 , P= 0.016) (Table 3)

Table 2. Total daily energy and dietary intakes at baseline and during the intervention

Nutrient	Group	baseline	WK5th	WK10th
Energy (kcal)	Control (n=21)	1779.95 ± 330.21	1723.35 ± 425.10	1775.38 ± 369.51
	Soy (n=21)	1730.12 ± 355.96	1738.12± 319.20	1769.39±406.10
Protein (g)	Control(n=21)	85.65 ± 20.05	88.14. ± 23.64	88.65 ± 11.46
	Soy (n= 21)	81.14 ± 16.26	84.10 ± 6.13	86.64 ± 71.40
Carbohydrate(g)	Control(n= 21)	274.33 ± 34.89	280.89 ± 29.16	276.06 ± 31.15
	Soy (n= 21)	263.35 ± 55.50	271.27 ± 31.31	274.49 ±45.81
Fat (gr)	Control(n= 21)	38.54 ± 6.02	40.84 ± 14.96	41.96 ± 10.21
	Soy (n=21)	38.22 ± 6.02	38.21 ± 9.05	37.98 ± 15.84
Ca (mg)	Control(n=21)	808.61 ± 49.67	790.65 ± 36.70	798.46 ±167.17
	Soy (n= 21)	776.86 ± 200.81	786.23 ± 145.09	802.67 ± 63.26
Fe(mg)	Control(n=21)	8.65 ± 4.50	9.28 ± 3.54	7.91 ± 2.68
	Soy (n= 21)	10.87± 4.26	9.41 ± 3.46	8.08 ± 3.25
Zn (mg)	Control(n=21)	11.43 ± 4.13	9.16 ± 4. 94	10.57± 2.34
	Soy (n= 21)	10.95 ± 3.20	11.19 ± 4.34	9.59± 2.30
PUFA(mg)	Control (n=21)	6.93 ± 1.96	6.12 ± 2.24	8.31 ± 2.45
	Soy (n= 21)	6.1 ± 4.8	6.1 ± 1.44	7.63 ± 3.80
MUFA (mg)	Control (n=21)	7.70 ± 2.18	9.43 ± 4.19	8.59 ± 2.04
	Soy (n=21)	6.8 ± 1.0	9.46 ± 2.71	7.55 ± 3.32
SFA (mg)	Control(n= 21)	23.89 ± 6.75	23.28 ± 8.53	24.06 ± 5.72
	Soy (n= 21)	25.22 ± 3.97	24.63 ± 4.8	22.79 ± 8.7

Table 3. PON1 Activity and serum lipoprotein concentrations before and after intervention in two groups

Variable	Group	Time	
		Baseline	**After Intervention**
PON1(U/Liter)	Control (n=21)	52.78 ± 7.56	51.07 ± 8.03
	Soy (n=21)	50.76 ± 7.58	55.69 ± 4.62 A.B
LDL-C(mg/dl)	Control (n= 21)	195.93 ± 15.25	192.06 ± 12.54
	Soy (n= 21)	191.77 ± 14.77	171.49 ± 13.12 C,D
HDL-C(mg/dl)	Control (n=21)	38.66 ± 9.35	38.73 ± 9.75
	Soy (n=21)	36.59 ± 6.57	40.94 ± 3.29 E
TG(mg/dl)	Control (n=21)	329.87 ± 32.37	325.96 ± 41.71
	Soy (n=21)	315.15 ± 40.68	301.17 ± 42.46
TC(mg/dl)	Control (n=21)	287.14 ± 16.26	290.56 ± 20.50
	Soy (n=21)	291.40 ± 13.95	274.28 ± 17.33 F,G
LDL-C/HDL-C	Control (n=21)	5.38 ± 1.51	5.39 ± 1.77
	Soy (n=21)	5.41 ± 1.31	4.21 ± 0.48 H,I
TG/HDL-C	Control (n=21)	9.13 ± 3.12	9.19 ± 3.62
	Soy (n=21)	8.89 ± 2.03	7.39 ± 1.20 J,K
TC/HDL-C	Control (n=21)	7.90 ± 2.31	8.07 ± 2.54
	Soy (n=21)	8.19 ± 1.44	6.73 ± 0.67 L,M

A: P= 0.029 Student's test
B: P= 0.015 Paired test
C: P= 0.0001 Student's test
D: P= 0.001 Paired test
E: P= 0.011 Paired test
F: P= 0.008 Student's test
G: P= 0.002 , Paired t test
H: P= 0.001 , Paired t test
I: P= 0.012 , Student's t test
J: P= 0.016 , Paired t test
K: P= 0.041 , Student's t test
L: P=0.001 , Paired t test
M:P=0.029 , Student's t test

DISCUSSION:

This study showed that cunsumption of 50 g soy protein containing 164 mg isoflavones for 10 weeks lead to a significant decrease in LDL-c , TC , LDL-c/HDL-c, TG/HDL-c and a significant increase in HDL-c and PON1 activity .

To our knowledge, the present study is the first to evaluate the relation between soy protein intake and PON1 activity.

There was a significant increase (+9.7%) in PON1 activity after consumption of soy protein in our study, but Oh et al reported no significant change in PON1 activity. The disparate findings concerning effects on PON1 between two studies may be due to:

1- Oh used purified isoflavones in his study [25] but we used soy protein in our study. Other reports indicated that soy protein with isoflavones but not isoflavones alone, could improve LDL metabolism and atherogenesis. [18] Nevertheless, it must be emphasized that isoflavones alone (without soy protein) have not been shown to lower serum lipids. [9]

2- Compared to our study, the sample size of Oh study is rather small and extrapolating the conclusion to the general population of interest may not be valid. [25]

PON1 is thought to attenuate the oxidation of LDL and also inhibited the accumulation of lipid peroxides in LDL. Aside from inhibition of LDL oxidation, there is evidence from animal and in vitro models that PON1 can protect the HDL particle from oxidation and preserve the integrity of HDL. [22]

As PON1 is a protective enzyme against atherogenesis, a significant increase of PON1 activity in soy group means lowering of atherogenesis, protection of HDL particle from oxidation, preservation of integrity of HDL, attenuation the oxidation of LDL, which leads to a lower occurance of CHD. So it seems that the control group is more prone to develop atherosclerosis than the soy group.

The most consistently reported benefical effect of soy has been on lipoproteins, and a meta – analysis of 38 controlled human clinical trials using on average of 47 g of soy protein daily showed significant reductions in TC (9%), LDL-c(13%) and TG(11%).[7] This analysis reported that initial serum cholesterol concentrations determined the extent of response, with mildly hypercholesterolemic exhibiting nonsignificant reductions,[7] however in our study all postmenopausal women had clear hypercholesterolemia and for this reason, we saw a significant decrease in TC and LDL-c. Hodgson reported no alteration in serum lipid concentration in postmenopausal women. It seems that normal cholesterol level of participants and using purified isoflavonoids in Hodgson's study leads to different results compared to our study. [26] In the present study, we observed significant improvements in TC, LDL-c, LDL-c/HDL-c, and TC/HDL-c, but no change in TG. These results are consistent with those of studies in perimenopausal and postmenopausal women in whom soy consumption had no significant effect on TG [10,27,28,30] and HDL-C [10,27,28,30] concentrations compared to the placebo, but resulted in reductions in TC [17,28,30] and LDL-C [10,17,28,30] concentrations. 21mg/dl decrease in LDL-c in our study could be associated with a 42% reduction in coronary artery disease (CAD) risk. [10] However, most studies showed decrease in TC but Baum et al [31] found no change in serum TC concentrations in hypercholesterolemic women which may be related to lower intake of soy protein (40g) compared to our study.

The decrease of serum TG in Teede [29] and Jayagopal [7] studies was possibly due to longer duration of study (3 months), larger sample size, higher baseline TG concentrations than our participants. However, some studies reported no significant difference in HDL-c [8, 30] but Baum et al [31] found an increase in HDL-c concentrations among postmenopausal women consuming soy protein.

In accordance with Jayagopal [7] and Teede [29], HDL-c was not affected by soy treatment compared to the control group. However, TG/HDL-c decreased significantly compared to control group and also compared to initial value in soy group in our study. TG/HDL-C has been identified as a stronger predictor of myocardial infarction than either TC/HDL-C or LDL-c/HDL-c. [3] Decrease of TG/HDL-C in our study leads to an increase in LDL particle size and lower formation of small dense LDL particles. [32] Small, dense LDL particles are associated with an increased risk CAD. [32] So it seems that, in our study CAD risk is lower in postmenopausal women in soy group compare to placebo group. The mechanisms of the lipid- lowering effect of soy is unclear. It has been suggested that soy may induce a hyperthyroid state; however, we didn't measure it. [7,14] Another possibility is due to amino acid composition of the diet (increases in arginine) are accompanied by decrease in serum cholesterol concentrations, alterations in bile acid or cholesterol absorption and alterations in the ratio of serum glucagon to serum insulin may affect hepatic cholesterol synthesis. [7,14] On the other hand, the LDL-receptor activity of monocyte was eight times greater in human subjects receiving soy protein than in those eating control diets. [7,14] However, these variables have not been measured, which were limitations in our present study. Obesity is a known important variable in lipid metabolism, and it is important to determine whether the effect of soy protein is influenced by the degree of obesity. In this study, not only did all of the subjects had 25 < BMI < 30, but also no significant change took place in BMI during the intervention. Thus, obesity was not a determinant factor in our results.

Initial serum cholesterol concentration had a powerful effect on changes in serum cholesterol and LDL-c concentration and accounted for approximately 77 percent of the variance among studies. [17] The amount of soy protein ingested had a significant effect on serum cholesterol concentration when the effects of the soy diet were examined alone, without the effects of the control diet. [9, 10] Soy protein intake averaged 47 g per day, and 37 percent of the studies used 31 g per day or less. [9,10]

These observations suggest that the daily consumption of 31 to 47 g of soy protein can significantly decrease serum cholesterol and LDL-c concentrations, [9,10,32] After adjustment for initial serum cholesterol concentrations and other variables, the ingestion of 25 or 50 g of soy protein per day was estimated to decrease serum cholesterol concentrations by 8.9 or 17.4 mg per deciliter, respectively, which was consistant to results of our study. [9,10,32] Persons with moderate or server hypercholesterolemia (> 250 mg per deciliter) should have even larger decreases in serum cholesterol concentrations when soy protein replaces animal protein in the diet. [30,34]

In our study, however, serum TC was higher than 250 mg/dl, but animal protein was not replaced by soy protein. Thus, we didn't see larger decreases in serum TC.

CONCLUSIONS:

In summary, this study is the first to show that consumption of soy protein leads to an increase of PON1 activity. We conclude that 50g soy protein had benefical effects on PON1 activity and serum lipoproteins and thus may improve risk factors for CAD in postmenopausal women. Further studies are required to determine whether all these benefits can be achieved by taking soy as a supplement or whether it is better to substitute soy for animal-protein foods. Is there an optimal ratio of vegetable to animal proteins to maximize the soy effects? Certainly the displacement of saturated fat and cholesterol in the diet with soy foods may results in the extra dietary advantage of cholesterol lowering. [10]

Acknowledgements:

We thank the study volunteers for their dedication and hard work throughout the study: The staff of the Endocrine Research Center, Iran University of Medical Sciences, Tehran, Iran

REFERENCES:

1. Hermansen k, Dinesen B, Hoie LH, Morgenstern E.Effects of soy and other natural products on LDL: HDL ratio and other lipid parameters: a literature review. Adv Ther.2003 Jan- Feb ; 20(1) : 50-78
2. Rosell MS, Appleby PN, Spencer EA, Key TJ.Soy intake and blood cholesterol concentrations : a cross-sectional study of 1033 pre-and postmenopausal women in the Oxford arm of the European Prospective Investigation into Cancer and Nutrition .Am J Clin Nutr, 2004 NOV; 80(5): 1391-6
3. Stark KD, Holub BJ. Differential eicosapentaenoic acid elevations and altered cardiovascular disease risk factor responses after supplementation with docosahexaenoic acid in postmenopausal women receiving and not receiving hormone replacement therapy. Am J Clin Nutr.2004 May; 79(5) : 765-773
4. Teede HJ, McGrath BP;Desilva L, Cehum M, Fassoulakis A. Isoflavones reduce arterial Stiffness: a placebo- controlled study in men and postmenopausal women. Arterioscler Thromb Vasc Biol. 2003 Jun; 23(6):1066-71
5. Bush NJ, Griffin- Sobel JP. Hormone replacement therapy in postmenopausal women Oncol Nurs Forum. 2003 Sep- Oct ; 30(5): 747-9
6. Rossouw GE, Anderson GL, prentice RL. Risks and benefits of estrogen plus progestin in Healthy postmenopausal women: Principal results from the women's Health Initiative randomized controlled trial. JAMA, 2002; 288:321-33
7. Jayagopal V, Albertazzi P, Kilpatrick ES, Howarth EM, Jennings PE. Benefical effects of soy phytoestrogen intake in postmenopausal women with type 2 diabetes. Diabetes Care,2002 Oct ; 25(10) : 1709-14
8. Tonstad S, Smerud K, Hoie L.A Comparison of the effects of 2 doses of soy protein casein on serum lipids, serum lipoproteins and plasma total homocysteine in hypercholesterolemic subjects. Am J Clin Nutr, 2002 July ; 76(1): 78-84
9. Jenkins DJA, Kendall CWC, Jackson CJA, Connelly PW.Effects of high- and low-isoflavone soy foods on blood lipids, oxidized LDL, homocysteine blood pressure in hyperlipidemic men and women. Am J Clin Nutr, 2002 Aug ; 76(2) : 365-372
10. Wangen KE, Duncan AM, Xu X , Kurzer MS. Soy isoflavones improve plasma lipids in normocholesterolemic and mildly hypercholesterolemic postmenopausal women . Am J Clin Nutr, 2001 Feb ; 73(2) : 225-231
11. Dent SB, Peterson CT, Brace LD, Swain JH, Reddy MB. Soy protein intake by perimenopausal women does not affect circulating lipids and lipoproteins or coagulation and fibrinolytic factors. J Nutr , 2001; 131: 2280-2287
12. Nestel P,Yamashita T, Sasahara T,Pomeroy S,Dart A.Soy isoflavones improve systemic arterial compliance but not plasma lipids in menopausal and perimemopausal women.
13. Arteriosclerosis , Thrombosis and Vascular Biology, 1997; 17:3392-3398
14. Engelman HM, Alekel DL, Hanson LN, Kanthasamy AG, Reddy MB.Blood lipid and oxidative stress responses to soy protein with isoflavones and phytic acid in postmenopausal women. Am J Clin Nutr, 2005 March ; 81(3) : 590-596

15. Hermansen K, Sondergaard M, Hoie L, Carstensen M. Benefical effects of a soy-based Dietary supplement on lipid levels and cardiovascular risk markers in type 2 diabetic subjects. Diabetes Care, 2001; 24: 228-233
16. 15- Dalias FS, Ebeling PR, Kotsopoulos D, McGrath BP, Teed HJ. The effects of soy protein containing isofalvones on lipids and indices of bone resorption in postmenopausal women. Clin Endocrinol , 2003 Jun : 58161 : 704-9
17. West SG, Hipert KF, Juturu V, Bordi; PL, Lampe JW. Effects of including soy protein in a blood cholesterol- lowering diet on markers of cardiac risk in men and in postmenopausal women with and without hormone replacement therapy. J Womens Health. 2005 April ; 14(3) : 253-62
18. Goodman Gruem D, Kritz- Silverstein D. Usual dietary isoflavone intake is associated with cardiovascular disease risk factors in postmenopausal women. J Nutr. 2001; 131: 1202-1206
19. Wanger JD, Schwenke DC, Greaves KA, Zhang L, Anthony MS. Soy protein with isoflavones, but not an isoflavone- rich supplement improves arterial low- density lipoprotein metabolism and atherogenesis. Arterioscler Thromb Vasc Biol, 2003 Dec; 23(12) : 2241-6
20. Yamane T,Matsumoto T, Nakae I, Takashima H, Tarutani Y. Impact of paraoxonase polymorphism (Q192R) on endothelial function in intact coronary circulation. Hypertens Res.2006 Jun ; 29(6): 417- 22
21. Aviram M, Rosenblat M, Bisgaier CL, Newton RS, Primo- Parmo SL. Paraoxonase inhibits high- density lipoprotein oxidation and preserves its function , A possible peroxidative role for paraoxonase. J Clin Invest ,1998 April; 1581-1590
22. Sutherland WHF, Manning PJ, DeJong SA,Allum AR, Jones SD. Hormone – replacement therapy increases serum paraoxonase arylesterase activity in diabetic postmenopausal women. Metabolism, 2001 March; 50(3): 319-324
23. Himbergen TMV, Van Tits LJH, Roest M, Stalenhoef AFH. The story of PON1: how an oranophosphatehydrolysing enzyme is becoming a player in cardiovascular medicine. The Netherlands Journal of Medicine , 2006 Feb; 64(2) : 34-36
24. Gan KN, Smolen A, Eckerson HW. Purification of human serum paraoxonase/arylesterase.Drug Metab Dispos, 1991; 19: 100-106
25. Jonnalegadda SS, Diwan S.Nutrient intake of first generation Gujarati Asian Indian immigrants in the U.S.J Am Coll Nutr, 2002 ; 21(5): 372-80
26. Oh Hy,Kim SS, Chung HY, Yoon S. Isoflavone Supplements exert hormonal and antioxidant effects in postmenopausal Korean women with diabetic retinopathy . J Med Food,2005 Spring ; 8(1) : 1-7
27. Hodgson JM, Puddly IB, Beilin LJ, Mori TA. Supplementation with isoflavonoid phytoestrogens does not alter serum lipid concentration: A randomized controlled trial in Humans. The Journal of Nutrition , 1998 Apr,128(4): 728-732
28. Chiechi LM, Secreto G, Vimercati A, Greco P, Venturelli E. The effects of soy rich diet on serum lipids: The Menfis randomized trial. Maturitas , 2002 Feb, 26;41(2):97-104
29. Vincent A, Fitzpatrick LA.Soy isoflavones: are they useful in menopause? Mayo Clin Pro, 2000 Nov; 75(11) : 1174-84
30. Teede HJ, Dalais FS, Kotsopulos D, Liang Y-L, Davis S.Dietary soy has both benefical and potentially adverse cardiovascular effects: A placebo – controlled study in men and postmenopausal women. The Journal of Clinical Endocrinology and Metabolism , 2001; 86(7) : 3053-3060

31. Anderson JW, Johnstone BM, Cook-Newell ME. Meta- analysis of the effects of soy protein intake on serum lipids .The New England Journal of Medicine , 1995; 333(5) : 276-282
32. Baum JA,Teng H,Erdman JW, Long-term intake of soy protein improves blood lipid profiles and increases mononuclear cell LDL receptor messenger RNA in hypercholesterolemic, postmenopausal women. Am J Clin Nutr, 1988;68:545-51
33. Mori TA, Burke V, Puddey IB, Watts GF, Oneal DN. Purified EPA and DHA have differential effects on serum lipids and lipoproteins, LDL particle size, glucose and insulin midly hyperlipidemic men. Am J Clin Nutr, 2000 ; 71(5): 1085-1094
34. Anderson JW, Smith BM, Washnock CS. Cardiovascular and renal benefits of dry bean and soybean intake. Am J Clin Nutr. 1999 ; 70(3) : 3645-4745
35. Puska P,Korpelainen V, Hoie LH, Skovlund E, Smerud K. Isolated soya protein with standard levels of isoflavones , cotyledon soya fibres and soya phospholipids improves plasma lipids in hypercholesterolaemia : a double blind , placebo-controlled trial of a yoghurt formulations .Br J Nutr , 2004 Mar ; 91(3) : 393-401

COMPLEX REDUCTION OF CARDIOVASCULAR PROGNOSTIC RISK WITH TIME-RELEASED GARLIC POWDER TABLETS ALLICOR

Igor A. Sobenin[1,2], Valentin V. Pryanishnikov[1], Alexander N. Orekhov[1,2]

Institute of General Pathology and Pathophysiology, Russian Academy of Medical Sciences[1]; Institute for Atherosclerosis Research[2], Moscow, Russia.

Corresponding author:
Igor A. Sobenin, MD, PhD,
Institute of General Pathology and Pathophysiology
Russian Academy of Medical Sciences
125315 Moscow, Russia
E-mail: sobenin@cardio.ru

Keywords: coronary heart disease, myocardial infarction, multivariate risk, primary prevention, atherosclerosis, garlic

Running title: Garlic tablets reduce prognostic cardiovascular risk

SUMMARY

Polyetiological nature of atherosclerosis allows the suggestion that simultaneous complex reduction of several risk factors appears to be valid and clinically effective way to primary prevention of cardiovascular diseases. To test this hypothesis, double-blinded placebo-controlled study was performed in 167 mildly hyperlipidemic patients to evaluate the effectiveness of garlic-based drug Allicor treatment in primary coronary heart disease (CHD) prevention and reduction of multifunctional cardiovascular risk. Ten-year prognostic multifunctional cardiovascular risk that was calculated using algorithms derived from Framingham and Muenster Studies. It has been demonstrated that 12-months treatment with Allicor lowers 10-years prognostic risk of CHD development in men by 10.7% from the baseline level ($p<0.05$), and 10-years prognostic risk of acute myocardial infarction (fatal and non-fatal) and sudden death by 22.7% from the baseline level ($p<0.05$). In women, Allicor treatment prevented the rise of cardiovascular risk with ageing ($p<0.05$). The effects of treatment on the dynamics of cardiovascular risk were mainly due to a decrease in LDL cholesterol and partially due to the changes in HDL cholesterol and systolic blood pressure. The most prominent changes in serum lipids was the reduction of total and LDL cholesterol in men by 27.9 mg/dl ($p<0.05$) and 22.5 mg/dl ($p<0.05$), respectively, and in women by 11.4 mg/dl ($p<0.05$) and 10.8 mg/dl ($p<0.05$), respectively. In the subgroup of 79 intermediate- and high-risk patients 10-years prognostic risk of CHD development lowered by 13.2% from the baseline level in men ($p=0.005$), and by 7.1% in women ($p=0.040$). Ten-year prognostic risk of acute myocardial infarction (fatal and non-fatal) and sudden death was lowered by 26.1% from the baseline level in men ($p=0.025$), but in women did not change significantly. The effects of treatment on the dynamics of cardiovascular risk were mainly due to the decrease in LDL cholesterol by 0.61±0.17 mmol/L in men, and the rise in HDL cholesterol by 0.07±0.04

mmol/L in women. The results of the study have demonstrated the effectiveness of complex treatment of different risk factors in the reduction of the overall multivariate cardiovascular risk. Being the remedy of natural origin, garlic-based drug Allicor is safe with the respect to adverse effects and allows even perpetual administration, which may be quite necessary for primary prevention of atherosclerosis and atherosclerotic diseases.

INTRODUCTION

Clinical manifestations of atherosclerosis and especially coronary heart disease remain the number one killer and a leading cause of illness and disability worldwide. The social, medical and economical burden of CHD and the other complications of advanced atherosclerosis continue to increase in spite of the evident drop in cardiovascular death rates in the United States of America and several European countries. This fact is mostly due to population ageing, the rapid increase in the prevalence of type 2 diabetes mellitus, metabolic syndrome, overweight and obesity, as well as static or even increasing smoking prevalence rates. These unfavorable tendencies are characteristic for many industrial and developing countries, therefore CHD remains the major cause of mortality and morbidity in the world. By now, the results of prospective epidemiological studies allow us to reveal a number of clinical and biochemical conditions tightly associated with the development of atherosclerotic diseases and are therefore called "risk factors". The traditional strategy for prevention of cardiovascular diseases is aimed at the reduction of risk factors.

Many factors appear to contribute to the development of atherosclerosis including alterations in plasma lipid and lipoprotein levels, blood pressure regulation, platelet function and clotting factors, etc. In recent years special algorithms for the estimation of overall cardiovascular risk based on the results of major epidemiological studies have been developed, which allow for the complex impact of different risk factors [1]. On the other hand, current approaches to the prevention of atherosclerotic diseases usually employ the effect on isolated risk factors. However, the polyetiological nature of atherosclerosis allows for the suggestion that simultaneous complex reduction of several risk factors appears to be the most valid and clinically effective way for primary prevention of cardiovascular diseases. Theoretically, such approaches may decrease overall integral cardiovascular risk. To test this assumption, double-blinded placebo-controlled randomized study was performed in mildly hyperlipidemic patients. Garlic powder tablets, Allicor, were used as the drug capable of reducing different risk factors. Garlic contains a number of biologically active compounds and is widely used as traditional medicine of many cultures. In recent years, antiatherosclerotic and cardiovascular-protective effects of garlic have been extensively evaluated. Evidence from numerous studies demonstrate that garlic-based preparations can bring about the normalization of plasma lipids, along with the enhancement of fibrinolytic activity, inhibition of platelet aggregation and reduction of blood pressure.[2;3] The present study was performed to evaluate the effectiveness of Allicor treatment in primary CHD prevention and its effects on the estimates of multifunctional cardiovascular risk.

PATIENTS AND METHODS

This study was kept in accordance with the Helsinki Declaration of 1975, as revised in 1983, and approved by the local ethical committee. All participants gave their informed consent prior to their inclusion in the study. Men and women aged 40-74 years who had serum cholesterol level above 200 mg/dl (5.2 mmol/L) upon primary examination were eligible for inclusion. The absence of documented CHD, high arterial hypertension (systolic blood pressure above 160 mm Hg or diastolic blood pressure above 95 mm Hg), or lipid-lowering drugs administration were also regarded as inclusion criteria. Patients were randomized according to gender, age, total cholesterol and smoking history as covariates into two groups, one who received Allicor (coated tablets containing 150 mg garlic powder, INAT-Farma, Moscow, Russia) one tablet twice a day for 12 months, and the other one who received a placebo in the same manner. Allicor and placebo looked identical.

Clinical and biochemical examination was performed upon the inclusion at the end of the study. Venous blood taken after overnight fasting was used for total cholesterol, triglycerides and high density lipoprotein (HDL) cholesterol measurements with commercial enzymatic kits (Boehringer Mannheim GmbH, Germany). Low density lipoprotein (LDL) cholesterol was calculated according to the Friedewald formula.

Ten-years prognostic risk of CHD development was calculated according to Weibull model derived from the results of Framingham study [4]. Following variables were used for risk determination: gender, age, systolic blood pressure, total cholesterol, HDL cholesterol, smoking status, diagnosed diabetes mellitus, left ventricular hypertrophy.

Ten-years prognostic risk of fatal or non-fatal myocardial infarction and sudden death was calculated with Cox proportional hazards model derived from PROCAM study [1], where following variables were used: gender, age, systolic blood pressure, total cholesterol, triglycerides, smoking status, diabetes mellitus, family history of acute myocardial infarction in 1st degree relative occurred before the age of 60 years. For this risk, the regional adjustment factor was applied. [5;6]

The significance of differences was estimated using SPSS 10.1.7 program package (SPSS Inc., USA). After examination of variable distribution, Mann-Whitney statistics or t-test was used for between-group comparisons, Wilcoxon statistics was used for within-group effect assessments, and Pearson's chi-square statistics was used for the comparison of nominal variables distributions. To estimate the relationship between changes in risk values and clinical and biochemical variables, Pearson's correlation analysis and regression analysis were used. The data are presented in terms of mean and S.E.M.; significance was defined at the 0.05 level of confidence.

RESULTS

Before the study, 278 patients were screened, and according to inclusion criteria 195 patients were recognized as eligible and randomized into two groups. The Allicor-treated group consisted of 98 patients (34 men, 64 women), and the placebo group consisted of 97 patients (33 men, 64 women). During the study, 12 patients (4 men, 8 women) dropped out of Allicor group, and 16 patients (6 men, 10 women) were lost in the placebo group. The reason for the exclusion from the study was the discontinuation of study medication. So, the retirement accounted for 12.2% and 16.5% in Allicor-treated and placebo groups, respectively, and the difference between groups was not statistically significant. By the end of the study, there were 86 evaluable patients in Allicor-treated group (30 men, 56 women) and 81 in the placebo group (27 men, 54 women).

The baseline clinical data on evaluable patients are presented in Table 1. It can be seen that groups of patients did not differ significantly in all clinical variables. In Allicor-treated patients, the mean age was higher in women than in men ($P<0.05$), and in placebo recipients, diastolic blood pressure was lower in women as well ($P<0.05$); with respect to other variables, men and women did not differ significantly within groups.

Allicor and the placebo groups did not differ significantly in mean total cholesterol, HDL and LDL cholesterol and triglycerides levels. Within groups, men and women differed significantly only with respect to HDL cholesterol ($P<0.05$). Serum triglycerides were higher in men as compared to women in both groups, but the difference did not reach statistical significance. The changes in lipid levels that occurred during the study are shown in Table 2. In the placebo group, no statistically significant changes were observed, except to triglycerides lowering in men by 21.2% ($P<0.05$). By the end of the study, the significant difference in HDL cholesterol between men and women was preserved ($P<0.05$).

Table 1. Clinical characteristic of patients at the baseline.

Variable	Allicor			Placebo		
	All (n=86)	Men (n=30)	Women (n=56)	All (n=81)	Men (n=27)	Women (n=54)
Age, years	54.5±1.0	51.3±1.9	56.1±1.0$^{\$}$	54.7±1.0	52.0±2.1	56.1±1.1
SBP, mm Hg	134.0±2.1	137.2±4.4	132.3±2.2	136.9±2.0	139.1±3.1	135.7±2.6
DBP, mm Hg	84.2±1.0	86.3±2.0	83.1±1.2	84.3±1.0	87.4±1.4	82.7±1.3$^{\$}$
BMI, kg/m^2	25.9±0.4	25.8±0.5	26.0±0.4	27.1±0.5	26.8±0.6	27.2±0.6
DM, n	1	1	0	1	0	1
LVH, n	9	4	5	7	2	5
Smoking, n	13	10	3	12	8	4

SBP – systolic blood pressure; DBP – diastolic blood pressure; BMI – body mass index; DM – diabetes mellitus; LVH – left ventricular hypertrophy.
$^{\$}$ - a significant within-group difference between men and women, *t*-test, $P<0.05$.

In both men and women, there was some reduction in total and LDL cholesterol that did not reach statistical significance (95% CI: 0.8, 16.6 mg/dl for total cholesterol, and 1.8, 12.3 mg/dl for LDL cholesterol).

In Allicor recipients, the substantial decrease in total and LDL cholesterol was observed both in men and women ($P<0.05$). Total cholesterol decreased by 17.2±4.4 mg/dl (95% CI: 8.4, 26.5), and in men the effect was more prominent than in women (27.9±5.1 mg/dl vs. 11.4±6.1 mg/dl). LDL cholesterol decreased by 14.9±3.8 mg/dl (95% CI: 7.5, 22.8 mg/dl) in total group, by 22.5±2.0 mg/dl in men and by 10.8±5.0 mg/dl in women. So, Allicor treatment resulted in the reduction of total cholesterol by 6.4±1.7% and LDL cholesterol by 8.4±2.1% from the baseline. Hypolipidemic effects differed significantly in total group as well as in women, as compared to the placebo group ($P<0.05$). HDL cholesterol levels did not change significantly and the difference between men and women still persisted (*P*<0.05). Triglycerides in Allicor-treated men lowered significantly by 24.9% (*P*<0.05), but the effect was similar to that observed in the placebo group.

At the baseline, 10-years prognostic risk of CHD in placebo group was low (below 5%) or moderate (from 5% to 10%) in 42 patients (51.9% of the total group), intermediate (from 10% to 20%) in 24 patients (29.6%), and high (above 20%) in 15 patients (18.5%). In Allicor recipients, the distribution of patients into the same cohorts of CHD 10-years risk was as follows: 46 patients (53.5% of the total group) vs. 27 patients (31.4%) vs. 13 patients (15.1%). Thus, placebo and Allicor-treated groups did not differ in the distribution of risk levels at the baseline (χ^2=0.353, *P*=0.950).

Baseline lipid values are presented in Table 2.

Table 2. The changes of serum lipid parameters.

Time	Allicor			Placebo		
	All (n=86)	Men (n=30)	Women (n=56)	All (n=81)	Men (n=27)	Women (n=54)
			Total cholesterol, mg/dl			
At the baseline	261.4±5.2	260.5±8.9	261.9±6.5	263.3±4.4	257.2±9.2	265.6±4.9
After 12 months	244.2±4.8*#	232.6±7.0*	250.4±6.3*#	254.1±4.1	241.6±8.3	260.4±4.8
			HDL cholesterol, mg/dl			
At the baseline	55.1±1.7	48.2±2.0	58.7±2.3$	54.7±1.7	47.2±2.5	58.4±2.1$
After 12 months	55.4±1.4	51.2±1.9	57.7±1.7$	53.9±1.8	48.9±2.2	56.5±2.4$
			Triglycerides, mg/dl			
At the baseline	141.7±11.0	167.2±23.0	128.0±11.4	139.3±9.9	161.7±19.8	128.2±10.8
After 12 months	128.6±8.0	125.6±10.3*	130.3±11.1	125.9±7.4	127.4±10.8*	125.2±9.8
			LDL cholesterol, mg/dl			
At the baseline	178.0±4.7	178.9±7.9	177.5±5.9	180.3±4.1	177.6±8.0	181.6±4.8
After 12 months	163.1±4.3*#	156.4±6.5*	166.7±5.5*#	175.0±4.6	167.3±8.8	178.9±5.3

* - a significant difference from the baseline, paired Wilcoxon test, *P* <0.05;
\# - a significant difference from placebo, Mann-Whitney test, *P* <0.05;
$ - a significant within-group difference between men and women, *t*-test, *P*<0.05.

After 12 months of the treatment, no significant changes in the distribution of patients into risk categories in placebo group were observed: 50.6% patients had low or moderate CHD risk, 29.6% - intermediate risk, and 19.8% - high risk. However, the dynamic of changes demonstrated that the category of CHD risk did not change in 53 patients (65.4%), while in 11 patients (13.6%) risk category decreased, and in the other 17 patients (21.0%), it increased. Thus, during the one-year follow-up, a significant redistribution of placebo recipients into different CHD risk categories occurred, and an overall increase of risk prevailed (χ^2=86.875, P<0.001).

In the Allicor-treated group, the amount of patients who had low or moderate 10-years prognostic risk of CHD increased up to 60.5%, while 25.5% patients had intermediate risk, and other 14.0% patients had high CHD risk. The redistribution of patients into risk categories was as follows: 68 patients (79.1%) had the same category of CHD risk as at the baseline, 13 patients (15.1%) have demonstrated the decrease in risk category, and only in 5 patients (5.8% of the group) the category of risk increased. Thus, under the Allicor treatment, the tendency of decrease of CHD risk prevailed (χ^2=143.748, P<0.001). However, by the end of the study, the Allicor-treated group did not differ significantly from placebo group with the respect to patients' distribution into CHD risk cohorts (χ^2=2.121, P=0.548), but the trends to redistribution of patients within groups differed significantly in Allicor and placebo recipients (χ^2=8.429, P=0.015).

The data on the changes of absolute 10-years CHD risk that are presented in Table 3 seem even more informative. It is necessary to note that men and women looked quite different with respect to CHD prognostic risk values. In men, the mean CHD risk was 1.9-fold higher, and 10-years prognostic risk of acute myocardial infarction and sudden death was 9.4-fold higher than in women. Thus, the analysis of absolute risk changes in total groups without subdivision according to gender would be incorrect. It can be seen that in the placebo group, both men and women did not demonstrate significant changes in 10-years CHD risk. On the contrary, in Allicor-treated men the mean level of CHD risk decreased by 10.7±4.3% from the baseline (P<0.05); in women the decrease of risk did not reach statistical significance.

Table 3. The dynamics of absolute cardiovascular risk.

Time	Allicor			Placebo		
	All (n=86)	Men (n=30)	Women (n=56)	All (n=81)	Men (n=27)	Women (n=54)
	10-years prognostic risk of CHD, %					
At the baseline	11.55±1.05	17.26±2.11	9.09±1.08$^{\$}$	12.54±1.09	18.15±2.01	9.73±1.12$^{\$}$
After 12 months	10.98±1.03*#	15.41±1.99*	9.00±1.08$^{\$}$	12.38±1.07	17.49±2.12	9.82±1.05$^{\$}$
	10-years prognostic risk of myocardial infarction and sudden death, %					
At the baseline	5.02±1.2	11.79±3.17	1.39±0.28$^{\$}$	5.40±1.01	13.58±2.30	1.32±0.27$^{\$}$
After 12 months	4.19±0.93*#	9.11±2.31*#	1.56±0.41#$^{\$}$	5.15±1.04	12.51±2.56	1.47±0.25*$^{\$}$

* - a significant difference from the baseline, paired Wilcoxon test, $P<0.05$;
# - a significant difference from placebo, Mann-Whitney test, $P<0.05$;
$ - a significant within-group difference between men and women, *t*-test, $P<0.05$.

The changes occurred with CHD prognostic risk correlated with the dynamics of systolic blood pressure (r=0.419; $P<0.001$), diastolic blood pressure (r=0.233; $P=0.003$), total cholesterol (r=0.432; $P<0.001$), triglycerides (r=0.394; $P<0.001$), HDL cholesterol (r= -0.419; $P<0.001$) and LDL cholesterol (r=0.469; $P<0.001$). The analysis of regression has revealed that the dynamics of CHD risk depended on the changes in systolic blood pressure ($P<0.001$), LDL cholesterol ($P<0.001$), HDL cholesterol ($P<0.001$) and triglycerides ($P=0.001$). The leading factors that determined CHD risk dynamics both in men and women were the changes in LDL and HDL cholesterol, and in women the changes in systolic blood pressure and serum triglycerides played an additional role.

The relative 10-years risk of CHD in women from placebo group increased by 10.7±5.4% (95% CI: 0.0, 21.6, $P<0.05$); in Allicor-treated women there was no significant changes of risk ($P<0.05$ vs. placebo). In placebo-treated men, the relative CHD risk did not change, while in Allicor-treated men the risk lowered significantly by 9.2±4.2% (95% CI: 0.6, 17.8, $P<0.05$).

The data on the changes in 10-years prognostic risk of fatal and non-fatal myocardial infarction and sudden death are presented in Table 3. In placebo-treated men, mean level of absolute risk did not change; in placebo-treated women there was an increase by 11.4% from the baseline ($P<0.05$). On the other hand, in Allicor-treated women, the increase of risk was insignificant, while in men the significant reduction of risk by 22.7±15.5% from the baseline was observed ($P<0.05$). Both in men and women, the dynamic of changes differed significantly from placebo recipients ($P<0.05$).

The changes of risk of myocardial infarction and sudden death correlated with the dynamics of systolic blood pressure (r=0.185; $P=0.017$), total cholesterol (r=0.530; $P<0.001$), serum triglycerides (r=0.295; $P<0.001$), HDL cholesterol (r= -0.173; $P=0.025$) and LDL cholesterol (r=0.544; $P<0.001$). Regression analysis revealed that the leading factors in men were the changes of LDL cholesterol ($P<0.001$) and HDL cholesterol ($P=0.027$), and in women, the changes of LDL cholesterol ($P<0.001$) and triglycerides ($P<0.001$).

The relative risk of myocardial infarction and sudden death increased in placebo-treated women by 45.1±13.9% (95% CI: 17.4, 72.9, $P<0.05$). In Allicor-treated women, the relative risk also increased by 14.2±11.8% (95% CI: -9.4, 37.8, $P<0.05$), but the increase was lower than in placebo group ($P<0.05$). In placebo-treated men, there was no significant increase in relative risk, while in Allicor recipients the relative risk of myocardial infarction and sudden death decreased by 11.9±11.2% ($P<0.05$).

Additional data set was obtained from this study based on the data on those patients with high (above 10%) 10-year prognostic risk of CHD development. There were 40 patients (20 men, 20 women) in Allicor-treated group, and placebo group consisted of 39 patients (20 men, 19 women) in this subgroup.

In placebo-treated high-risk men, serum triacylglycerols tended to decrease by 25.0±7.2%, or by 0.49±0.26 mmol/L from the baseline (p=0.054), and HDL cholesterol increased significantly by 9.0±5.4%, or 0.10±0.04 mmol/L from the baseline (p=0.037). The changes in total and LDL cholesterol in placebo-treated men did not reach statistical significance (95% CI: -

1.16; 0.28 and -0.93; 0.23 mmol/L for total and LDL cholesterol, respectively). At the same time, no significant changes in lipid parameters in placebo-treated women were observed.

In Allicor-treated high-risk patients, all lipid parameters changed significantly. Total cholesterol decreased by 6.1±2.9%, or 0.42±0.21 mmol/L from the baseline ($p<0.001$). Serum triacylglycerols lowered by 20.1±7.2%, or 0.46±0.17 mmol/L from the baseline ($p=0.001$). HDL cholesterol level increased by 8.2±3.5%, or 0.10±0.04 mmol/L from the baseline ($p=0.005$). Accordingly, LDL cholesterol level lowered by 4.4±3.7%, or 0.31±0.18 mmol/L ($p=0.002$). The most prominent changes were observed in men: total cholesterol decreased by 0.82±0.17 mmol/L ($p=0.001$), LDL cholesterol decreased by 0.61±0.17 mmol/L ($p=0.004$), serum triacylglycerols decreased by 0.72±0.28 mmol/L from the baseline ($p=0.004$), and HDL cholesterol tended to increase by 0.12±0.07 mmol/L ($p=0.062$). In Allicor-treated women, total and LDL cholesterol levels did not change significantly, serum triacylglycerols tended to decrease by 0.20±0.18 mmol/L ($p=0.067$), and only HDL cholesterol increased significantly by 6.5±4.3%, or 0.07±0.04 mmol/L from the baseline ($p=0.040$).

The data on the changes in prognostic cardiovascular risk levels are presented in Table 4. It is notable that 10-year prognostic risk of acute myocardial infarction and sudden death in women was 8.7-fold lower than in men. At the same time, 10-year prognostic risk of CHD development in women was only 1.3-fold lower. Thus, further estimation of changes in the risk of myocardial infarction had to be gender-oriented, whereas the changes in CHD prognostic risk could be also analyzed without subdivision into men and women.

In placebo-treated patients, no significant changes in cardiovascular prognostic risks were observed. On the opposite, Allicor treatment resulted in a significant decrease in CHD prognostic risk calculated by systolic blood pressure algorithm, in spite of gender differences. In Allicor-treated high-risk men CHD prognostic risk decreased by 13.2% from the baseline ($p=0.005$), in women – by 7.1% ($p=0.040$), and in total group – by 10.7% from the baseline ($p<0.001$). When diastolic blood pressure algorithm was used, the decrease of CHD prognostic risk by 6.7% from the baseline was observed ($p=0.010$), but the subdivision into men (CHD risk decrease by 7.1%) and women (CHD risk decrease by 6.3%) resulted in the loss of significance ($p=0.058$ and $p=0.057$, respectively), obviously due to the insufficient sample size.

Ten-year prognostic risk of myocardial infarction (fatal and non-fatal) and sudden death did not change in placebo-treated men and women (Table 4). On the opposite, in

Table 4. The dynamics of absolute cardiovascular risk in high-risk subgroup.

Time	Allicor recipients			Placebo recipients		
	All (n=40)	Men (n=20)	Women (n=20)	All (n=39)	Men (n=20)	Women (n=19)
	10-year prognostic risk of CHD, % (systolic blood pressure algorithm)					
At the baseline	19.6±1.6	22.8±2.3	16.4±2.1	20.4±1.4	22.5±1.9	18.3±1.9
After 12 months	17.5±1.6*	19.8±2.4*	15.1±2.1*	19.5±1.5	21.5±2.2	17.5±1.8
	10-year prognostic risk of CHD, % (diastolic blood pressure algorithm)					

At the baseline	19.4±1.7	22.6±2.6	15.9±1.8	20.7±1.4	22.6±2.0	18.7±1.9
After 12 months	18.1±1.7*	21.0±2.5	14.9±2.0	20.2±1.6	21.9±2.4	18.3±2.0
10-year prognostic risk of myocardial infarction and sudden death, %						
At the baseline	9.7±2.4	16.5±4.4	3.0±0.6	10.3±1.8	17.3±2.6	2.9±0.6
After 12 months	7.8±1.8*#	12.2±3.2*	3.4±1.0	9.5±1.9	15.7±3.1	3.0±0.5

* - a significant difference from the baseline, paired Wilcoxon test, $p < 0.05$;
[#] - a significant difference from placebo, Mann-Whitney test, $p < 0.05$.

Allicor-treated high-risk men the risk of myocardial infarction and sudden death decreased by 26.1% from the baseline (p=0.025). In Allicor-treated women there were no significant changes in prognostic risk level.

DISCUSSION

The probability of the development of cardiovascular diseases is determined by the mutual action of different risk factors that can be classified into non-modifiable (such as gender, age, family history of CHD, diabetes mellitus, etc.) and modifiable ones (dyslipidemia, arterial hypertension, smoking status, abdominal obesity, etc.) Furthermore, a number of other factors have been attributed to the development of CHD, such as elevated plasma fibrinogen, lipoprotein(a), homocysteine, and microalbuminuria in patients with diabetes mellitus or insulin resistance. It is possible that physical inactivity, heavy alcohol consumption, hyperglycemia, fasting hyperinsulinemia, insulin resistance, social and psychologic stress, different thrombogenic factors etc. may also play a role [7]. Some of these are not definitely shown to be independent risk factors, and for some of them, measurements and analysis are not widely performed or considered to be useful. The strategy of primary prevention of cardiovascular diseases is fairly based on the modification of independent major risk factors. In the Framingham Study, the 44% decrease in the CHD rate in men between 1950 and 1989 could be attributed to improvements of risk factors by one third to one half, at least [8]. The reductions of risk factors in USA could account for 50% of the drop in CHD death rate between 1980 and 1990, while the improvements in other treatments could account for 43% of the decline [9]. In the Netherlands, 44% of the decline in CHD mortality rate between 1978 and 1985 could be attributed to effective primary prevention [10]. The decrease in plasma cholesterol and blood pressure along with smoking prevalence resulted in 48% effect on the reduction of CHD death rate across 20 years of the North Karelia Project [11].

The main approach to CHD primary prevention is traditionally based on blood cholesterol lowering. Indeed, the increased level of LDL cholesterol possesses the highest prognostic significance as compared to other isolated risk factors with the respect of its sensitivity and specificity. However, modern algorithms of calculation of the prognostic risks based on several risk factors measurements provide much better estimates,[12] since different factors may cluster and interact synergistically, increasing the probability of CHD development.[13;14] For example, the presence of three major risk factors triples the risk for CHD event and nearly doubles the risk of dying of any cause; having four or five risk factors increases the risk for CHD event by 5-fold and risk for death by 3-fold during the 20 years of follow-up [15]. Thus, no CHD risk factor can be judged in isolation, and if any risk factor is present in patient, it is necessary to assess for other risk components and, more importantly, to treat all of them vigorously [7]. Presently, multivariate cardiovascular risk estimation is developed from findings of Prospective Cardiovascular Münster Study (PROCAM) and the Framingham Study data sets. Such approach, to a certain extent, emphasizes the polyetiological nature of atherosclerotic diseases, although the algorithms do not include the assessment of all definitely known risk factors. Quite naturally, algorithms derived in one population may provide incorrect estimates of absolute risk when applied in another geographical region.[16;17] Ideally, the solution lies in the performance of similar prospective studies in different countries and regions, but the more pragmatic approach is the recalibration of existing algorithms based on cross-sectional observational data. So, the MONICA project has provided recalibration of the PROCAM algorithm using the observed CHD morbidity, mortality and case fatality data from many countries [5;6]. Thus, the relevant recalibration coefficients which are characteristics of Russia have been used in our study.

The results of the given study have demonstrated that 12-months treatment with garlic-based drug Allicor resulted in significant reduction of multivariate prognostic risk of cardiovascular diseases or, at least, prevents its rise with aging. These effects were observed as well for absolute and relative risk estimates. In placebo-treated women risk values increased during the follow-up, and Allicor treatment obviously prevented this rise. In men, risk values remained stable in placebo group, and Allicor treatment resulted in significant risk reduction. So, the men differed from women not only by risk level at the baseline, but also by its dynamics during the one-year follow-up. The favorable dynamics of lipid profile observed not only in Allicor recipients but also in the placebo group may be attributed, at least partially, to the placebo-controlled design of the study, since the patients were motivated to give more attention to the improvement of diet and lifestyle. The similar reduction of serum triglycerides observed in men both in Allicor-treated and placebo group indirectly confirms this suggestion. It is notable that much better beneficial changes in cardiovascular risk estimates were observed in the subgroup of patients who were at a high risk at the baseline.

The main effect that underlies the dynamics of multivariate cardiovascular risks in Alicor-treated patients is a hypolipidemic action of the drug. Cholesterol lowering potential of garlic-based preparations is studied rather well [2;18;19]. In spite of several controversial data, [20] the meta-analysis has demonstrated moderate lipid-lowering effect of garlic that is comparable with rational dietary improvement.[2;21] However, the association of beneficial effects of Allicor exclusively with its action on LDL cholesterol is incorrect, since the regression analysis has revealed the relation of cardiovascular risk dynamics with the changes in arterial blood pressure, HDL cholesterol and even triglycerides. Although the last effects did not reach statistical significance, they have provided benefits upon the calculation of individual multivariate risks. It is known that garlic-based preparations possess moderate hypotensive action, and in some studies the increase in HDL cholesterol has been demonstrated.[3;22-25] Additionally, it is necessary to note that garlic-based drugs can affect the processes of platelet aggregation and fibrinolysis substantially, that is especially important in the prevention of CHD events, such as myocardial infarction.[19;26] The last parameters are not considered in the algorithms of multivariate cardiovascular risk estimation, but independently influence on the probability of fatal and not-fatal events.

Garlic contains a variety of organosulfur compounds, amino acids, vitamins and minerals.[27] Some of the sulfur-containing compounds such as allicin, ajoene, S-allylcysteine, S-methylcysteine, diallyl disulfide and sulfoxides may be responsible for antiatherosclerotic activity of garlic that can be realized through different mechanisms of action.[20] Allicor contains dehydrated garlic powder that is thought to retain the same biologically active ingredients as raw garlic, both water-soluble and organic-soluble.[28;29] On the other hand, it possesses a prolonged mode of action, as its biological effect lasts for 12-16 hours after a single dose is administered.[30] Being the remedy of natural origin, Allicor is safe with the respect to adverse effects and allows even perpetual administration, which may be quite necessary for primary prevention of atherosclerosis and atherosclerotic diseases.

REFERENCES

1. Assmann G, Cullen P, Schulte H. Simple scoring scheme for calculating the risk of acute coronary events based on the 10-year follow-up of the prospective cardiovascular Münster (PROCAM) study. Circulation 2002;105(3):310-315.
2. Ackermann RT, Mulrow CD, Ramirez G, Gardner CD, Morbidoni L, Lawrence VA. Garlic shows promise for improving some cardiovascular risk factors. Arch Intern Med 2001;161(6):813-24.
3. Agarwal KC. Therapeutic actions of garlic constituents. Med Res Rev 2003;16:111-24.
4. Odell PM, Anderson KM, Kannel WB. New models for predicting cardiovascular events. J Clin Epidemiol 1994;47(6):583-92.
5. Tunstall-Pedoe H, Kuulasmaa K, Mahonen M, Tolonen H, Ruokokoski E, Amouyel P. Contribution of trends in survival and coronary-event rates to changes in coronary heart disease mortality: 10-year results from 37 WHO MONICA project populations. Monitoring trends and determinants in cardiovascular disease. Lancet 1999;353(9164):1547-57.
6. International Task Force for Prevention of Coronary Heart Disease. Pocket guide to prevention of coronary heart disease. Grunwald, Germany: Borm Bruckmeier Verlag GmBH, 2003.
7. Gotto AM, Jr., Assmann G, Carmena R et al. The ILIB lipid handbook for clinical practice. Blood lipids and coronary heart disease. New York: International Lipid Information Bureau, 2000.
8. Sytkowski PA, Kannel WB, Agostino RB. Changes in risk factors and the decline in mortality from cardiovascular disease. The Framingham Heart Study. N Engl J Med 1990;322(23):1635-41.
9. Hunink MG, Goldman L, Tosteson AN et al. The recent decline in mortality from coronary heart disease, 1980-1990. The effect of secular trends in risk factors and treatment. JAMA 1997;277(7):535-42.
10. Bots ML, Grobbee DE. Decline of coronary heart disease mortality in The Netherlands from 1978 to 1985: contribution of medical care and changes over time in presence of major cardiovascular risk factors. J Cardiovasc Risk 1996;3(3):271-76.
11. Ginter E. Prevention of cardiovascular diseases: the Finnish experience and current situation. Bratisl Lek Listy 1997;98(2):67-72.
12. Voss R, Cullen P, Schulte H, Assmann G. Prediction of risk of coronary events in middle-aged men in the Prospective Cardiovascular Münster Study (PROCAM) using neural networks. Int J Epidemiol 2002;31(6):1253-62.

13. Kannel WB, Neaton JD, Wentworth D et al. Overall and coronary heart disease mortality rates in relation to major risk factors in 325,348 men screened for the MRFIT. Multiple Risk Factor Intervention Trial. Am Heart J 1986;112(4):825-36.
14. Stamler J, Dyer AR, Shekelle RB, Neaton J, Stamler R. Relationship of baseline major risk factors to coronary and all-cause mortality, and to longevity: findings from long-term follow-up of Chicago cohorts. Cardiology 1993;82(2-3):191-222.
15. Yusuf HR, Giles WH, Croft JB, Anda RF, Casper ML. Impact of multiple risk factor profiles on determining cardiovascular disease risk. Prev Med 1998;27(1):1-9.
16. Empana JP, Ducimetiere P, Arveiler D et al. Are the Framingham and PROCAM coronary heart disease risk functions applicable to different European populations? The PRIME Study. Eur Heart J 2003;24(21):1903-11.
17. Hense HW, Schulte H, Lowel H, Assmann G, Keil U. Framingham risk function overestimates risk of coronary heart disease in men and women from Germany - results from the MONICA Augsburg and the PROCAM cohorts. Eur Heart J 2003;24(10):937-45.
18. Berthold HK, Sudhop T. Garlic preparations for prevention of atherosclerosis. Curr Opin Lipidol 1998;9(6):565-69.
19. Bordia A, Verma SK, Srivastava KC. Effect of garlic (Allium sativum) on blood lipids, blood sugar, fibrinogen and fibrinolytic activity in patients with coronary artery disease. Prostaglandins Leukot Essent Fatty Acids 1998;58(4):257-63.
20. Yeh YY, Liu L. Cholesterol-lowering effect of garlic extracts and organosulfur compounds: human and animal studies. J Nutr 2001;131(3s):989S-93S.
21. Silagy C, Neil A. Garlic as a lipid lowering agent - a meta-analysis. J R Coll Physicians Lond 1994;28(1):39-45.
22. Bordia A. Effect of garlic on blood lipids in patients with coronary heart disease. Am J Clin Nutr 1981;34(10):2100-2103.
23. Ernst E. Cardiovascular effects of garlic (Allium sativum): a review. Pharmatherapeutica 1987;5(2):83-89.
24. Auer W, Eiber A, Hertkorn E et al. Hypertension and hyperlipidaemia: garlic helps in mild cases. Br J Clin Pract Suppl 1990;693-6.
25. Silagy CA, Neil HA. A meta-analysis of the effect of garlic on blood pressure. J Hypertens 1994;12(4):463-68.
26. Bordia A, Verma SK, Srivastava KC. Effect of garlic on platelet aggregation in humans: a study in healthy subjects and patients with coronary artery disease. Prostaglandins Leukot Essent Fatty Acids 1996;55(3):201-5.
27. Block E. The chemistry of garlic and onions. Sci Am 1985;252(3):114-19.
28. Iberl B., Winkler G., Müller B., Knobloch K. Quantitative determination of allicin and alliin from garlic by HPLC. Planta Med 1990;56320-326.
29. Amagase H, Petesch BL, Matsuura H, Kasuga S, Itakura Y. Intake of garlic and its bioactive components. J Nutr 2001;131(3s):955S-62S.
30. Orekhov AN, Tertov VV, Sobenin IA, Pivovarova EM. Direct anti-atherosclerosis-related efects of garlic. Ann Med 1995;27(1):63-65.

NUTRITION AND CANCER: CAUSES AND PREVENTION

Parviz Ghadirian

Department of Nutrition, Faculty of Medicine, University of Montreal, Montreal, Canada

Corresponding author: Parviz Ghadirian, PhD, E-mail: parviz.ghadirian@umontreal.ca

Keywords: nutrition, colon cancer, pancreatic cancer, dietary fiber, vitamin C and beta-carotene

Cancer is the second leading cause of death in North America, after cardiovascular disease. In North America, 1 person out of 3 will develop cancer during his or her lifetime and 1 in 4 individuals will die from it. In North America, the top 3 cancers among males are cancer of the prostate, lung and colon-rectum (Table 1). Among females, breast cancer is followed by lung and colorectal cancer. In regard with the mortality due to cancer, lung cancer is the first cause of death in both male and female.

Table 1. The Most Frequent Cancers in North America

Males		Females	
Incidence	**Mortality**	**Incidence**	**Mortality**
1. Prostate	Lung	1. Breast	Lung
2. Lung	Prostate	2. Lung	Breast
3. Colorectal	Colorectal	3. Colorectal	Colorectal

In North America, cancer is the principal cause of mortality among women aged 35 to 74 years.

It has been estimated that 1 woman out of 2.9 will develop cancer during her lifetime, and 1 in 4.5 will die because of cancer. The incidence is 1 out of 9.3 for breast cancer, 1 out of 21.4 for lung cancer and 1 out of 17.9 for colorectal cancer (Table 2).

Among males, 1 man out of 2.4 (Table 3) will develop cancer during their lifetime and 1 out of 3.7 will die from it. The probability of developing prostate cancer is around 1 out of 8, while the rates for lung and colorectal cancers are1 out of 11 and 15.8, respectively.

Table 4 shows the high-risk regions of certain types of cancer in the world. It indicates the environmental, lifestyle and genetic factors causing certain types of cancer in different regions.

Table 2. Lifetime Probability of Females Developing and Dying from Cancer in North America

CANCER	INCIDENCE		MORTALITY	
	1 out of:	%	1 out of:	%
ALL CANCERS	**2.9**	35.0	**4.5**	22.4
BREAST	**9.3**	10.8	**25.3**	4.0
LUNG	**21.4**	4.7	**23.6**	4.2
COLORECTAL	**17.9**	5.6	**37.3**	2.7

Source: Statistics Canada, 2005

For example, the incidence of lip and skin cancer in Australia is very high, while cancer of the oral cavity is extremely high in France (among males) and India (among females). We know that the incidence rate of esophageal cancer in Turkman Sahra of North-eastern Iran, particularly among women, is the highest in the world, while USA has the highest rates for cancer of the breast, colon-rectum and pancreas. Japan is on top of the list regarding the incidence of stomach and gallbladder cancer.

Table 3. Lifetime Probability of Males Developing and Dying from Cancer in North America

CANCER	INCIDENCE		MORTALITY	
	1 out of:	%	1 out of:	%
ALL CANCERS	**2.4**	40.9	**3.7**	26.9
PROSTATE	**8.1**	12.3	**26.1**	3.8
LUNG	**11.0**	9.1	**12.2**	8.2
COLORECTAL	**15.8**	6.3	**34.8**	2.9

Source: Statistics Canada, 2005

Scientific investigations during the past few decades have shown that around 90% of cancers are due to environmental risk factors such as nutrition, occupation, lifestyle, etc., with nutrition playing a major etiologic role. In general, about 35% of cancer incidence and mortality are due to food habits and nutrition.

Table 4. High-risk Regions of Certain Types of Cancer in the World

Cancer	Men	Both	Female
Lip	Australia		
Skin		Australia	
Tongue		Bermuda	
Salivary gland		N.E. Scotland	
Mouth	**France**		**India**
Nasopharynx		Hong Kong	
Œsophagus		Iran	
Breast		USA	
Colon-rectum		USA	
Pancreas		USA	
Liver		Thaïland	
Gallbladder		Japan	
Stomach		Japan	
Lung		New Zealand	

Source: WHO, 1992

It is interesting to know when and how hypotheses on nutrition and cancer were porposed and developed. In 960 AD, Young-He-Yan thought that poor nutrition was a cause of the condition we now call cancer of the esophagus.

In 1396 AD, Hakim Gorgani observed that the main cause of esophageal cancer in Turkman Sahra of North-eastern Iran (where the incidence of this disease is the highest in the world) was partly due to malnutrition and food habits.

In 1676, Wiseman suggested that cancer might arise from an error in diet, high intake of meat and alcoholic beverages, while Lambe in 1815 believed that excess consumption of food in general and meat in particular may cause cancer.

In 1914, Peyton Raus observed that the restriction of food consumption delayed the development of tumor metastases in mice. In the 1920s and 1930s, vital statistics accumulated by insurance companies revealed an association between obesity and mortality from cancer in different organs.

Exploration of the role of diet in human cancers began in the 1930s, and, even at that stage, evidence emerged that high intake of plant foods may reduce the risk of cancer. In the 1940s, the protective effects of under-feeding on tumor formation were recognized in experimental animals.

In the 1980s, the possible role of diet and nutrition in the etiology of several cancer sites in humans was suggested. About a decade later, in 1990-92, Cancer Prevention Trial II showed a lower risk of colon cancer among people who ate diets rich in fruits and vegetables. And, finally, in 1996, American Cancer Society guidelines on diet, nutrition and cancer affirmed that one-third of all cancer deaths could be prevented through healthy eating and physical activity.

There are several hypotheses regarding why and how diet and nutrition may cause cancer.

1) Carcinogens (cancer-causing agents) in natural foodstuffs: Some foods, such as bracken fern (a spinach-like wild vegetable), are powerful, direct-acting carcinogens. This vegetable has been related to cancer in humans.

2) Carcinogens produced by cooking: Humans are the only species which cook their food. It has been known for many years that benzo(a)pyrene and other polycyclic hydrocarbons can be produced by pyrolysis when meat and fish (animal proteins) are broiled, barbecued, or smoked, or when any food is fried in repeatedly-used fat.

3) Carcinogens produced by microorganisms in stored foods: Carcinogens may be produced in stored foods by the action of microorganisms. Aflatoxin, a product of the fungus *aspergillus flavus* which commonly contaminates peanuts and other staple carbohydrate foods stored in hot and humid climates, is a major factor in the etiology of liver cancer in certain tropical countries.

4) **Formation of carcinogens in the body:** N-nitroso compounds are powerful carcinogens formed in the body. They are present in small amounts in the gastric juice, and may be formed in the digestive tract or possibly in the infected bladder by reactions between nitrites and various nitrosable compounds (secondary amines). Their formation can be inhibited by antioxidants (e.g. vitamin C).

5) **High or low consumption of calories, fat, protein, carbohydrates, fiber, vitamins and minerals:** A high intake of some nutrients, such as protein, caloric carbohydrates and fat, may play a causative role in the etiology of some cancers. The protective effects of vitamin C and A on the occurrence of different cancers have been suggested. Among minerals, an inverse relationship between cancer and selenium, iron, iodine and molybdenum has been reported from different parts of the world. Excess amounts of zinc, copper and cadmium in food and water may also play an important role in the etiology of some cancers.

The protective influence of dietary fiber on cancer, particularly cancer of the colon and pancreas, has been suggested. Despite this information, self-prescription of vitamins, particularly fat-soluble vitamins such as A, D, E and K, as well as minerals should be avoided or taken only after consultation with a physician.

Alcohol intake will facilitate the absorption of carcinogens in the digestive tract. Several studies suggest the important role of alcohol in the etiology of several cancers, particularly breast cancer among women. This habit should also be eliminated for well-being and the prevention of chronic diseases such as cancer.

Although extensive epidemiological studies have been carried out on the etiology of cancer, very little is known about the relationship between nutrition and cancer. As mentioned above, the effect of nutrition on cancer could be due to either specific nutritional deficiencies, unbalanced metabolism from dietary excesses or foods contaminated by carcinogens, certain food additives and preservatives, which may increase the risk of cancer. Currently, more than 500 chemical food additives and preservatives are used in the food industry, of which only 50 are legal and harmless. Therefore, the consumption of prepared and preserved foods containing chemical carcinogens should be avoided. Several studies have shown that overnutrition (overeating) and obesity play major roles in the etiology of certain cancers, such as of the breast, cervix, prostate, colon and gallbladder.

It has been reported by several investigators that drinking tea and food at high temperatures may cause esophageal cancer by thermal irritation. This risk could be reduced or avoided by changing these food habits.

The assessment of diet and its relationship to disease is complicated by time. In general, it is not yet known when diet is most relevant in the cancer process. This factor probably varies from cancer site to site. For some cancers, diet may be important during childhood, even though the disease occurs decades later. For other cancers, diet may act as a late-stage promoting or inhibiting factor; thus, intake over a continuous period up to just before diagnosis may be important.

Finding the causes of cancer is extremely difficult, because cancer often develops very slowly over many years. It may be 20-30 years after a number of people are exposed to a cancer-causing agent (carcinogen) before there is a significant increase in cancer among them.

We know that around 35% of cancer incidence and mortality are due to nutrition, while the possible role of tobacco is around 30%. The third highest risk factor is infection, which adds another 10% to the overall risk. Although alcohol plays a 5% role in the etiology of cancer, its effect rises to about 5-fold when its consumption is combined with tobacco.

For a further explanation regarding nutrition and cancer, two cancers namely colon and pancreatic cancer, will be discussed below.

Table 5. Proportion of Cancer Deaths Attributed to Different Factors

Factor or class of factors	*% of all cancer deaths*
1. Nutrition	35
2. Tobacco	30
3. Infection	10
4. Reproductive & sexual behavior	7
5. Alcohol	5
6. Occupation	4
7. Geophysical factors	3
8. Pollution	2
9. Medicine and medical procedures	1
10. Industrial products	< 1
11. Food additives	< 1
12. Unknown	3-4

Source: Doll and Peto, 1981

Colon cancer

Internationally-based correlational studies suggest that, *per capita,* fat from animal sources, but not vegetable sources, correlates with colon cancer rates. Most epidemiological investigations have found a positive association between the per capita consumption of red meat intake and the risk of colon cancer. It has been suggested that energy intake, regardless of whether the source is fat, protein, or carbohydrates, is associated with colon cancer risk, with an overall relative risk of 1.56 (56% increased risk).

The most consistent finding is the negative association between frequent consumption of vegetables and fruits and the risk of developing colorectal cancer. Table 6 reports the results of one of our studies on dietary fiber and colon cancer in Canada.

Table 6. Odds Ratios and 95% Confidence Intervals of Colon Carcinoma Risk Associated with Intake of Dietary Fiber

NUTRIENT	QUARTILE				*P* VALUE
	1 (low)	2 RR (CI)	3 RR (CI)	4 (high) RR (CI)	
Dietary fiber	1.00	**0.62** (0.42 – 0.89)	**0.72** (0.50 – 1.04)	**0.50** (0.34 – 0.74)	**0.0018**
Vegetable fiber	1.00	0.91 (0.63 – 1.33)	0.95 (0.65 – 1.38)	**0.57** (0.39 – 0.84)	**0.0096**
Fruit fiber	1.00	0.92 (0.63 – 1.33)	0.74 (0.51 – 1.08)	**0.74** (0.51 – 1.08)	**0.0687**
Cereal fiber	1.00	0.81 (0.55 – 1.18)	0.66 (0.45 – 0.96)	0.78 (0.53 – 1.13)	0.1046

Odds ratios and 95% confidence intervals from a logistic regression model adjusted for gender, age, marital status, history of colon carcinoma in first-degree relatives, and total energy intake.

Source: Ghadirian et al. Cancer, 80(5) :858-864, 1997

In this case-control study (Ghadirian et al, 1997), it was found that total dietary fiber may reduce the risk of colorectal cancer to 50%. Comparing fiber types, it was observed that vegetable fiber was the most important type, reducing the risk by 43%. Although fruit fiber showed 26% risk reduction, it was not statistically significant, whereas cereal fiber seemed to exert no protective role in the etiology of this disease.

Another promising hypothesis is that relatively high calcium intakes may reduce the risk of colorectal cancer, perhaps by forming complexes with secondary bile acids in the intestine. In other words, high dietary calcium intake (Table 7) can decrease the risk of colon cancer by about around 31% (Ghadirian et al, 1997).

We also studied the possible role of certain vitamins in the etiology of colorectal cancer (Table 8). It appears that total vitamin A, retinol and alpha-tocopherol can reduce the risk of colorectal cancer while beta-carotene plays no role. For the first time, to our knowledge, this study found that high vitamin E intake can reduce the risk of colorectal cancer, by around 47%.

Table 7. Odds Ratios and 95% Confidence Intervals of Colon Carcinoma Risk Associated with Dietary Calcium Intake

NUTRIENT	QUARTILE 1 (low)	 2 RR (CI)	 3 RR (CI)	 4 (high) RR (CI)	*P* VALUE
Calcium	1.00	**0.76** (0.52 – 1.27)	**0.68** (0.46 – 0.99)	**0.69** (0.47 – 1.00)	**0.0411**

Odds ratios and 95% confidence intervals from a logistic regression model adjusted for gender, age, marital status, history of colon carcinoma in first-degree relatives, and total energy intake.

Source: Ghadirian et al. Cancer, 80(5),858-864,1997

Table 8. Odds Ratios and 95% Confidence Intervals of Colon Carcinoma Risk Associated with the Intake of Certain Dietary Vitamins (402 cases and 668 controls)

NUTRIENT	QUARTILE 1 (low)	 2 RR (CI)	 3 RR (CI)	 4 (high) RR (CI)	*P* VALUE
Total Vitamin A	1.00	1.18 (0.82 – 1.72)	0.89 (0.61 – 1.29)	**0.67** (0.46 – 0.98)	**0.0162**
Retinol	1.00	0.81 (0.55 – 1.18)	0.71 (0.48 – 1.04)	**0.69** (0.48 – 1.00)	**0.0409**
Beta-carotene	1.00	0.95 (0.65 – 1.37)	1.14 (0.79 – 1.67)	0.72 (0.49 – 1.06)	0.2036
Other carotenes	1.00	1.29 (0.89 – 1.88)	1.01 (0.69 – 1.47)	0.76 (0.52 – 1.11)	0.0740
Vitamin E	1.00	0.54 (0.37 – 0.80)	0.52 (0.35 – 0.78)	**0.53** (0.36 – 0.78)	**0.0026**
Alpha-tocopherol	1.00	0.74 (0.50 – 1.09)	0.68 (0.45 - 1.02)	**0.63** (0.43 – 0.94)	**0.0256**

Odds ratios and 95% confidence intervals from a logistic regression model adjusted for gender, age, marital status, history of colon carcinoma in first-degree relatives, and total energy intake.

Source : Ghadirian et al. Cancer, 80(5),858-864,1997

We undertook a case-control study of toenail selenium, the best representative of nutritional selenium intake and colon cancer in Canada. We observed that higher selenium intake significantly reduced the risk of colon cancer (OR=0.42 or around 58%), and this was particularly significant among females (62%), but not in males (Ghadirian et al, 2000).

Pancreatic cancer

Because of its increasing incidence and poor prognosis, cancer of the pancreas is one of the most important malignant diseases in humans. Several studies have been carried out on the relations between dietary pattern, nutrition and cancer of the pancreas. We have carried out a multi-center case-control study (Ghadirian et al, 1991) with some interesting results. In this study it was found that an increased risk of pancreas cancer associated with high levels of reported energy intake, particularly energy derived from carbohydrate (Table 9). The highest quintiles of caloric intake showed 2-fold increase in the risk of pancreatic cancer.

It is interesting that total fat intake increased the risk of pancreatic cancer by 3-fold and the rate for saturated fat was more than 5-fold.

Table 9. Relative Risks by Quintiles of Daily Intake of Kilocalories, All Studies Combined

Relative risks (95% confidence intervals)[1] for quintiles					*p*(trend)[2]
1	2	3	4	5	
1.00	1.22 (0.89-1.68)	1.20 (0.87-1.65)	2.00 (1.45-2.76)	2.07 (1.47-2.91)	< 0.0001

[1]The model includes kilocalories (indicators) and lifetime cigarette consumption (continuous).

[2]The model includes kilocalories (categorical) and lifetime cigarette consumption (continuous).

Howe, G.R., Ghadirian, P. et al., 51:365-372,1992

We also found (Table 10) that dietary fiber (66%), vitamin C (59%) and beta-carotene (63%) significantly reduced the risk of pancreatic cancer.

Before this study was designed, it was speculated that high coffee intake may increase the risk of pancreatic cancer. Therefore, the World Health Organization (IARC) sponsored our multi-centre study. We found that coffee drinkers were collectively at lower risk (45%) than non-drinkers, particularly when coffee was consumed with meals, but not on an empty stomach (Ghadirian et al, 1991).

A strong positive association (around 4-fold) was observed between total cigarette smoking and the risk of pancreatic cancer (OR=3.76).

We also found that high salt intake increased the risk of pancreatic cancer to more than 4-fold (OR=4.28). The rate for smoked food was 4.68, while eating dehydrated foods (OR=3.10) and refined sugar (OR=2.81) was also a major risk factor, raising the risk by around 3-fold. It is interesting that the consumption of raw foods decreased the risk of pancreatic cancer by around 72%, and the risk reduction for natural foods was 77%, which means that foods with additives and preservatives as well as foods cooked at a high temperature may heighten the risk of this cancer.

Recently, we observed that lycopene, mostly from tomato, may reduce the risk of pancreatic cancer (around 31%), which is good news (Nkondjock, Ghadirian et al, 2005).

Table 10. Relative Risk by Quintiles of Daily Intake of Some Nutrients (Kilocalorie-Adjusted)

Nutrient	Relative risks[1] for quintiles					*p*(trend)[2]
	1	2	3	4	5	
Cholesterol	1.00	1.37	1.60	1.96	2.34	<0.0001
Dietary fiber	1.00	0.61	0.51	0.43	0.34	<0.0001
Vitamin C	1.00	0.81	0.50	0.55	0.41	<0.0001
Beta-carotene	1.00	0.50	0.61	0.42	0.37	<0.0001
Pre-formed vitamin A	1.00	1.37	1.30	1.68	1.47	0.027

[1]The model includes single nutrient (indicators), kilocalories (categorical), and lifetime cigarette consumption (continuous). — [2]This model includes single nutrients (categorical), kilocalories (categorical), and lifetime cigarette consumption (continuous).

Howe, G.R., Ghadirian, P. et al., 51:365-372,1992.

So, for cancer prevention through dietary habits and nutrition, the following recommentations are made:

1. Avoid obesity.
2. Cut down on total fat intake.
3. Eat more high-fiber foods such as whole grain cereals, fruits and vegetables.
4. Include foods rich in beta-carotenes, vitamins A and C in your daily diet.
5. Include some cruciferous vegetables, such as cabbage, in your daily diet.
6. Avoid alcoholic beverages.
7. Be moderate in the consumption of salt-cured, smoked and nitrite-cured foods.

REFERENCES

1. Ghadirian, P., Simard, A., Baillargeon, J., Maisonneuve, P. and Boyle, P.: Nutritional factors and pancreatic cancer in the francophone community in Montreal, Canada. International Journal of Cancer, 47: 1-6, 1991.
2. Ghadirian, P., Simard, A., and Baillargeon, J.: Tobacco, Alcohol, Coffee and Cancer of the Pancreas: a population-based case-control study in Québec, Canada. Cancer, 67: 2664-2670, 1991.
3. Ghadirian, P., Maisonneuve, P., Lacroix, A., Perret, C., Potvin, C., Gravel, D., Bernard, D.and Boyle, P.: Nutritional factors and colon cancer: A case-control study involving French Canadians in Montreal, Quebec, Canada. Cancer 80:858-864, 1997.
4. Ghadirian, P., Maisonneuve, P., Perret, C., Kennedy, G., Boyle, P., Krewski, D., Lacroix, A.: A case-control study of toenail selenium and cancer of the breast, colon and prostate. Cancer Detection and Prevention, 24(4):305-313, 2000.
5. Nkondjock, A., Ghadirian, P., Johnson, K.C., Krewski, D., and the Canadian Cancer Registries Epidemiology Research Group: Dietary intake of lycopene is associated with reduced pancreatic cancer risk. Journal of Nutrition. 135:592-597, 2005.
6. Statistics Canada. Canadian Cancer Statistics. NCIC, 2005.
7. Doll, R., and Peto, R.: The causes of cancer. Oxford University Press, 1981.

FOOD CONSUMPTION PATTERN AND OBESOGENIC FACTORS OF OBESE ADOLESCENT SCHOOL GIRLS IN OWERRI MUNICIPALITY

[1]Asinobi, C.O. and [2]Akeredu, M

[1]Department of Nutrition and Dietetics, Faculty of Agricultural and Veterinary Medicine, Imo State University, Owerri, Nigeria
[2]Safe/Pr/IFRA, University of Mali Owerri, Nigeria

Corresponding author: Asinobi C.O. E-mail: coasinobi@yahoo.com

Keywords: food consumption, overweight, obesity, body mass index, and cholesterol

ABSTRACT

This study focused on high risk/obese adolescent school girls in Owerri Municipality with a view of determining the relationship between food consumption patterns and obesity, and other obesogenic factors among adolescent girls. The anthropometric measurements of 500 adolescent high risk/obese and non-obese girls aged 10 -19 years in three schools in Owerri

Municipality were measured. Body mass index (BMI) was calculated from the weight and height measurements.

The information on food consumption patterns and the obesogenic factors of the adolescent girls were obtained by structural questionnaire and 24 hour recall methods. The adolescent girls were classified into high risk/obese and non-obese using the BMI values calculated. Student T-test and chi-squared test were used for the determination of significance of difference. The mean BMI values of the high risk/obese were significantly higher ($P<0.05$) than non-obese adolescent girls. Frequency of consumption of risk foods and energy intake were more in high risk/obese than non-obese adolescent school girls. Associations of food consumption pattern and energy intake with the BMI values for the adolescent girls were significant and higher in high risk/obese than in non-obese adolescent girls ($p<0.05$). The differences in the obesogenic factors between the high risk/obese and non-obese adolescent girls were significantly higher in high risk/obese than in the non-obese. Food consumption pattern as well as the obesogenic factors could have contributed to obesity among the adolescent girls. The changing dietary patterns that are usually attributed to rapid socio-economic transition, in particular, the excessive consumption of various nutrients and dietary components especially fats and oils, sugar should be controlled at the local/(starting at family level) and national levels.

INTRODUCTION

Food and Nutrition issues are perceived in developing societies as those reciting to inadequate food or nutrient deficiency diseases.

However, diet and nutrition along with life-style changes are recognized as the principal; environmental component affecting a wide range of diseases of public health importance in developing countries. Prevalence of over weight and obesity among children and adolescents has increased dramatically in the last half of the 20th century in virtually every country of the world. The rise is particularly apparent in the last ten years, and by 2002 some 155 million school-age children worldwide were estimated to be overweight or obese (Lobstein et al 2004). The emerging epidemic of overweight and obesity is adding to the burden of malnutrition and, unlike what was previously believed, are no longer a problem restricted to affluent, industrialized countries but is increasingly also affecting developing countries. Thus, in developing societies, diseases caused by calorie inadequacy and deficiency continue to persist, and co-exist with the growing presence of diet related chronic diseases among adults; hence contributing to the double burden of malnutrition. The public health problems are the result of changes in diet and life styles that characterize the nutrition transition, which accompanies economic development, the increasing urbanization of societies, and the globalization of food systems.

The last few decades have shown fundamental changes in food consumption patterns around the world. The changes are characterized not only by an increase in the amounts of food consumed, but also by a shift in the composition of the diet towards more fat and oils (Smil, 2000).

In poorer countries, diet quality has traditionally been equated with sufficient intakes of energy and essential nutrients. But with dietary patterns undergoing radical transformation (becoming more concentrated in sugars, saturated fats and salt while being low in fruits, vegetables and whole grain cereals) brought about a rapid increase in the prevalence of overweight, obesity and related non-communicable diseases (Deckelbaun, 2001). Obesity is associated with increases in nearly all chronic diseases, morbidity and mortality. It has long been recognized that genetic and behavioural predisposition, excess caloric intake and insufficient exercise are directly associated with obesity Deckelbaun 2001). Overweight and obese children are at a raised risk of co-morbidities including type 2 diabetics, fatty liver disease, and endocrinal and orthopedic disorders (Cole et al 2000). Adult obesity in turn carries an increased likelihood of metabolic and cardiovascular disease, certain cancers and a range of other disorders including psychiatric problem (WHO, 2000). Even if subsequent weight loss is achieved and maintained, there is evidence that mortality rates are higher among those adults who have been obese as adolescents (Must 1992).

Growing health concerns have also given rise to an intense debate about possible remedies to stop and reverse the obesity epidemic in developed countries and perhaps even more importantly, to prevent similar developments in developing countries.

This paper examines the relationship between food consumption patterns and obesity as well as identifying obesogenic factors among the obese adolescent girls in Owerri Municipality.

METHODOLOGY

This study was conducted in Owerri Municipality of Imo State, Nigeria. A random selection of two secondary schools; namely; Girls secondary School Ikenegbu and Alvana Secondary School and one primary school; Imo State University Primary staff school, was obtained from a list of the secondary and primary schools in Owerri Municipality. All girls between the age of 10-19 years who were willing to participate were measured for height and weight parameters according to the standard procedures of WHO (1976). A total sample of 500 subjects who were measured and the age obtained from school records were classified into high risk/obese (174) and non-obsese (326) after calculation of the adolescent girls body mass index (BMI); Wt/Ht2 = BMI (WHO 1987). The subjects whose BMI were below $25kg/m^2$ were grouped as “non-obese while those whose BMI were between $26kg/m^2$ – $3Ikg/m^2$ and above were grouped as “high risk/obese”. The combination of “high risk/obese” adolescents was obtained due to a small sample size of those that were obese as compared to the large number of those that were overweight and normal.

The frequency of consumption per day of some selected foods namely; low and high risks foods, as well as the obesogenic factors among the adolescent girls were obtained by structured questionnaire.

The mean frequency of consumption per day of those foods with high content of saturated fatty acids, sugar, cholesterol (termed “high risk foods”) was calculated for each groups of the adolescent girls by obtaining the number of times of consumption per day for each food item per subjects and the average per food for all the adolescent girls calculated. A similar calculation was done to obtain the mean daily frequency of consumption of “low risk foods” of the adolescent girls.

A 24 hour- recall method was used to obtain the type and amount of food consumed using food models of known weights for staple foods. The subjects were asked to recall all the food eaten and drunk at home and outside the home in the previous day. Food items such as stew, porridge, soups consumed by the adolescent girls were prepared as were normally prepared at home. Cemical analysis (AOAC, 2002) for protein and fats, of the prepared samples and other local recipes for which no values were available in the food composition tables (Platts, 1975, Oguntona and Akenyele, 1995) was done. Carbohydrate content was determined by differences.

Calorie was calculated using Atwater factors (4 x protein, 9 x fat, 4 carbohydrates). The calorie intake for each subject was calculated.

The student T-test was used to determine the significance difference in BMI and energy intake between the obese. Chi-square (x2) test was used to determine whether any relationship exists between the types of food, energy intakes and obesity among the adolescent girls.

RESULTS AND DISCUSSIONS

Table 1 shows that there was no significant difference in the mean BMI between high risk/obese and non-obese adolescent girls (P> 0.05). However, the difference was higher in high risk/obese than non-obese adolescent girls. The prevalence of high risk/obese was 34.8% and 65.2% for non-obese adolescent girls.

Table 1: Mean body mass index and prevalence of Obesity among adolescent girls by group.

Group	BMI (kg2/m2			Prevalence
	N	Mean	SD	%
High risk/Obese	174	26.1	1.02	34.8
Non-obese	326	24.9	2.81	65.2

P < 0.05, t-ratio = 2.804

This could be due to the fact that BMI does not account for the wide variation in body fat distribution and may not correspond to the same degree of fatness or associated health risk in different individuals and population (SCN, 2005). Although BMI provides the most useful, though crude, population-level measure of over weight and obesity, it can be used to estimate the prevalence of over weight and obesity with a population and the risks associated with them.

Table 2 shows that higher percent of high risk/obese consumed high-risk foods less than 3 times or more daily than non-obese adolescent girls.

Table 2: Percentage of adolescent girls consuming high and low risk foods daily by numbers of times of consumption.

Food Items	High risk/obese (n = 174)		Non-obese (n = 326)	
	% 1 - 3X	% 4 - 6X	% 1 - 3X	% 4 - 6X
High Risk/Foods				
Sugary foods	56	40	39	21
Butter/veg. oils	59	24	43	75
Animal source foods and product	69	30	50	18
Low Risk Foods				
Fruits	41	-	53	-
Vegetables	61	19	72	24
Whole grain products	10	-	14	-
Staple Foods	53	14	65	9

Note: 1 -3 = 1 to 3 times, 4 -6x = 4 – 6 times

Conversely, the percentage of high risk/obese that consumed low risk foods was lower than the non-obese adolescent girls. This could be explained by the regulating mechanism of energy intake that is controlled by the hypothalamus in the brain by steroids, a fatty agent in the body. Findings suggested that the hypothalamus centre controlling feeding behaviour receive information about the size of the store of energy in the adipose tissue by natural steroid, (Hervey 1975) a humoral agent with high fat water partition that determines the energy intake in the body. The humoral agent with a high fat water coefficient has been reported to inform the brain of changes in the energy reserves in adipose tissue. With increased quantity of steroids in the body the food intake is increased. Hervey (1975) has shown that in rats' progesterone a humoral agent, increased food intake, decreased energy expenditure in certain circumstances and led to gain weight and fat. This could be the case of high risk/obese adolescent girls who could have more fats than the non-obese. Also during fasting or feelings of hunger pangs, the sensation of hunger can increase the level of free fatty acids in the blood, which could act as stimulus to the centres in the hypothalamus which could be more in higher risk/obese than among obese adolescent girls, that resulting to the highest consumption of high risks foods observed among high risk/obese adolescent girls. More so, fundamental changes in the social environment and family life-styles; have resulted to a population-wide shift in food consumption; with overweight replacing wasting and making the heaviest children becoming even heavier (de Onis and Blossner, 2000) as many countries undergo the nutrition transition (Wang and Dietz 2002).

The mean daily energy intakes of the high risk/obese were significantly higher than the non-obese ($p < 0.05$), Table 3.

Table 3: Mean energy intake of the adolescents by group

Group	Energy Intake Kcal/day		
	N	mean	SD
High risk/Obese	174	2600	431
Non-obese	326	2350	573

$P < 0.05$, t-ratio = 2.024

The values were 2600 kcal/day and 2350 kcal/day for high risk/obese and non-obese adolescent girls, respectively. The significant differences observed could be partly as a result of higher intake of high-risk foods and partly as a result of increased concentration of humoral natural steroids that could be in the blood of high risk/obese adolescent girls that are conveying information to the hypothalamus centre controlling feeding behaviour. High fat humoral agents such as steroid, increases food intake and decrease energy expenditure (Hervey, 1975).

Table 4 shows that the associations between the BMI and energy intakes, as well as frequency of consumption of selected "risk foods" were significant ($P < 0.05$).

The association was significantly lower among the high risk/obese adolescent girls that consumed, equal to or more than 24000 kcal/day of energy, those in high risk/obese group whose frequency of consumption of low risk foods was equal to or more than 3 times per day, and also for those in high risk/obese group whose frequency of consumption of high risk foods was equal to or more than 3 times per day. This observation has been attributed to rapid socio-economic transition (Popkin, 2001 & Swinburn, 2004) in particular the excessive consumption of various nutrients and energy components (such as fats and oils, sugars) on the one hand and low levels of intake of nutrients (such as complex carbohydrates and fibres) on the other. The percentage of the adolescent girls indulging in practices that contribute to obesity was higher among high risk/obese than among non-obese adolescent girls in most cases (Table 5) except in food preference through the peer groups influence and excessive parental control.

Table 4 Association between BMI and energy intake, and frequency of consumption of foods among the adolescents

Factor	BMI Level (kg/-m2 > 30 (high risk/Obese)		BMI Level (kg/-m2 < 30 (non-Obese)	
	n	%	n	%
a) Energy Intake (kcal/day				
Below Average Recc. < 2400	63	36.2	252	77.3
Above average Recc. > 2400	111	63.7	74	22.6
b) Frequency of Consumption of low risk foods				
< 3 times	104	59.7	113	34.5
> 3 times	70	40.2	213	65.3
c) Frequency of Consumption of high risk foods				
< 3 times	54	31.0	169	51.8
> 3 times	120	689	137	48.1

Note: Recc = Recommended daily Allowance

a) X2 = 23.80; P< 0.05; df = 2
b) X2 = 11.93; P< 0.05; df = 2
c) X2 = 10.83; P< 0.05; df = 2

However, the differences in the factors were higher in "high risk/obese than among non-obese. The practices that contribute to obesity that was significantly among high risk/obese than among control obese adolescent girls excepting peer group pressure and parental control factors suggest that external influences encourage weight again (Swinburn, 2004). Gluttony or slot that was believed by the traditional medical to cause obesity has been replaced by new and powerful environmental factors. The fast foods commonly found in developing countries are undergoing radical transformations; becoming more concentrated in sugars, saturated fats and salts while being low in fruits, vegetables and whole grain cereals. This practice could lead to over weight and obesity. It is recognized that the reduction or even elimination of children's ability to play safely outside the home, the increased use of motorized transportation, and the marked increase in the proportion of both parents working outside the home have greatly contributed to increase in children obesity. Television and computer games help confine children and reduce their physical activity. Social pressure from peers lead to purchase of certain foods, or to undertaking of sedentary activities. Food advertising and labeling policies, use of vending foods in schools and other factors will also contribute to the list of potential obesogens (Hill et al, 2003, Murata 2000, Swinburn 1999).

Table 5: Percentage of adolescent girls indulging in practices contributing to obesity

Factor	(High risk/Obese) (n = 174)		(Non-Obese) n = 326	
	n	%	n	%
1) Gluttony/Sloth	133	76.4	103	53.3*
2) preference for fast and sugary food	127	72.9	211	64.7*
3) T.V Commercial	94	58.0	150	46.0*
Peer groups	40	22.9	65	19.9**
Others	40	22.9	127	38.9*
4) Patronizing food vendors in schools	123	70.6	171	52.4*
5) Excessive Parental Control	54	310	93	28.5**
6) Playing computer games more than 24 hour	98	56.3	92	28.2*
7) Watching T.V more than 3 hrs	86	49.4	94	28.8*
8) Use of motorized transportation in all major activities	89	51.1	118	36.1*
9) Both parents working outside	149	85.6	213	65.3*

Note: *P < 0.05, ** P> 0.05

CONCLUSION

The food consumption pattern as well as food practices have brought about an increase in the prevalence of overweight and obesity. These changes are characterized not only by an increase in the amount of food consumed but also a shift in the composition of the diet towards more meat, eggs, diary products, as well as more fats and oils and sugars. The result was an increase in the calories consumed in tandem with a shift towards diets that are much richer in saturated fats and cholesterol.

In view of its rapid development in the developing countries, there is a general agreement among experts that the environments, rather than biology, are driving the epidemics. Early recognition of excessive weight gain relative to linear growth is essential throughout childhood. Changes on growth patterns, such as upward crossing of weight-for-weight or BMI percentages, should be recognized using the appropriate reference data and addressed before children become severely overweight. Underlying predisposing factors should be discussed with parents and other caregivers and dietary and physical activity interventions initiated right away after an increase in weight-for-height or BMI percentages has been observed. Although data is extremely limited, it is likely that intervention before overweight has become severe will be more successful. The pediatric community should take a leadership role in the prevention and early recognition of childhood overweight and obesity by incorporating into routine clinical practice assessment and anticipatory guidance about weight, diet and physical activity.

REFERENCES

1. Lobstin. T. Baur, LB, Uauy. Obesity in Children and Young people: A Crisis in Public health. Report to the World health organization by the International Obesity task Force. 2004, Obesity Reviews 5 (Suppl) 5-104.
2. Smil. V. Feeding the World: A Challenge for the Twenty-First Century. The MIT Press Cambridge, Massachusetts and London, England, 2000
3. Deckelbaun, R.J CL, Williams. Childhood Obesity: The Health Issue. Obesity Reviews, 2001, 9 (Suppl 4) 239S – 243S.
4. Cole, T. J, Bellizzi M.C, Flegal KM. WH, Dietz. Establishing a standard definition for Child Overweight an Obesity Worldwide: International Survey. British Medical Journal, 2000, 320: 140 – 1343
5. World Health Organization. Obesity: Preventing and Managing the Global Epidemic Report of a WHO Consultation. WHO Technical report series No 894. WHO: Geneva. 2000.
6. Must. A. Jacques, PF, Dalla IGE, Bajema CJ, W.H Dietz. Long term Morbidity and Mortality of Over weight Adolescents. A Follow-up of the Harvard Growth Study of 1922 to 1935. New English Journal of medicine, 1992. 327: 1350-1355.
7. Word health Organization (WHO). Methodology of Nutrition Requirements- World Health Organization, Geneva, 1976.
8. World Health Organization. Global Nutritional Status. Anthropometric Indications. Nutrition Unit, Division of family Health, Geneva, 1987, 932-941.
9. AOAC, Official Methods of Analysis of the Association of Official Analytical Chemists 16th Education, Washington, D.C Association of Official Analytical Chemistry, 2002
10. Platt, B. S. Tables of Representative values of Food Commonly used in tropical Countries. Pergamon Press, Oxford press, London, 1975
11. Oguntona, E.B, I. O Akinyele. Nutrient composition of commonly eaten foods in Nigeria-Raw, Processed and prepared. Food Basket Foundation Publication Series. ISBN 987-31106-3-2, 1995
12. SCN, Overweight and Obesity. A New Nutrition Emergency? A Periodical Review of Developments in International Nutrition. United Nations System. Numbers 29, ISSN 1564-3743, 2005
13. de Onis M. M. Blossner. The World Health Organization Global Database on Child growth and malnutrition: Methodology and Apllications. International Journal of Epidermiology, 2000, 32: 518 – 526
14. Wang. G, WH Dietz Economic Burden of Obesity in Youths Aged 6 to 17 years: 1979-1999. Pediatrics 109/5/e81, 2002
15. Popkin, B. M. The Nutrition Transition and Obesity in the Developing World. Journal of Nutrition, 2001, 131:8715-8735.
16. Swinburn, B.A. Diet Nutrition and the Prevention of Excessive Weight gain and Obesity. Public health Nutrition. 2004, 7(IA): 123-146.

17. Hill J. O. Wyatt, HR, Reed, GN, J.C Peters. Obesity and Environment: Where Do We Go From Here? Science, 2003, 299-853.
18. Murata, M. Secular Trends in Growth and Changes in the Eating pattern of Japanese Children, American Journal of Clinical Nutrition, 2000, 72 (Suppl) 13795-13835.
19. Swinburn, B. Egger, G. F. Raza. Dissection Obesogenic Environment: The Development and Application of a framework for identifying and Prioritizing Environmental; Interventions For Obesity. Preventive Medicine 1999, 29: 563-570.

QUALITATIVE ANALYSIS OF PROPOLIS FOR THE TREATMENT OF CHRONIC DISEASES

[1]Kristina Ramanauskienė, [1]Arūnas Savickas, [1]Liudas Ivanauskas, [1]Zenona Kalvėnienė, [1]Giedrė Kasparavičienė, [1]Algirdas Amšiejus, [1]Valdas Brusokas, [1]Asta-Marija Inkėnienė and [2]Danik M. Martirosyan,

[1]Kaunas University of Medicine, LT – 44307 Kaunas, Lithuania
[2]Functional Food Center, Richardson, TX, USA

Corresponding author: Professor Arunas Savickas, E-mail: farmkurs@med.kmu.lt

Keywords: propolis, ferulic acid, antioxidant activity, caffeic acid

INTRODUCTION

Most vascular plants can withstand little desiccation except during the later stages of seed development and the early stages of germination. There is increasing evidence that the development of desiccation intolerance is correlated with increased free radical attack and damage during water loss [Hendry GAF et al,1992]. Dried, desiccation-intolerant tissues generally exhibit extensive peroxidative damage to cellular membranes, resulting in loss of semi-permeability, changes in their microviscosity and lipid phase properties [Leprince O et al, 1994]. The origin of such peroxidative damage occurring under stress conditions is likely to be the formation of transient but highly reactive oxygen species [Leprince O et al, 1994]. Endogenous and exogenous antioxidants prevent free radical-induced damage by preventing the formation of free radicals, scavenging them or by promoting their decomposition. One of the roles of phenolic compounds is the scavenging of free radicals and the protection against excess oxidation caused by UV irradiation, chemical oxidants or pathogen attack, or other kinds of stress. Propolis ranks among natural raw materials that can be used as an antioxidant both in medicine and in agriculture. According to literature, propolis – due to its antimicrobial, anti-inflammatory, anti-oxidative, anti-cancer, immunostimulant and antiviral characteristics [Pietta et al, 2002; Bankova, 2005; Banskota et al, 2001, Menezes et al, 1997] – can be used as an active substance in the manufacturing of pharmaceutical preparations. The phenolic acids in propolis – such as coumaric, ferulic, caffeic gallic, and chlorogenic acids – have strong antioxidant activity [Kanski et al, 2001; Gulcin, 2006; Yeh et al, 2004]. However, it is known that the composition of the active substances in propolis is influenced by the geographic region where propolis has been harvested [Watson et al, 2006; Kosalec et al, 2003], as well as by the quality of the process of the harvesting of the raw material. For this reason, the application of propolis as the active and the complementary substance in the manufacturing of natural products in medicine requires the examination of the chemical composition of Lithuanian propolis, including the identification of the active substances and their amounts, as well as the evaluation of the antioxidant activity of the product.

MATERIALS AND METHODS

The object of the study was propolis harvested in Lithuania. The specimens of propolis were harvested in late July of 2006. The characteristics of the harvested raw material are presented in Table 1.

Table 1. Characteristics of the harvested propolis

Series	Location	Predominating plants in the region	The layer of the harvested propolis in the hive, mm
No -1	Kėdainiai region, Lithuania	Rape fields, linden	20
No-4	Kėdainiai region, Lithuania	Rape fields, linden	4
No-2	Varėna region, Lithuania	Buckwheat field, pine forest	4
No-3	Kaunas region, Lithuania	Leafy forest, coniferous wood, raspberry-canes	4

Methanol for HPLC analysis was of HPLC grade and was purchased from Carl Roth GmbH (Karlsruhe, Germany). Distilled water used for the preparation of solvents was filtered through the Millipore HPLC grade water preparation cartridge (Millipore, Bedford, USA) and 0.45μm pore size membrane filter. Standards of phenolic acids and phenylpropanoids were purchased from ChromaDex (Santa Ana, USA).

The experimental fluid extracts were produced from propolis by maceration. Ethanol at the concentration of 80% (v/v) was used for the extraction. All samples of propolis (30 g) were extracted for 5 days with 80% ethanol up to the 100 ml at room temperature. After extraction, the propolis extract was filtered through a paper filter.

Detection of phenolic acids via high-pressure liquid chromatography (HPLC). HPLC analysis with UV/PDA detection was carried out using the Waters 2690 chromatography system model (Waters, Milford, USA), equipped with a Waters 2487 UV/Vis detector and Waters 996 PDA detector. For separation, a Hichrom column Hypersil H5ODS-150A 150×4.6 mm (Hichrom Ltd., Berkshire, UK) and a H5ODS-10C guard-cartridge were used. The data were collected and analyzed using a personal computer and the Waters Millennium 2000® chromatographic manager system (Waters Corporation, Milford, USA). The mobile phase of the method consisted of solvent A (methanol) and solvent B (0.5% (v/v) acetic acid in water). The elution profile was: 0 min 10% A in B, 28 min 60% A in B, and 30 min 10% A in B. All gradients were linear. The flow rate was 1 ml/min, the column temperature was ambient, and the injection volume was 10 micro liters. UV detection was effected at 290 nm. The eluted components were identified on the basis of the retention time by comparison with the retention time of the reference standard. The identity of constituents was also confirmed with PDA detector by comparison with UV spectra of reference standard in the wavelength range of 190-400 nm.

Determination of the antioxidant activity.

Antioxidant activity of the tested propolis extracts was assessed by measuring 2.2-diphenyl-1-picryhydrazyl (DPPH•) free radical binding by antioxidant species. In its radical form, DPPH• has an absorption band at 515 nm, which disappears upon reduction by an antiradical compound. 50 μl of propolis extract (No-1, No-2, No-3, and No -4) was added to 2 ml of daily prepared DPPH• solution (0,025 g/l in methanol) in a 1 cm-wide cuvette. Absorbance at 515 nm was measured on UNICAM Helios α UV spectrophotometer (Unicam, Cambridge, UK) 16 min after starting the reaction. The comparative solution was methanol. The percent inhibition of the DPPH• radical was calculated according to the following formula:

$$\text{DPPH}\bullet\ \%\ \text{inhibition} = \frac{A_0 - A_1}{A_0},$$

A^0 – the absorbance of the control at t = 0 min;

A^1 – the absorbance of the reaction solution at t = 16 min.

RESULT AND DISCUSSION

High-pressure liquid chromatography (HPLC) showed that all the collected propolis specimens contained coumaric, ferulic, gallic, cinnamic and caffeic acids (Fig. 1), which rank among essential phenolic acids in propolis. The study found that ferulic and coumaric acids were the predominant in the investigated specimens.

Figure 1. The chromatogram of the identification of phenolic acids in propolis using the HPLC technique

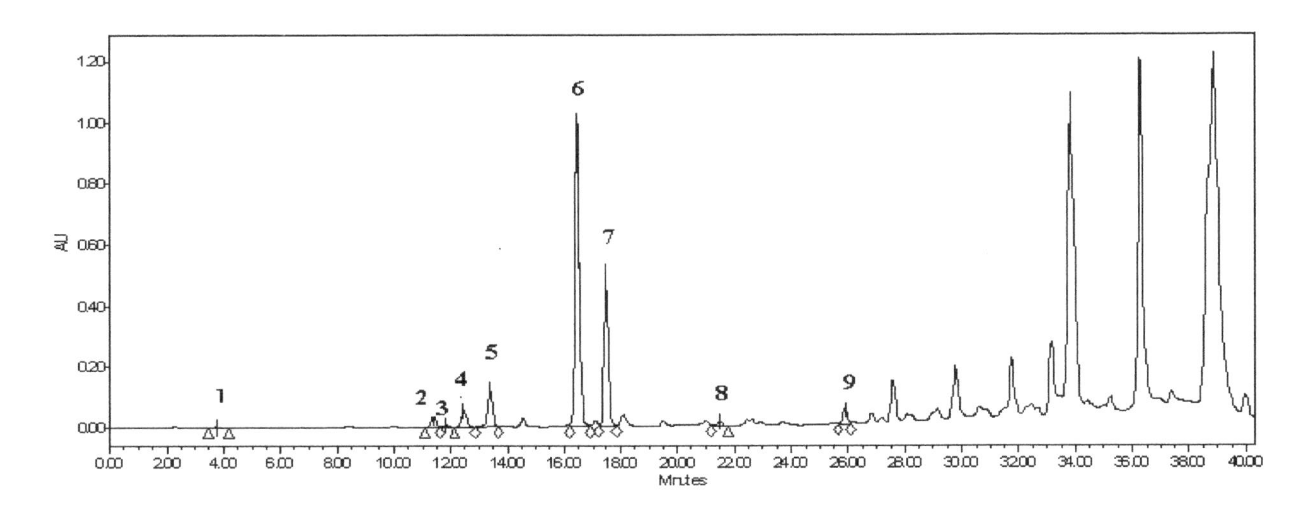

1- gallic acid, 2- chlorogenic acid, 4- caffeic acid, 5- vanillin, 6- coumaric acid, 7- ferulic acid, 8-rosmarinic acid, 9 - cinnamic acid.

Table 2. The quantitative identification of active compounds in propolis

Phenolic acid mkg/ml	No 1	No 2	No 3	No 4
Gallic acid	9,7	7,86	6,59	10,68
Caffeic acid	115,45	141,11	67,97	160,60
Coumaric acid	508,38	1025,51	623,59	1600,60
Ferulic acid	461,46	1085,18	526,94	1470,13
Cinnamic acid	49,49	60,23	43,62	80,21

The investigation showed that the amounts of caffeic, cinnamic and gallic acids in all specimens were negligible and did not differ significantly (Table 2). However, significant differences in the amounts of coumaric and ferulic acids were detected in the propolis specimens (Table 2).

Figure 2. Evaluation of the amounts of coumaric and ferulic acid using the HPLC technique in the harvested specimens of propolis

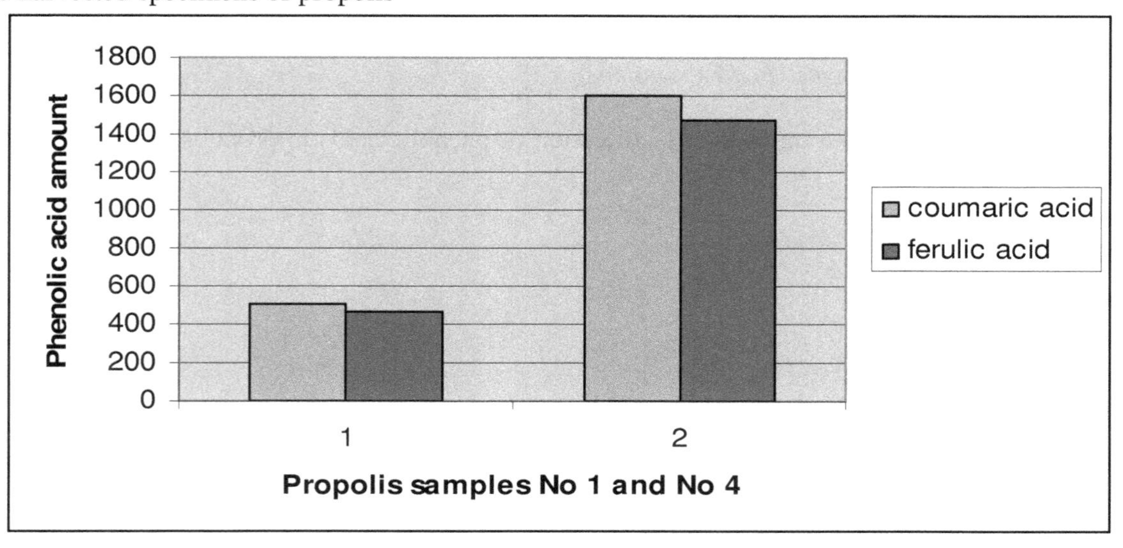

The comparison of the results of the examination of propolis (No 1 and No 4) collected in the same geographical region but using different collection techniques showed that the amounts of ferulic and coumaric acids were significantly higher in sample No 4, which was harvested using special propolis collectors, thus ensuring not greater than 4-mm layer of the raw material in the hive (Fig. 3).

Figure 3. Evaluation of the antioxidant activity of propolis specimens

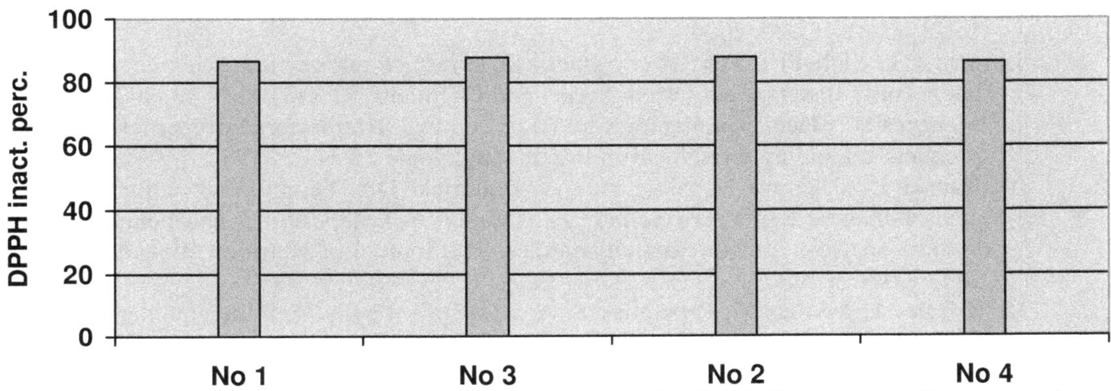

Antioxidant activity was evaluated by DPPH radical binding assay. Data on the percentage inhibition of the DPPH (Fig. 2.) showed that antioxidant activities of propolis samples decreased in the following order: No 3>No 2>No 1> No 4. The results showed high antioxidant activity for all samples of propolis.

CONCLUSIONS

We identified phenolic compounds in Lithuanian propolis and determined that the predominant phenolic acids were ferulic acid and coumaric acid. The results of the study confirmed that the concentration of the active substances in propolis depend not only on the geographical location, but also on the quality of the harvesting of the raw material. Scientific research has proven that propolis has antioxidant properties, and thus can be used as both the active and the complementary substance in both medicine and agriculture for the creation of ecological and environment-friendly products.

REFERENCES

1. Hendry GAF, Finch-Savage WE, Thorpe PC, Atherkon NM, Buckland SH, Nilsson KA, Seel WE (1992) Free radical processes and loss of seed viability during desiccation in the recadcitrant species Quercus robur L. New Phytol122 273-299.
2. Leprince O, Neil M. Atherton, Roger Deltour, and Ceorge A. F. Hendry. The Involvement of Respiration in Free Radical Processes during Loss of Desiccation Tolerance in Germinating Zea mays L. Plant Physiol. (1994) 104: 1333-1339.

3. Pietta P.G. , Gardana C., Pietta A.M. Analytical methods for quality of propolis. Fitoterapia. 73: 7-20, 2002.
4. Watson D.G., Peyfoon E., Zheng L, Lu D, et al. Application of principal components Analysis to H-NMR data obtained from propolis samples of different geographical origin. Phytocem.Anal. 17: 323-331, 2006.
5. Bankova V. Recent trends and important developments in propolis research.Evidence Based Compl Altern. Med. 2: 29-32, 2005.
6. Banskota AH, Tezuka Y, Kadota S, Recent progress in pharmacological research of propolis. Phytother. Res. 15: 561-571, 2001.
7. 8. Gulcin I. Antioxidant activity of caffeic acid (3,4-dihydrocinnamic acid.Toxicology. 16:217(2-3):213-20, 2006.
8. Yeh CT, Shih PH, Yeh GC. Synergistic effect of antioxidant phenolic acids on human phenolsulftransferase activity. J Agric Food Chem. 30; 52 (13):4139-43, 2004.
9. Menezes H., Bacci Jr, Oliveira S.D. Pagnocca F.C. Antibacterial properties of propolis and products containing propolis from Brazil. Apidologie 28:71-76, 1997.
10. Kanski J, Aksenova M, Stoanova A, Buttefield DA. Ferulic acid antioxidant protection against hydroxyl and peroxyl radical oxidation in synaptosomal and neuronal cell culture systems in vitro: structure-activity studies. The Journal of Nutrition Biochemistry. 13: 273-281, 2002.
11. Kosalec I, Bakmaz M, Pepeljnjak S. Analysis of propolis from the continental and Adriatic regions of Croatia. Acta. Pharm. 53: 275-287, 2003.

FUNCTIONAL FOODS IN CHRONIC KIDNEY DISEASE

Jaime Uribarri, MD

Mount Sinai School of Medicine, NY, USA

Corresponding author: Jaime Uribarri, E-mail: jaime.uribarri@mssm.edu

Kaywords: Chronic kidney disease, advanced glycation end products

ABSTRACT

The worldwide prevalence of chronic kidney disease (CKD) is increasing. Therefore the creation and commercialization of functional foods for CKD patients will become important items in the food industry in the near future. We propose to create functional foods for CKD patients by decreasing the food content of advanced glycation end products (AGEs), well-known pro-inflammatory and pro-oxidant compounds, which may contribute significantly to the high cardiovascular disease and other complications of CKD.

It has long been known that AGEs generate during cooking of food using heat, but only recently it has been appreciated that these dietary compounds undergo sufficient absorption from the gastrointestinal tract to make them major contributors to the body AGE pool, in both health and disease. In vitro experiments demonstrate that food-derived AGEs share the same pro-inflammatory and pro-oxidant properties as their endogenous counterparts. Experimentally, food-derived AGEs have been shown to produce and/or intensify several inflammatory processes including atherosclerosis, kidney disease, diabetes, etc in different animal models. More importantly, recent clinical studies support the hypothesis that these exogenous compounds play an important pathogenic role in health and disease.

The food AGE content is mostly determined by the cooking methods, especially the amount and duration of applied heat. Therefore we suggest reducing the complications of chronic diseases like CKD and diabetes in a cost-effective way by simply modifying (less heat application) the way we cook. Any food prepared commercially for CKD patients should take this principle under consideration.

INTRODUCTION

Chronic kidney disease (CKD) is a major public health issue in the United States with as many as 20 million adults affected (1). As the US population ages and the incidence of diabetes and hypertension increases, the incidence and prevalence of CKD will keep increasing at an alarming rate (1). Cardiovascular events are the leading cause of death in both the general US population and in CKD patients (1). Data from the USRDS in 1997 show a 50% prevalence of coronary artery disease, myocardial infarction or peripheral vascular disease in patients starting dialysis (2). The excess cardiovascular risk and mortality, however, is already shown in early renal dysfunction. In the Valsartan in Acute Myocardial Infarction Trial (VALIANT), each reduction of the estimated GFR by 10 units below 81 ml/min/1.73 m2 was associated with a hazard ratio for death and nonfatal cardiovascular outcomes of 1.10 (95 percent confidence interval, 1.08 to 1.12), which was independent of the treatment assignment (3). In a community-based population of 1,120,295 adults with a median follow-up of 2.84 years and a mean age of 52 years, an independent, graded association was observed between a reduced estimated GFR and the risk of death, cardiovascular events, and hospitalization (4). This strong association between CKD and CVD has also been shown in other populations, including studies from Japan (5) and China (6).

The high risk of CVD in CKD most likely results from the additive effect of multiple risk factors, some of them considered traditional such as hyperlipidemia and hypertension and others, more specific of renal disease, such as volume overload, anemia, secondary hyperparathyroidism and hyperhomocystinemia (1). However, even after adjusting for all these factors, an unexplained strong association between CKD and CVD remains. Recently, a new pathogenetic construct of CVD has evolved in which the interplay of inflammation, oxidative stress (OS) and endothelial dysfunction plays a major role (7-9). CKD has been described as a "micro-inflammatory state" (10), but it is unclear which factor(s) initiate the acute phase response in this population. Since renal failure is associated with increasing circulating levels of advanced glycation end products (AGEs) and these compounds are known to have pro-inflammatory, pro-oxidant and anti-endothelial actions, it is likely that they play an important role in the pathogenesis of CVD in this population. Figure 1 depicts our interpretation of the role of AGEs in the pathogenesis of CVD.

In this review we will discuss our current understanding on dietary advanced glycation end products (AGEs) and how it could be incorporated in the design of functional foods for CKD patients. These dietary modifications would appear to be appropriate as part of a multifactorial approach to improve health among CKD patients.

Advanced glycation end products (AGEs)

AGEs are a heterogeneous group of compounds resulting from the spontaneous reaction of reducing sugars and reactive carbonyl groups with the free amino groups of proteins, lipids or nucleic acids (11). AGEs, together with the interrelated processes of oxidative stress and inflammation, may account for many of the complications of diabetes (11). Evidence for this emerges from increasing numbers of in vitro and in vivo studies exploring the role of AGEs in different pathologies, as well as from studies demonstrating significant improvement of features of diabetic complications by the use of inhibitors of the glycation process (12-18).

AGEs form continuously in the body at a slow rate under physiological conditions, but this rate of formation is markedly increased in diabetes. AGEs also form externally during heat cooking of food and a portion of them is absorbed from the gastrointestinal tract (19,20). Recently, it has become apparent that dietary AGEs contribute significantly to the body AGE pool, in health and diseases such as diabetes and CKD (19-23).

Formation of AGEs in food

AGEs form spontaneously in foods, during storage at room temperature, but their formation rate accelerates markedly as temperature increases during heat cooking (13, 18). Figure 2 illustrates a simplified scheme of AGE formation in foods. Factors other than heat, which are known to affect AGE generation in foods include nutrient composition, amount of water, pH and duration of cooking (19,24). Many AGEs have been identified in foods, including εN-carboxymethyl-lysine

(CML) and methylglyoxal (MG)-derivatives (25). Recently, a large database with the contents of CML in more than 200 food items has become available (19). This database allows estimating daily dietary AGE intake as well as designing diets with variable AGE content. In general, foods high in lipid and protein content exhibit the highest AGE levels; for example, the fat and meat groups have 30- and 12-fold higher AGE content than the carbohydrate group, respectively (19). Temperature and method of cooking are more critical to AGE formation than time of cooking; for example, meat samples broiled or grilled at 230°C for shorter cooking times have higher AGE content when compared to samples boiled in liquid media at 100°C for longer periods (19). Analysis of the data (19) shows that meat and meat -derived products, processed by high and dry heat, such as in broiling, grilling, frying and roasting are major sources of AGEs. Since these foods are commonly consumed in the USA, most people are constantly ingesting a high AGE diet. Using the above database to analyze the 3-day food records from a group of 90 healthy subjects we estimated an average dietary AGE intake of 16,000 $\pm$ 5,000 AGE kilounits/day (mean $\pm$ SD) (19). Some representative values of food AGE content are shown in Figure 1, modified from reference 19.

Figure 1: Pathogenesis of CVD in CKD

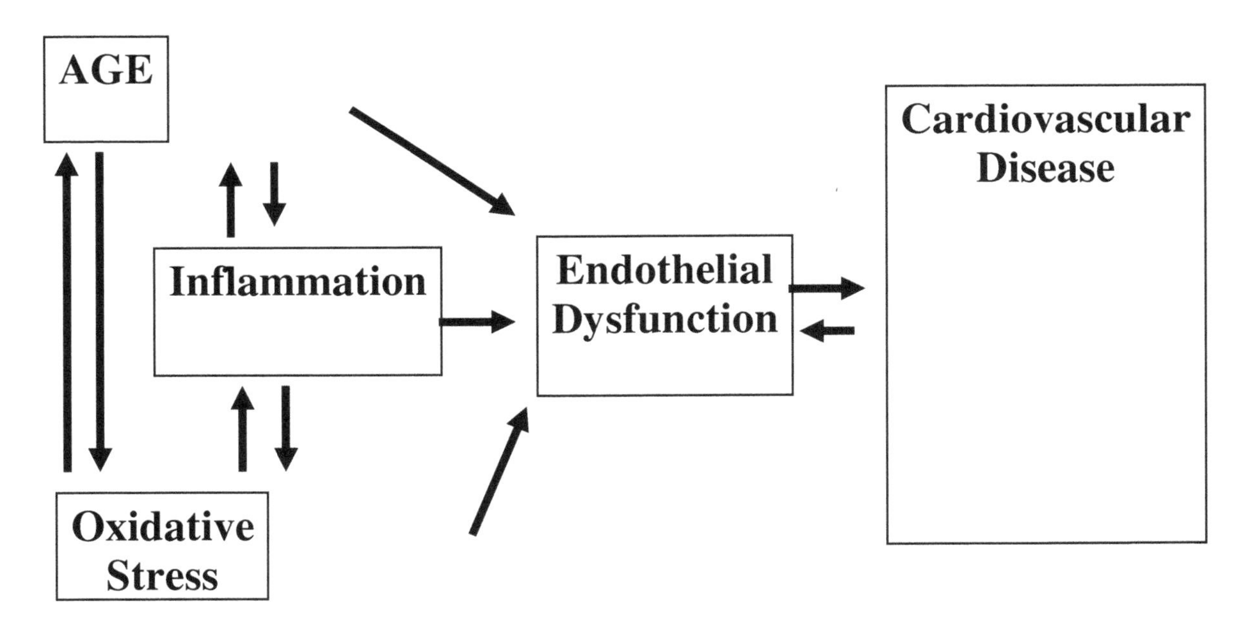

Knowledge about the effect of cooking temperature, duration of cooking, use of water and pH of the solution in the generation of food AGEs is of great importance to develop simple cooking modifications that will reduce the food AGE content without necessarily changing too much the type and amount of foods consumed. For example, stewing, by increasing the water content during cooking, will generate fewer AGEs than broiling or frying a similar piece of steak.

Marinating may have a beneficial effect by reducing the local pH of food (pH effect of lemon or vinegar).

Intestinal absorption of AGEs

Gastrointestinal absorption of AGEs has long been demonstrated, but largely ignored because of its small magnitude (24,25). Several animal studies have shown a correspondence between dietary AGE levels and serum and tissue AGE levels (26-28). The appearance of serum peaks of radiolabeled AGE-modified proteins within a few hours of feeding these radiolabeled AGEs to rats (29) is a more direct demonstration of the intestinal absorption of these compounds.

Studies on human subjects have shown a significant increase in plasma AGE levels within 2 hours following the oral administration of a single AGE-rich meal (30). In a study on 7 healthy subjects urinary pyrraline excretion decreased from 4.8 + 1.1 mg/day on a regular diet to 1.6 + 0.6, 0.4 + 0.2 and 0.3 + 0.1 mg/day, respectively on days 1, 2 and 3 after being placed on a very low AGE diet (31).

Figure 2: In vitro formation of AGEs

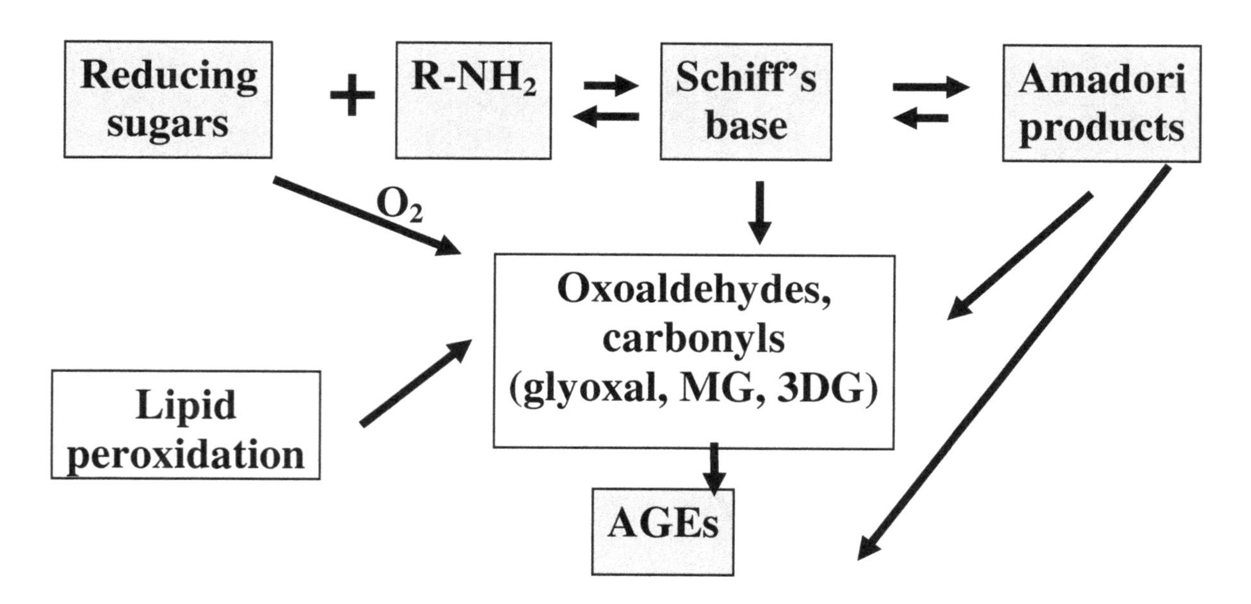

After cessation of dietary restriction by day 5, urinary pyrraline excretion increased immediately to 3.2 + 1.4 mg/day (31). Moreover, studies in healthy subjects, in diabetic patients and in renal failure patients have shown a significant association between dietary and circulating AGE levels (20-23). All the above data demonstrate that food-derived AGEs are absorbed into the circulation and contribute to the body AGE pool. The exact mechanism for this absorption, however, remains undefined at present.

Biological effects of food-derived AGE

Food-derived AGEs induce protein cross-linking and intracellular oxidant stress similar to their endogenous counterparts when tested in vitro using human-derived endothelial cells (Figure 3) (32).

These pro-oxidant and pro-inflammatory properties are also found in the circulating AGE fractions derived from dietary AGEs. LDL samples obtained from diabetic subjects exposed to a high AGE diet for several weeks produced a marked increase in MAPK phosphorylation, NF-κB activity and VCAM-1 secretion when added to cultured human endothelial cells (33). These effects did not happen in response to LDL extracted from diabetic subjects of similar glycemic control, but exposed to the low AGE diet (33).

Experimentally, a low AGE diet has been shown to have a significant protective effect against development of insulin resistance (26), atherosclerosis (27) and diabetic nephropathy (28) in animals. Dietary AGE restriction in a group of stable diabetic patients was associated with parallel reduction of serum AGEs as well as significant reduction of two markers of inflammation (CRP and TNF-α) and a marker of endothelial dysfunction (VCAM-1) (21). These observations were later extended to non-diabetic uremic patients on maintenance peritoneal dialysis (22), another clinical group with high circulating AGE levels. These studies illustrate the fact that a relatively short period of restriction of AGEs in the diet can lower AGEs and AGE-induced inflammatory changes in patients with especially high levels of AGEs, some of which were of endogenous origin. One can speculate that if this diet were to be combined with other therapies known to lower AGE levels, it may be possible to derive a greater therapeutic effect. Importantly, in a large cross-section of healthy subjects, dietary AGE intake – not calories, nutrients, glucose or lipids - was found to be an independent predictor of serum AGEs as well as of CRP (34).

Figure 3: Food-derived AGE have same biological effects as endogenous AGE (Cai W et al. Mol Med 8:337, 2002)

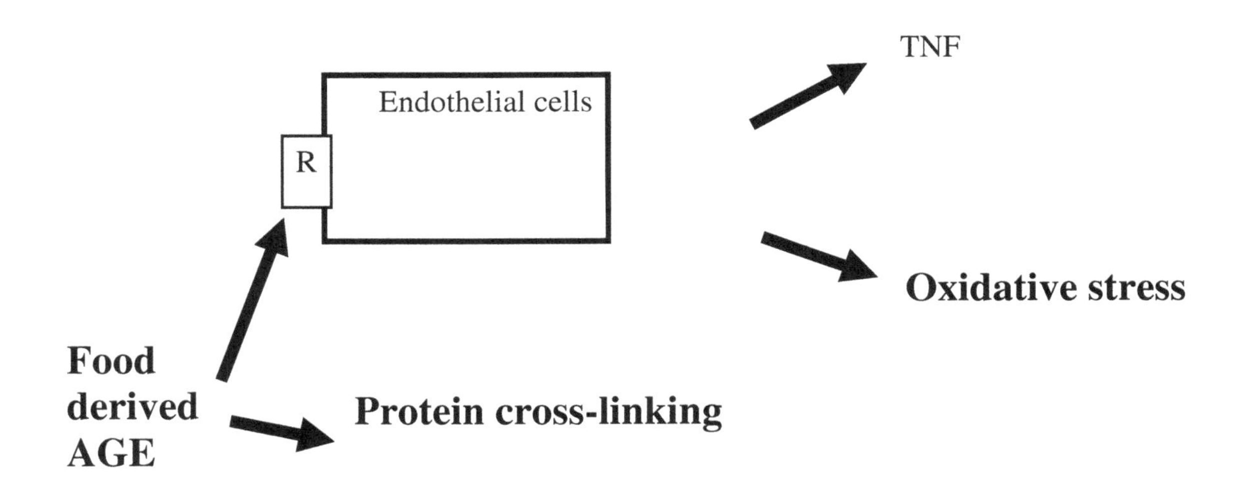

Recently, a single AGE-rich full meal was shown to produce acute impairment of brachial artery vasodilatation in response to ischemia (35), an early marker of atherosclerosis. A similar vasodilatory defect has been previously described after an oral load of glucose or fatty acids (36-37). Since the AGE-rich meal used in those studies (35) also contained carbohydrates and fatty acids it was difficult to distinguish a purely AGE-related effect on this impaired vasodilation. In order to further isolate the effect of dietary AGEs on endothelial function, a group of diabetic and healthy subjects drank over 20 minutes a specially prepared AGE-rich beverage without carbohydrates or fatty acids. Ninety minutes later, serum AGE increased while vasodilatory response (FMD) was markedly impaired in both groups (38). These findings demonstrate an acute harmful effect of dietary AGE on endothelial function. We postulate that repeated insults to endothelial function over the years in subjects ingesting high AGE containing foods may eventually lead to overt endothelial dysfunction and cardiovascular disease (35).

These data have generated a new paradigm of disease suggesting that excessive consumption of AGEs and related oxidants secondary to a "western lifestyle" represents an independent risk factor for inappropriate chronic oxidative stress, inflammation and endothelial dysfunction during the healthy adult years, which over time may facilitate the emergence of diseases, such as diabetes and cardiovascular disease (11). This hypothesis will only be proven by prospective interventional studies in which dietary AGE intake is reduced while following specific clinical outcomes. In the meantime, however, reducing the AGE content of common foods may prove a feasible, safe and easily applicable intervention in both health and disease.

Functional foods

We believe that there are enough data to support a role for dietary AGEs as a risk factor for chronic illness in humans. Thus the time might have come to create functional food for patients with chronic diseases such as CKD and diabetes by simply following the rule of decreasing the intensity and/or time of heat application during cooking.

The currently available information on dietary AGE content can be easily used to recommend meal patterns, similar to those currently recommended against cardiovascular disease and cancer in the general population. A person can achieve a reduced dietary AGE intake by decreasing the intake of high fat cheese, meats and highly processed foods and increasing the consumption of low fat milk products, fruits and vegetables. These recommendations are consistent with those of The Dietary Approaches to Stop Hypertension (39) the American Heart Association (40) and the American Cancer Society (41). We are also saying that a person can reduce the dietary AGE intake, without changing the types of food consumed, by modifying the culinary techniques from those which require high heat applied on dried conditions (grilling, roasting) to those utilizing more humidity (stewing or team-cooking).

When initiating any dietary intervention one should keep in mind that real-life, dietary patterns are complex and include the simultaneous presence of many factors other than AGEs. Some of these dietary factors are pro-inflammatory, such as fats and AGEs, while others are clearly anti-inflammatory, such as polyphenols and a variety of natural antioxidants. Most likely it is the combined influence of all these factors, each one of them acting in different directions and with different intensity, what explains the final biological effect of any dietary pattern.

Interestingly, the only study measuring circulating AGE levels in vegans found them higher than those of omnivores (42), suggesting that something about vegetarian diets may promote endogenous AGE production. Unresolved explanations for this phenomenon include the relatively high-fructose content of vegetarian diets (42) as well as taurine deficiency associated with vegetarian diets (43).

REFERENCES

1. Weiner DE, Tabatabai S, Tighiouart H, Elsayed E, Bansal N, Griffith J, Salem DN, Levey AS, Sarnak MJ. Cardiovascular Outcomes and All-Cause Mortality: Exploring the Interaction Between CKD and Cardiovascular Disease. Am J Kid Dis 48:392-401, 2006
2. U.S. Renal Data System, USRDS 2005 Annual Report, National Institute of Health, National Institute of Diabetes and Kidney Diseases, Bethesda, MD, 2005
3. Anavekar NS,McMurray JJ, Velazquez EJ, Solomon SD, Kober L, Rouleau JL, White HD, Nordlander R, Maggioni A, Dickstein K, Zelenkofske S, Leimberger JD, Califf RM, Pfeffer MA. Relation between Renal Dysfunction and Cardiovascular Outcomes after Myocardial Infarction. NEJM 351:1285-1295
4. Go AS, Chertow GM, Fan D, McCullough CE, Hsu C. Chronic Kidney Disease and the Risks of Death, Cardiovascular Events, and Hospitalization. NEJM 351:1296-1305
5. Najamura K, Okamura T, Hayakawa T, Kadowaki T, Kita Y, Ohnishi H, Saitoh S, Sakata K, Oyakama A, Ueshima H. Chronic kidney disease is a risk factor for cardiovascular death in a community-based population in Japan: NIPPON DATA90 Research Group. Circ J 70:954-956, 2006
6. Zhang L, Zuo L, Wang F, Wang M, Wang S, Lv J, Liu L, Wang H. Cardiovascular disease in early stages of chronic kidney disease in a Chinese population. J Am Soc Nephrol 17:2617-2621, 2006
7. Ross R. Atherosclerosis – an inflammatory disease. N Engl J Med 340:115-126, 1999
8. Himmelfarb J, Stenvinkel P, Ikizler TA, Hakim RM: The elephant in uremia: oxidant stress as a unifying concept of cardiovascular disease in uremia. Kidney Int 62:1524-1538, 2002
9. Stenkinvel P, Pecoits-Filho R, Lindholm B: Coronary artery disease in end-stage renal disease: no longer a simple plumbing problem. J Am Soc Nephrol 14:1927-1939, 2003
10. Kaysen GA: The microinflammatory state in uremia: causes and potential consequences. J Am Soc Nephrol. 12:1549-57, 2001
11. Vlassara H, Uribarri J. Glycoxidation and diabetic complications: modern lessons and a warning? Rev Endocr Metab Disord 2004; 5:181-8
12. Vlassara H, Striker LJ, Teichberg S, Fuh H, Li YM, Steffes M: Advanced glycation end products induce glomerular sclerosis and albuminuria in normal rats. Proc Natl Acad Sci U S A 1994; 91: 11704–11708
13. Yamamoto Y, Kato I, Doi T, Yonekura H, Ohashi S, Takeushi M, Watanabe T, Yamagishi S, Sakurai S, Takasawa S, Okamoto H, Yamamoto H. Development and prevention of advanced diabetic nephropathy in RAGE-overexpressing mice. J Clin Invest 2001; 108:261-268
14. Wendt TM, Tanji N, Guo J, Kislinger TR, Qu W, Lu Y, Bucciarelli LG, Rong LL, Moser B, Markowitz GS, Stein G, Bierhaus A, Liliensiek B, Arnold B, Nawroth PP, Stern DM, D'Agati VD, Schmidt AM. RAGE drives the development of glomerulosclerosis and implicates podocyte activation in the pathogenesis of diabetic nephropathy. Am J Pathol 2003; 162:1123-37
15. Soulis T, Cooper ME, Sastra S, Thallas V, Panagiotopoulos S, Bjerrum OJ, Jerums G: Relative contributions of advanced glycation and nitric oxide synthase inhibition to aminoguanidine-mediated renoprotection in diabetic rats. Diabetologia 1997; 40: 1141–1151

16. Nakamura S, Makita Z, Ishikawa S, Yasumura K, Fujii W, Yanagisawa K, Kawata T, Koike T .Progression of nephropathy in spontaneous diabetic rats is prevented by OPB-9195, a novel inhibitor of advanced glycation. Diabetes 1997; 46:895-9
17. Forbes JM, Thallas V, Thomas MC, Founds HW, Burns WC, Jerums G, Cooper ME. The breakdown of preexisting advanced glycation end products is associated with reduced renal fibrosis in experimental diabetes. FASEB J 2003; 17:1762-64
18. Thomas MC, Tikellis C, Burns WM, Bialkowski K, Cao, Z, Coughlan MT, Jandeleit-Dahm K, Cooper ME, Forbes JM: Interactions between renin angiotensin system and advanced glycation in the kidney. J Am Soc Nephrol 2005; 16:2976-2984
19. Goldberg T, Cai W, Peppa M, Dardaine V, Uribarri J, Vlassara H. Advanced glycoxidation end products in commonly consumed foods. J Am Diet Assoc 2004; 104:1287-1291
20. Uribarri J, Cai W, Sandu O, Peppa M, Goldberg T, Vlassara H. Diet-derived glycation end products are major contributors to the body's AGE pool and induce inflammation in healthy subjects. Ann N Y Acad Sci 2005; 1043:461-6
21. Vlassara H, Cai W, Crandall J, Goldberg T, Oberstein R, Dardaine V, Peppa M, Rayfield E. Inflammatory mediators are induced by dietary glycotoxins, a major risk factor for diabetic angiopathy. Proc Natl Acad Sci 2002; 99:15596-601
22. Uribarri J, Peppa M, Cai W, Goldberg T, Lu M, He C, Vlassara H. Restriction of dietary glycotoxins markedly reduces AGE toxins in renal failure patients. J Am Soc Nephrol 2003; 14: 728-731
23. Uribarri J, Peppa M, Cai W, Goldberg T, Baliga S, Lu M, Vassalotti JA, Vlassara H. Dietary glycotoxins correlate with circulating advanced glycation end products in renal failure patients. Am J Kid Dis 2003; 42:532-38
24. Somoza V. Five years of research on health risks and benefits of Maillard reaction products: an update. Mol Nutr Food Res 2005; 49:663-72
25. Chuyen NV. Toxicity of the AGEs generated from the Maillard reaction: on the relationship of food-AGEs and biological-AGEs. Mol Nutr Food Res. 2006; 50:1140-9
26. Hofmann, SM, Dong, HJ, Li, Z, Cai W, Altomonte J, Thung SN, Zeng F, Fisher EA, Vlassara H. Improved insulin sensitivity is associated with restricted intake of dietary glycoxidation products in the db/db mouse. Diabetes 2002; 51:2082-9
27. Lin RY, Reis ED, Dore AT, Lu M, Ghodsi N, Fallon JT, Fisher EA, Vlassara H. Lowering of dietary advanced glycation endproducts (AGE) reduces neointimal formation after arterial injury in genetically hypercholesterolemic mice. Atherosclerosis 2002; 163:303-11
28. Zheng F, He C, Cai W, Hattori M, Steffes M, Vlassara H. Prevention of diabetic nephropathy in mice by a diet low in glycoxidation products. Diabetes Metab Res Rev 2002; 18:224-37
29. He C, Sabol C, Mitsuhashi T, Vlassara H. Dietary glycotoxins. Inhibition of reactive products by aminoguanidine facilitates renal clearance and reduces tissue sequestration. Diabetes 1999; 48:1308-15
30. Koschinsky T, He CJ, Mitsuhashi T, Bucala R, Liu C, Buenting C, Heitmann K, Vlassara H. Orally absorbed reactive advanced glycation end products (glycotoxins): an environmental risk factor in diabetic nephropathy. Proc Natl Acad Sci 1997; 94:6474-79
31. Foerster A, Henle T. Glycation in food and metabolic transit of dietary AGEs (advanced glycation end-products): studies on the urinary excretion of pyrraline. Biochem. Soc. Trans 2003; 31:1383–1385
32. Cai W, Cao Q, Zhu L, Peppa M, He CJ, Vlassara H. Oxidative stress-inducing carbonyl compounds from common foods: Novel mediators of cellular dysfunction. Molecular Medicine 2002; 8:337-46

33. Cai, W, He, C, Zhu, L, Peppa M, Lu C, Uribarri J, Vlassara H. High Levels of Dietary Advanced Glycation End Products Transform Low-Density Lipoprotein Into a Potent Redox-Sensitive Mitogen-Activated Protein Kinase Stimulant in Diabetic Patients. Circulation 2004; 110:285-291
34. Uribarri J, Cai W, Peppa M, Goodman S, Ferrucci L, Striker G, Vlassara H. Circulating glycotoxins and dietary AGEs; two links to inflammatory response, oxidative stress and aging. J Gerontology. In press.
35. Stirban A, Negrean M, Stratmann B, Gawlowski T, Horstmann T, Gotting C, Kleesic K, Mueller-Roesel M, Koschinsky T, Uribarri J, Vlassara H, Tschoepe D. Benfotiamine prevents macro- and microvascular endothelial dysfunction and oxidative stress following a meal rich in advanced glycation endproducts in people with Type 2 Diabetes Mellitus. Diabetes Care 2006; 29:2064-2071
36. Kawano H, Motoyama T, Hitashima O, Hirai N, Miyao Y, Skamoto T, Kugiyama K, Ogawa H, Yasue H. Hyperglycemia rapidly suppresses flow-mediated endothelium-dependent vasodilation of brachial artery. J Am Coll Cardiol 34:146-154, 1999
37. Vogel RA, Corretti MC, Plotnick GD. Effect of a single high-fat meal on endothelial function in healthy subjects. Am J Cardiol 79:350-354, 1997
38. Uribarri J, Striban A, Sander D, Cai W, Negrean M, Buenting C, Koschinsky T, Vlassara H. Single oral challenge by advanced glycation en dproducts acutely impairs endothelial function in diabetic and nondiabetic subjects. Sumitted for publication.
39. L.J. Appel, T.J. Moore, E. Obarzanek, W.M. Vollmer, L.P. Svetkey, F.M. Sacks, G.A. Bray, T.M. Vogt, J.A. Cutler, M.M. Windhauser, P.H. Lin and N. Karanja, A clinical trial of the effects of dietary patterns on blood pressure. DASH Collaborative Research Group. *N Engl J Med.* 336 (1997), pp. 1117–1124
40. R.M. Krauss, R.H. Eckel, B. Howard, L.J. Appel, S.R. Daniels, R.J. Deckelbaum, J.W. Erdman, Jr, P. Kris-Etherton, I.J. Goldberg, T.A. Kotchen, A.H. Lichtenstein, W.E. Mitch, R. Mullis, K. Robinson, J. Wylie-Rosett, S. St Jeor, J. Suttie, D.L. Tribble and T.L. Bazzarre, AHA Dietary Guidelines: Revision 2000: A statement for healthcare professionals from the Nutrition Committee of the American Heart Association. *Circulation.* 102 (2000), pp. 2284–2299
41. W.C. Willet, Goals for nutrition in the year 2000. *CA Cancer J Clin.* 49 (1999), pp. 333–352
42. Sebekova k, Krajcoviova-Kudlackova M, Schinzel R, Faist V, Klvanova J, Heidland A. Plasma levels of advanced glycation end products in healthy, long-term vegetarians and subjects on a western mixed diet. Eur J Nutr 2000; 40:275-81
43. McCarty MF. The low-AGE content of low-fat vegan diets could benefit diabetics - though concurrent taurine supplementation may be needed to minimize endogenous AGE production. Med Hypotheses 2005; 64:394-8

THE ROLE OF CURCUMIN IN INSULIN ACTION AND GLUCOSE METABOLISM

Jessica Davis, Teayoun Kim, & Suresh T. Mathews

Department of Nutrition and Food Science, Auburn University, Auburn, AL 36849

Running Title: Curcumin and glucose metabolism

Corresponding author: Suresh T. Mathews, PhD

Cellular & Molecular Nutrition, Dept.of Nutrition & Food Science
260 Lem Morrison Dr., 101 Poultry Sci Bldg,
Auburn University, Auburn, AL 36849.
Email: smathews@auburn.edu

Keywords: curcumin, functional foods, obesity, diabetes

ABSTRACT

Diabetes is characterized by dangerously elevated blood glucose concentration resulting from decreased glucose uptake into skeletal muscle and increased endogenous glucose production by the liver. Though several pharmaceutical drugs are currently available to control blood glucose levels in diabetes, a more holistic approach leading to its prevention and management, with lesser side effects and lower costs, is preferred. One approach involves incorporating functional foods into the diet. Curcumin is an anti-inflammatory and antioxidant bioactive component of the curry spice, turmeric (*Curcuma longa*). Current research suggests that curcumin may act as a functional food lowering blood glucose animal models of diabetes. This study examines potential mechanisms of how curcumin may modulate glucose metabolism. Our data indicate that curcumin suppresses the expression of hepatic gluconeogenic genes, primarily glucose 6-phosphatase, without affecting insulin receptor activation or insulin-stimulated glucose uptake. Further, we demonstrate that curcumin activates AMP-activated kinase resulting in the phosphorylation of its downstream target, acetyl-CoA carboxylase in H4IIE rat hepatoma cells. Taken together, our results suggest that curcumin may play a role in the regulation of glucose metabolism by inhibiting hepatic gluconeogenesis, lending further support to its role as a functional food in the prevention and treatment of diabetes.

INTRODUCTION

Type 2 diabetes, currently affecting nearly 21 million people in the United States and 246 million worldwide, is characterized by impaired insulin action on skeletal muscle glucose uptake, adipocytes lipolysis, and suppression of endogenous glucose production by the liver. This escalating incidence of diabetes is paralleled with a spiraling increase in the prevalence of obesity in the United States, with current estimates indicating that a third of the adult population in the U.S. is obese [1]. Both obesity and diabetes increase the risk for cardiovascular disease. While the etiology of obesity and diabetes is complex, diet plays an important role in the development and management of these "*twin epidemics*" [2]. Functional foods, defined as foods or food ingredients that provide a health benefit beyond traditional nutrients may offer an alternative and complementary approach in the treatment and/or prevention of chronic conditions such as obesity and type 2 diabetes.

Several botanical, herbal, and biological products claim to lower blood glucose or decrease complications of diabetes. However, very little is known about their mechanism of action, efficacy, and/or safety. Two recent reviews summarize the literature on botanicals, herbal products and dietary supplements for glucose control among patients with diabetes [3,4]. Examples of herbs that may regulate glucose homeostasis include *Coccinia indica* (ivy gourd), *Gymnema sylvestre*, ginseng species, *Ocimum sanctum* (holy basil), *Trigonella foenum graecum* (fenugreek), and allium species (garlic). Claims have also been made for aloe, bilberry, milk thistle, and *Curcuma longa* (turmeric).

Curcumin is the natural yellow pigment in turmeric, isolated from the rhizomes of the plant *Curcuma longa*. It constitutes about 3-5% of the composition of turmeric. The use of turmeric in the treatment of different inflammatory diseases has been described in Ayurveda and in traditional Chinese medicine. The yellow pigmented fraction of turmeric is rich in phenolics, structurally identified as curcuminoids. The three main curcuminoids isolated from turmeric are curcumin, demethoxy curcumin, and bisdemethoxy curcumin (Fig.1). Curcumin has been shown to possess potent antioxidant and anti-inflammatory properties [5,6], as well as anticancer and cardioprotective effects [7,8]. The molecular effects of curcumin have been shown to be mediated by: transcription factors including nuclear factor kappa B (NF-κB) and peroxisome proliferator-activated receptor-γ (PPAR-γ); enzymes such as cyclooxygenase-2 (COX-2) and inducible nitric oxide synthase (iNOS); and cytokines including tumor necrosis factor alpha (TNF-α, interleukin-1 and 6 (IL-1, IL-6) [9]. In the Indian sub-continent, for centuries, curcumin has been consumed as a dietary spice at doses up to 100 mg/day. Further, recent Phase I clinical trials indicate that a dose as high as 8 g/day, was tolerable in humans, with no side effects [10].

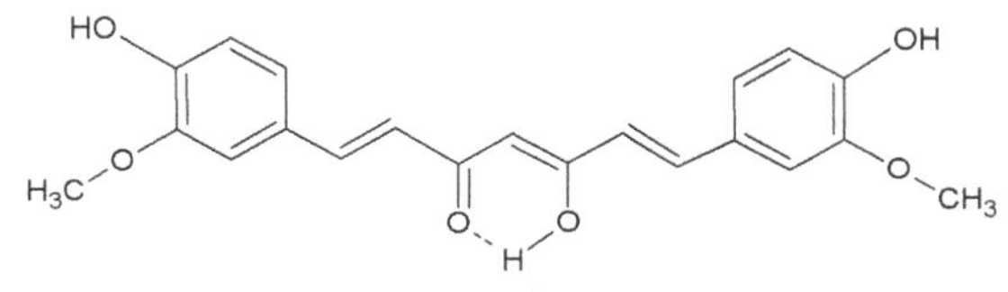

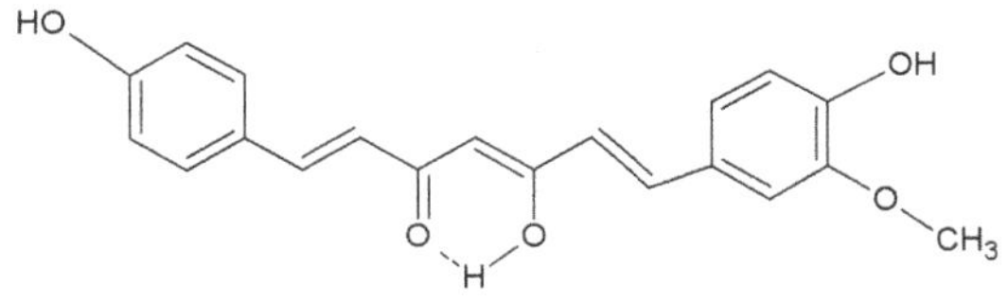

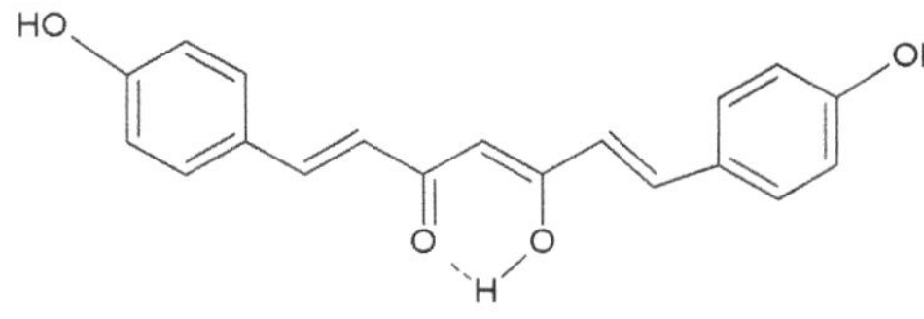

Fig. 1: Structures of three major curcuminoid compounds of turmeric; from top to bottom curcumin, demethoxycurcumin, and bis-demethoxycurcumin.

Several lines of evidence suggest that curcumin may play a beneficial role in diabetes, both in lowering blood glucose levels and in ameliorating the long-term complications of diabetes (Table 1). In streptozotocin-induced diabetic rats, administration of curcumin resulted in the lowering of blood glucose [11-13]. Further, in the KK/AY mice, a genetic model of type 2 diabetes, an ethanolic extract of curcuminoids added to the diet lowered blood glucose significantly [14,15]. Additionally, a recent study by Pari and coworkers demonstrated that tetra-hydrocurcumin THC), a synthetic analog of curcumin, normalizes blood glucose levels in STZ-treated rats [16]. A recent study by Kowluru *et al* demonstrated that curcumin administration decreases retinal oxidative stress and levels of pro-inflammatory markers, both of which are implicated in the development of diabetic retinopathy [17].

***Table 1**: Hypoglycemic effects of curcumin in animal models of diabetes*

Curcuminoid	**Animal model**	**Treatment period**	**Diabetic: Blood glucose (mg/dl or mM‡)**	**Curcumin-treated: Blood glucose (mg/dl or mM‡)**	**Reference**
Bis-o-hydroxy cinnamolylmethane (15 mg/kg BW)	STZ-treated Wistar rats	45 days	17.09 ± 1.46‡	5.65 ± 0.6‡	Srinivasan et al [11]
Tetrahydrocurcumin (80mg/kg BW)	STZ-treated Wistar rats	45 days	285.0 ± 15.0	110 ± 7.0	Pari et al [16]
Turmeric (1g/kg BW)	Alloxan-treated Wistar rats	21 days	204.34 ± 4.32	142.7 ± 1.68	Arun et al [13]
Curcumin (80mg/kg BW)	Alloxan-treated Wistar rats	21 days	204.32 ± 4.32	140.1 ± 1.40	Arun et al [13]
Curcuminoid mix (2g/kg BW)	KK-Ay/Ta mice	28 days	22.0 ± 2.5‡	14.0 ±3.5‡	Kuroda et al [14]
Curcumin (60 mg/kg BW)	Laka albino mice	28 days	412 ± 12.3	280.5 ± 5.6	Sharma et al [18]
Curcumin (0.5g/kg diet)	STZ-treated Lewis rats	42 days	469 ± 81	451 ± 123	Kowluru et al [17]

These studies suggest that curcumin or dietary curcuminoids may have beneficial effects in diabetes, with the potential to lower or normalize blood glucose levels. However, our understanding of curcumin's molecular mechanisms that modulate insulin action is limited. Therefore the goal of this study was to analyze potential mechanisms that mediate curcumin's effects on insulin action and glucose metabolism.

MATERIALS AND METHODS

Reagents and antibodies: Curcumin was purchased from Cayman Chemical Company (Ann Arbor, MI, USA) and dissolved in DMSO. Recombinant human insulin was purchased from Roche Diagnostics (Indianapolis, IA); dexamethasone from Sigma (Milwaukee, WI); phospho-AMPKalpha (Thr172) antibody from Cell Signaling Technology (Beverly, MA), phospho-ACC2 (Ser79) antibody from Upstate Biotechnology (Charlottesville, VA), and GAPDH antibody from AbCam (Cambridge, MA).

Cell culture: Hep3B human hepatoma cells and H4IIE rat hepatoma cells were purchased from American Type Culture Collection (Manassas, VA, USA). Hep3B cells were cultured in Improved Modified Eagle's Medium (IMEM) media and H4IIE cells in α-MEM media. Both media were supplemented with 10% fetal bovine serum and 1% Penicillin/Streptomycin. L6 muscle cells stably expressing myc-tagged GLUT4 (L6-GLUT4myc cells) were cultured as described previously [19].

Insulin receptor tyrosine kinase assay: IRs were partially purified from rat liver using wheat germ agglutinin, as described previously [20]. Partially purified rat liver IRs were preincubated with curcumin in the assay buffer (25mM HEPES, pH 7.4, 0.1% Triton X-100, 0.05% BSA) for 30 minutes prior to 100 nM insulin stimulation for 10 minutes. An exogenous substrate, poly (Glu80Tyr20) was added and phosphorylation was carried out as described [21,22].

Glucose uptake assay: Confluent L6-GLUT4myc myoblasts in 24-well microplate were serum-starved by addition of serum-free DMEM containing 0.1% BSA for 3 hours. The media was then changed to glucose-free Krebs-Ringer HEPES buffer for 30 minutes. Cells were pre-incubated with curcumin for 30 minutes. Insulin (100 nM) treatment was for 30 minutes followed by the addition of [3H]2-deoxy-D-glucose (2-DOG), and incubation at 37oC for 10 minutes. After washing the cells four times with ice-cold PBS, the cells were solubilized in 0.2N NaOH, and radiolabeled 2-DOG uptake was measured by liquid scintillation counting. Non-specific transport was determined using 10 μM cytochalasin, and this was subtracted from all values.

Gluconeogenic gene expression: Hep3B human hepatoma cells or H4IIE rat hepatoma cells were treated with insulin, dexamethasone, or curcumin, overnight at 37oC in a CO2 incubator. The next morning total RNA was isolated using RNeasy mini-kit (QIAGEN, Valencia, CA). cDNA was synthesized using iQ SYBR Green Supermix (BioRad, Hercules, CA). Quantitative real-time PCR was conducted on an iCycler real-time PCR detection system (BioRad) using the following human gene specific primers (Invitrogen, Carlsbad, CA, USA): Fructose 1,6 bisphosphatase forward 5'-TCA TCG CAC TCT GGT CTA CG-3', Fructose 1,6 bisphosphatase reverse 5'-GCC CTC TGG TGA ATG TCT GT-3'; PEPCK forward 5'-GGG TGC TAG ACT GGA TCT GC-3', PEPCK reverse 5'-GAG GGA GAA CAG CTG AGT GG-3'; glucose-6 phosphatase forward 5'-GGG TGT AGA CCT CCT GTG GA-3', glucose-6 phosphatase reverse 5'-GAG CCA CTT GCT GAG TTT CC-3'; β-actin forward 5'- CCT CTA TGC CAA CAC AGT GC-3', β-actin reverse 5'-CAT CGT ACT CCT GCT TGC TG-3'. Reaction conditions were as follows: 95.0◦C, 3:0 min; 95.0◦C, 0:15 min, 60.0◦C, 0:30 min, 72.0◦C, 0:30 min, repeated 40x; 55.0◦C, 0:10 min, repeated 80x. Expression levels were first normalized to β-actin by calculating the ΔΔCt (ΔΔCt = Target gene Ct - β-actin Ct). Next, relative mRNA expression was calculated as 2ΔΔCt and expressed as fold change, as set forth by Pfaffl [23]; this value was used for statistical analyses (Real-Time PCR Applications Guide, BioRad, Hercules, CA). All assays were carried out in triplicate.

Western blot analysis of phospho-AMPK and phospho-ACC: Confluent 6-well dishes of Hep3B or H4IIE cells were treated with 20μM curcumin for various lengths of time. AICAR (0.5mM) treatment for 30 minutes was used as a positive control of AMPK activation. Cells were washed with ice-cold PBS three times and lysed in 200μl of cell lysis buffer [50mM HEPES pH7.4, 100mM sodium pyrophosphate, 100mM EDTA, 20mM sodium orthovanadate, 2mM phenylmethylsulfonyl fluoride, 0.01mg/ml aprotinin, 0.01mg/ml leupeptin, 1% Triton X-100, and protease inhibitor cocktail tablet (Roche Diagnostics, Indianapolis)]. Cell lysate proteins were separated by SDS-PAGE followed by Western blot analysis using antibodies against phospho-AMPKα (Thr172) and phospho-ACC2 (Ser79). GAPDH was used as a loading control.

Chemiluminescence detection and image analysis were performed with UVP-Biochimie Bioimager and LabWorks software (UVP, Upland, CA).

RESULTS AND DISCUSSION

Plant derivatives with purported anti-diabetic properties have been used in folk medicine, traditional healing systems, and as complementary and alternative therapy. The discovery and synthesis of the hypoglycemic drug, metformin (Glucophage), can be attributed to early findings of hypoglycemic activity of guanidine in French lilac (*Galega officinalis*) extracts [24]. A wide range of plant-derived principles belonging to compounds, mainly alkaloids, glycosides, galactomannan gum, polysaccharides, hypoglycans, peptidoglycans, guanidine, steroids, phenolics, glycopeptides and terpenoids, have demonstrated bioactivity against hyperglycaemia [25]. Curcumin has been widely used for centuries in indigenous medicine for the treatment of a variety of inflammatory conditions and other diseases. Several studies have shown that curcumin exhibits hypoglycemic effects in animal models of diabetes [11-13]. In this study, we examined curcumin'effects on glucose metabolism in intact skeletal muscle and liver cell lines, in an effort to characterize potential mechanisms that may contribute to improvement of insulin action.

Insulin exerts its pleiotropic effects on several target tissues. In the skeletal muscle, insulin stimulates glucose uptake and glycogen synthesis. In the fat tissue, insulin has an anti-lipolytic effect which results in decreased circulating free fatty acids, and in the liver, insulin suppresses hepatic glucose production. Under normal conditions, the combined effects of insulin action in these insulin-sensitive tissues maintain normal glucose homeostasis. Insulin action begins by the binding of insulin to its plasma membrane receptor, which activates the ß-subunit tyrosine kinase (for a review, [26]). This induces tyrosine phosphorylation of insulin receptor substrate (IRS) family members (IRS1, 2, 3 and/or 4). The phosphorylated IRS proteins recruit phosphatidylinositol 3-kinase (PI3K) through its association with the regulatory subunit p85α. Central to the actions of insulin is the lipid kinase, PI 3-kinase, which generates phosphatidylinositol-3,4,5-triphosphate (PIP3). Insulin-stimulated PIP3 leads to the recruitment of Akt/protein kinase B (PKB) to the plasma membrane where PIP3 dependent kinase 1 (PDK1) fully activates the serine/threonine kinase Akt. The activation of Akt ultimately results in the translocation of the insulin-sensitive glucose transporter GLUT4 to the plasma membrane facilitating the entry of glucose into the cell.

Curcumin does not affect either insulin receptor activation or glucose uptake of skeletal muscle cells: Of the 3 critical nodes of insulin signaling that Kahn and coworkers have proposed [26], the first node is the activation of insulin receptor tyrosine kinase (IR-TK) and the phosphorylation of its substrate, the IRS proteins. We assayed the effect of curcumin on insulin receptor activation by assessing IR-TK activity in partially-purified IRs. Insulin induced a 4-fold increase in TK activity *(Fig.2A)*. Curcumin treatment along with insulin had no additive effect. Similarly curcumin alone, at the highest concentration of 100 µM had no effect on stimulating IR-TK activity. Next, we tested the distal effect of insulin action, i.e., glucose uptake into skeletal muscle cells. As demonstrated earlier, insulin treatment induced a 2-fold increase in 2-DOG uptake in L6-GLUT4myc myoblasts, compared to basal levels [19] (Fig. 2B). Curcumin treatment had no effect on basal glucose uptake levels. However, insulin-stimulated 2-DOG uptake was decreased ~30% by curcumin treatment (2.5µM). Increasing curcumin concentrations did not further alter insulin-stimulated glucose uptake. These results suggest that curcumin does not have a direct effect on either the proximal or distal steps of insulin action, viz., IR activation and glucose uptake in skeletal muscle cells. These results are consistent with earlier findings of Yang et. al. [27], who show that curcumin does not alter IR autophosphorylation. Further, Ikonomov et.al [28] demonstrated that curcumin is an inhibitor of insulin-stimulated GLUT4 translocation and glucose transport in 3T3-L1 adipocytes. These findings indicate that curcumin's hypoglycemic effects are not mediated by an insulinomimetic activity acting via the insulin receptor or through upregulation of glucose transport.

Curcumin modifies the expression levels of gluconeogenic genes: One of the hallmarks of diabetes is the inability of insulin to inhibit hepatic glucose production. Normally, insulin acts to suppress the expression of the genes for the key gluconeogenic enzymes phosphoenolpyruvate carboxykinase (PEPCK), glucose-6-phosphatase (G6Pase) and fructose 1,6 bisphosphatase (F1,6bP) [29]. Further, the rate of transcription of the hepatic PEPCK and G6Pase genes are increased by several hormones, including glucocorticoids, retinoic acid, and glucagon (via its second messenger, cAMP). Therefore, it was of interest to test curcumin's effects on the expression of hepatic gluconeogenic genes. We demonstrate that insulin treatment significantly decreased the expression of G6Pase in Hep3B human hepatoma cells (Fig.3A). Similarly, curcumin treatment also decreased G6Pase expression in a dose-dependent manner (Fig.3A).

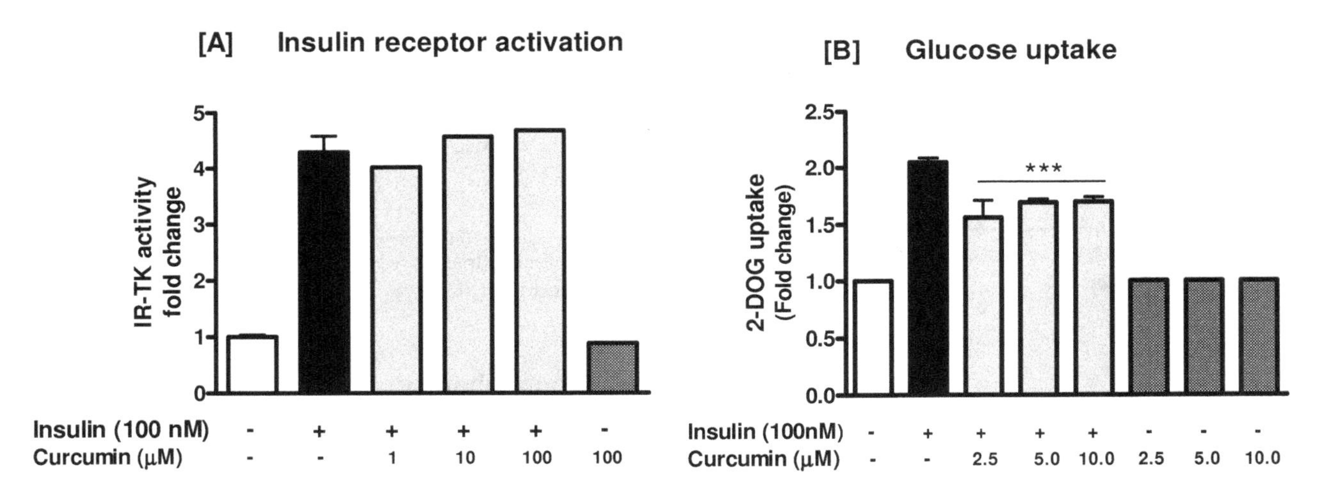

*Fig.2: Effect of Curcumin on insulin receptor tyrosine kinase activity (A) and glucose uptake in L6-GLUT4myc myoblasts (B), *** $p < 0.001$, compared to insulin-stimulated glucose uptake.*

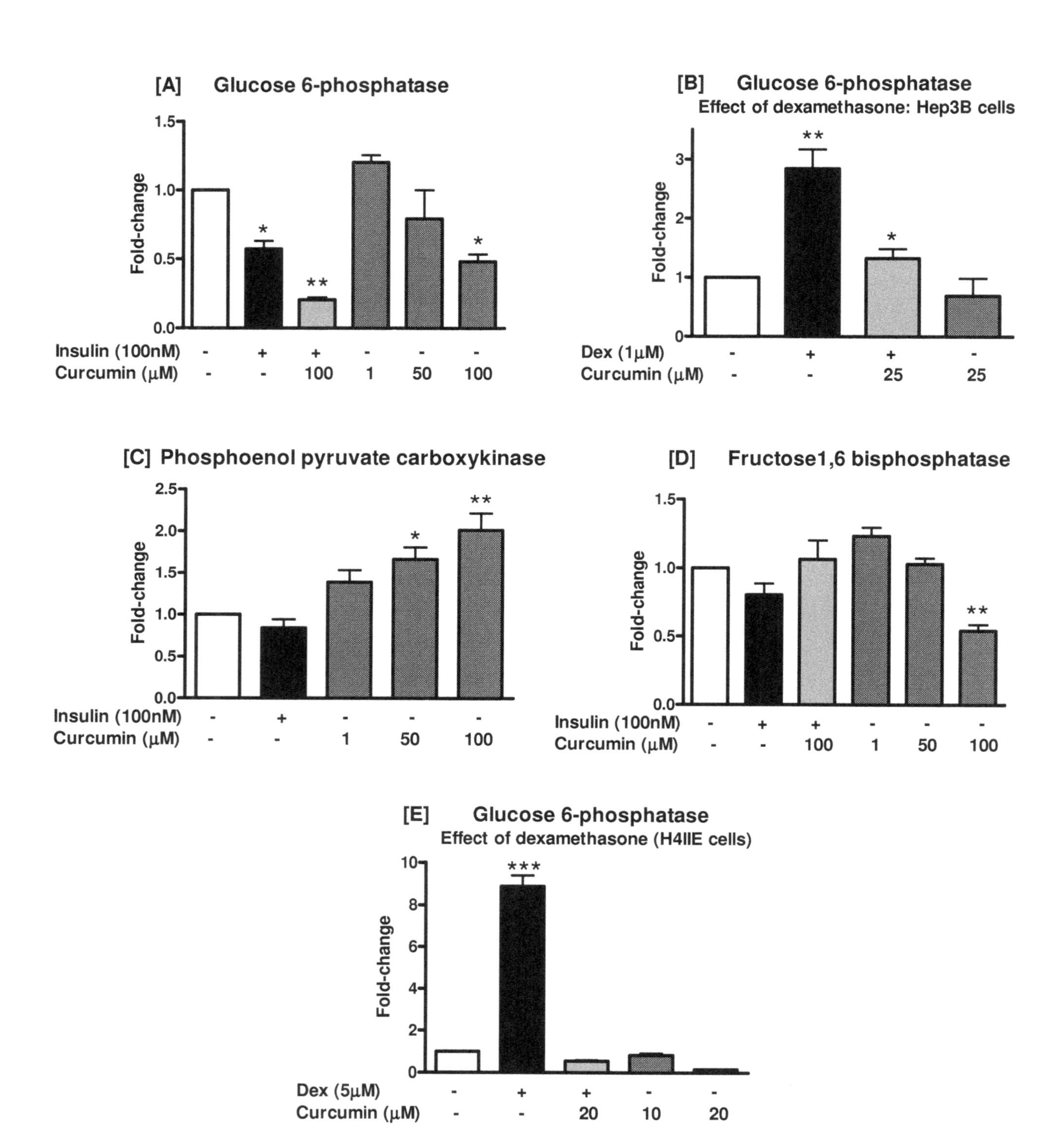

Fig. 3: *Effect of curcumin on gluconeogenic gene expression in hepatoma cell lines treated with insulin or dexamethasone. Gene expression for glucose-6-phosphatase (G6Pase) in Hep3B human hepatoma cells (A,B) or H4IIE rat hepatoma cells (E), and phosphoenolpyruvate carboxykinase (PEPCK) (C),and fructose-1,6-bisphosphatase (F16BPase) (D) in Hep3B cells were analyzed by quantitative real-time PCR. Expression was normalized to β-actin gene and expressed as fold-change, using the ΔΔCt method (* $p < 0.05$, ** $p < 0.01$, *** $p < 0.001$, compared to control)*

Simultaneous treatment of Hep3B cells with insulin and curcumin resulted in a significant suppression of G6Pase gene expression. This effect of curcumin was additive to insulin-mediated inhibition of G6Pase expression. This suggests that curcumin's effects on G6Pase expression may be mediated by pathways distinct from insulin signaling. Since dexamethasone is shown to increase the gene expression of G6Pase, we tested curcumin's ability to repress this effect. Treatment of Hep3B cells with 25 μM curcumin repressed the ability of dexamethasone to stimulate G6Pase gene expression. Next, we assayed curcumin's effects on the expression of PEPCK and F1,6BPase. Insulin-mediated suppression of PEPCK and G6Pase was modest. Curcumin treatment decreased the expression of F1,6BPase (Fig.3D) but failed to suppress PEPCK gene expression, and was stimulatory at higher doses (Fig.3C). G6Pase catalyzes the final step in both the gluconeogenic and glycogenolytic pathways, namely the hydrolysis of glucose-6-phosphate to glucose and inorganic phosphate while PEPCK is the first committed step in hepatic gluconeogenesis. To confirm our findings, we tested curcumin's effects in the H4IIE rat hepatoma cell line. This cell line has been used successfully to study the regulation of gluconeogenic gene expression. We demonstrate that dexamethasone (5 μM) induces an 8-fold increase in G6Pase gene expression. Curcumin treatment (20 μM) suppresses basal G6Pase gene expression. Further, curcumin treatment represses dexamethasone-mediated increase in G6Pase gene expression. These results are consistent with our findings in Hep3B cells. Another human hepatoma cell line, HepG2, fail to express PEPCK efficiently. However, our studies indicate that PEPCK expression is fairly robust in Hep3B cells. It is possible that in Hep3B cells, insulin's inability to suppress gene expression of PEPCK may be attributed to lower levels of transcription factors (PGC1α, SRC1, and PCAF) or idle regulatory factors (HNF4α) [30-32].

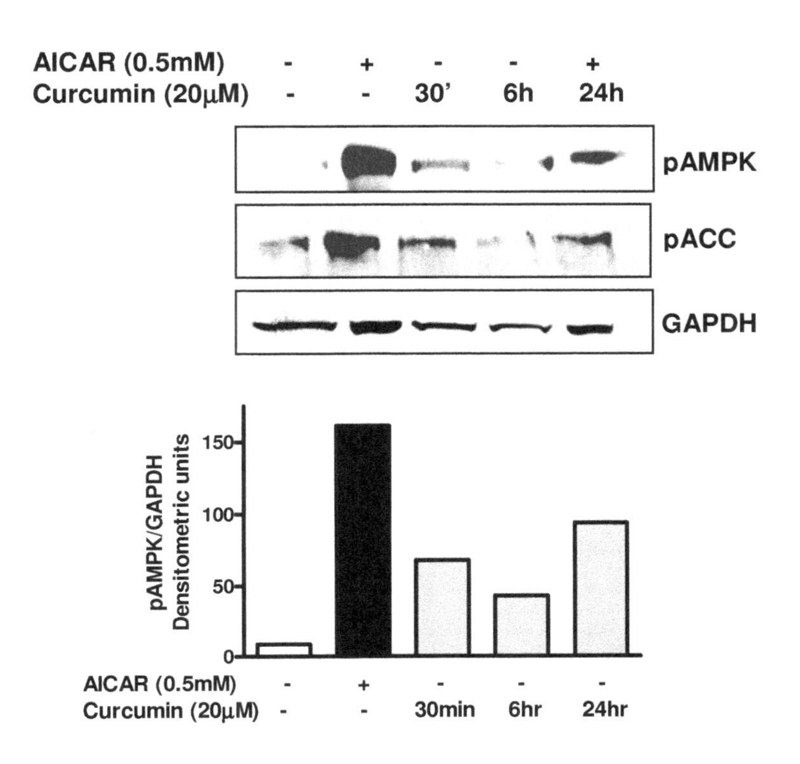

Fig. 4: *Effect of curcumin on the phosphorylation of 5'AMP-activated protein kinase (AMPK) and acetyl-coenzyme A carboxylase (ACC) in H4IIE rat hepatoma cell line. GAPDH is used as loading control. Densitometric analysis of phospho-AMPK levels compared to GAPDH, represented as a bar diagram, is shown in the bottom panel.*

Curcumin stimulates the phosphorylation of AMPK and ACC: AMP-activated protein kinase (AMPK) has been implicated as a key regulator of physiological energy dynamics, including glucose transport, gluconeogenesis, and lipolysis [33]. Further, it has been shown that polyphenols such as apigenin, resveratrol, and S17834 (a synthetic polyphenol) stimulate AMPK activity [34]. Therefore, it was of interest to examine the effect of curcumin, a polyphenol, in

AMPK activation. We treated H4IIE cells with curcumin for 30 min, 6 h and 24 h and assayed AMPK activation by Western blot using phospho-AMPK antibody. 5-aminoimidazole-4-carboxamide 1-beta-D-ribofuranoside (AICAR) was used as a positive control. AICAR treatment stimulated pAMPK/GAPDH ratio up to 20-fold, curcumin treatment (20 μM) for 24 hours also stimulated as much as 10-fold. Curcumin also stimulated the downstream target of AMPK, acetyl-coenzyme A carboxylase (ACC). Many compounds have been shown to activate 5'AMP-activated protein kinase (AMPK) including the anti-diabetic drug metformin [35,36]. Further, activation of AMPK has been shown to suppress the transcription of PEPCK and G6Pase in hepatoma cells [37] indicating a role of AMPK in gluconeogenesis. Our data demonstrating that curcumin activates AMPK is consistent with our findings that it also suppresses hepatic gluconeogenic gene expression. Recent data have provided evidence that inhibition of the gluconeogenic program by AMPK involves a transcriptional coactivator, transducer of regulated CREB activity 2 (TORC2) [38]. Additionally, curcumin's effects on gene expression of PGC-1a and FOXO-1 transcription factors, both implicated in hepatic gluconeogenesis, need to be further evaluated.

CONCLUSIONS

In conclusion, we have examined the molecular mechanisms mediating curcumin's role in insulin action and glucose metabolism. We demonstrate that curcumin may exert an effect on hepatic gluconeogenesis by modulating the expression of the key gluconeogenic gene G6Pase through the activation of AMPK. These studies offer a mechanistic basis for curcumin's hypoglycemic effects observed in animal models of diabetes, and suggest a role for curcumin as a functional food in the treatment of chronic conditions such as obesity and diabetes.

ACKNOWLEDGEMENTS

This work was supported by the Alabama Agricultural Initiative on Natural and Human Resources (awarded to S.T.M), the Auburn University Undergraduate Research Fellowship (awarded to J.D) and the Auburn University Cellular and Molecular Biosciences Undergraduate Summer Research Scholarship (awarded to J.D).

REFERENCES

1. Ogden, C.L., Carroll, M.D., Curtin, L.R., McDowell, M.A., Tabak, C.J. and Flegal, K.M. (2006) Prevalence of overweight and obesity in the United States, 1999-2004. Jama 295, 1549-55.
2. Smyth, S. and Heron, A. (2006) Diabetes and obesity: the twin epidemics. Nat Med 12, 75-80.
3. Yeh, G.Y., Eisenberg, D.M., Kaptchuk, T.J. and Phillips, R.S. (2003) Systematic review of herbs and dietary supplements for glycemic control in diabetes. Diabetes Care 26, 1277-94.
4. Mentreddy, S.R. (2007) Review - Medicinal plant species with potential antidiabetic properties. Journal of the Science of Food and Agriculture 87, 743-750.
5. Khopde, M.S., Priyadarsini, K.I., Venkatesan, P. and Rao, M.N. (1999) Free radical scavenging ability and antioxidant efficiency of curcumin and its substituted analogue. Biophys Chem 80, 85-91.

6. Jobin, C., Bradham, C.A., Russo, M.P., Juma, B., Narula, A.S., Brenner, D.A. and Sartor, R.B. (1999) Curcumin blocks cytokine-mediated NF-kappa B activation and proinflammatory gene expression by inhibiting inhibitory factor I-kappa B kinase activity. J Immunol 163, 3474-83.
7. Aggarwal, B.B., Kumar, A. and Bharti, A.C. (2003) Anticancer potential of curcumin: preclinical and clinical studies. Anticancer Res 23, 363-98.
8. Aggarwal, B.B., Kumar, A., Aggarwal, M.S. and Shishodia, S. (2005) Curcumin derived from turmeric (*Curcuma longa*): a spice for all seasons. In: Phytochemicals in Cancer Chemoprevention (Debasis Bagchi, P.D. and Preuss, H.G., eds.), pp. 349-387, CRC Press, New York.
9. Shishodia, S., Sethi, G. and Aggarwal, B.B. (2005) Curcumin: getting back to the roots. Ann N Y Acad Sci 1056, 206-17.
10. Cheng, A.L., Hsu, C.H., Lin, J.K., Hsu, M.M., Ho, Y.F., Shen, T.S., Ko, J.Y., Lin, J.T., Lin, B.R., Ming-Shiang, W., Yu, H.S., Jee, S.H., Chen, G.S., Chen, T.M., Chen, C.A., Lai, M.K., Pu, Y.S., Pan, M.H., Wang, Y.J., Tsai, C.C. and Hsieh, C.Y. (2001) Phase I clinical trial of curcumin, a chemopreventive agent, in patients with high-risk or pre-malignant lesions. Anticancer Res 21, 2895-900.
11. Srivivasan, A., Menon, V.P., Periaswamy, V. and Rajasekaran, K.N. (2003) Protection of pancreatic beta-cell by the potential antioxidant bis-o-hydroxycinnamoyl methane, analogue of natural curcuminoid in experimental diabetes. J Pharm Pharm Sci 6, 327-33.
12. Pari, L. and Murugan, P. (2005) Effect of tetrahydrocurcumin on blood glucose, plasma insulin and hepatic key enzymes in streptozotocin induced diabetic rats. J Basic Clin Physiol Pharmacol 16, 257-74.
13. Arun, N. and Nalini, N. (2002) Efficacy of turmeric on blood sugar and polyol pathway in diabetic albino rats. Plant Foods Hum Nutr 57, 41-52.
14. Kuroda, M., Mimaki, Y., Nishiyama, T., Mae, T., Kishida, H., Tsukagawa, M., Takahashi, K., Kawada, T., Nakagawa, K. and Kitahara, M. (2005) Hypoglycemic effects of turmeric (Curcuma longa L. rhizomes) on genetically diabetic KK-Ay mice. Biol Pharm Bull 28, 937-9.
15. Nishiyama, T., Mae, T., Kishida, H., Tsukagawa, M., Mimaki, Y., Kuroda, M., Sashida, Y., Takahashi, K., Kawada, T., Nakagawa, K. and Kitahara, M. (2005) Curcuminoids and sesquiterpenoids in turmeric (Curcuma longa L.) suppress an increase in blood glucose level in type 2 diabetic KK-Ay mice. J Agric Food Chem 53, 959-63.
16. Pari, L. and Murugan, P. (2007) Changes in glycoprotein components in streptozotocin--nicotinamide induced type 2 diabetes: influence of tetrahydrocurcumin from Curcuma longa. Plant Foods Hum Nutr 62, 25-9.
17. Kowluru, R.A. and Kanwar, M. (2007) Effects of curcumin on retinal oxidative stress and inflammation in diabetes. Nutr Metab (Lond) 4, 8.
18. Sharma, S., Chopra, K. and Kulkarni, S.K. (2007) Effect of insulin and its combination with resveratrol or curcumin in attenuation of diabetic neuropathic pain: participation of nitric oxide and TNF-alpha. Phytother Res 21, 278-83.
19. Ueyama, A., Yaworsky, K.L., Wang, Q., Ebina, Y. and Klip, A. (1999) GLUT-4myc ectopic expression in L6 myoblasts generates a GLUT-4-specific pool conferring insulin sensitivity. Am J Physiol 277, E572-8.
20. Freidenberg, G.R., Klein, H.H., Cordera, R. and Olefsky, J.M. (1985) Insulin receptor kinase activity in rat liver. J Biol Chem 260, 12444-12453.

21. Mathews, S.T., Chellam, N., Srinivas, P.R., Cintron, V.J., Leon, M.A., Goustin, A.S. and Grunberger, G. (2000) α2-HSG, a specific inhibitor of insulin receptor autophosphorylation, interacts with the insulin receptor. Mol Cell Endrocrinol 264, 87-98.
22. Zick, Y., Grunberger, G., Rees-Jones, R.W. and Comi, R.J. (1985) Use of tyrosine containing polymers to characterize the substrate specificity of insulin and other hormone-stimulated tyrosine kinase. Eur. J. Biochem 148, 177-182.
23. Pfaffl, M.W. (2001) A new mathematical model for relative quantification in real-time RT-PCR. Nucleic Acids Res 29, e45.
24. Watanabe, C.K. (1918) Studies in the metabolic changes induced by the administration of guanidine bases. V. The change of phosphate and calcium content in serum in guanidine tetany and the relation between the calcium content and sugar in the blood. Journal of Biological Chemistry 36, 531-546.
25. Marles, R.J. and Farnsworth, N.R. (1995) Antidiabetic plants and their active constituents. Phytomedicine 2, 137-189.
26. Taniguchi, C.M., Emanuelli, B. and Kahn, C.R. (2006) Critical nodes in signalling pathways: insights into insulin action. Nat Rev Mol Cell Biol 7, 85-96.
27. Yang, X., Thomas, D.P., Zhang, X., Culver, B.W., Alexander, B.M., Murdoch, W.J., Rao, M.N., Tulis, D.A., Ren, J. and Sreejayan, N. (2006) Curcumin inhibits platelet-derived growth factor-stimulated vascular smooth muscle cell function and injury-induced neointima formation. Arterioscler Thromb Vasc Biol 26, 85-90.
28. Ikonomov, O.C., Sbrissa, D., Mlak, K. and Shisheva, A. (2002) Requirement for PIKfyve enzymatic activity in acute and long-term insulin cellular effects. Endocrinology 143, 4742-54.
29. O'Brien, R.M. and Granner, D.K. (1996) Regulation of gene expression by insulin. Physiol Rev 76, 1109-61.
30. Martinez-Jimenez, C.P., Gomez-Kechon, M. J., Castell, J. V., and Jover, R. (2006) Underexpressed Coactivators PGC1 AND SRC1 Impair Hepatocyte Nuclear Factor 4 Function and Promote Dedifferentiation in Human Hepatoma Cells. J. Biol. Chem. 281, 29840-29849.
31. Wang, N.D., Finegold, M. J., Bradley, A., Ou, C. N., Abdelsayed, S. V., Wilde, M. D., Taylor, L. R., Wilson, D. R., and Darlington, G. J. (1995) Impaired energy homeostasis in C/EBP alpha knockout mice Science 269, 1108-1112.
32. Yeagley, D., Moll, J., Vinson, C. A., and Quinn, P. G. (2000) Characterization of Elements Mediating Regulation of Phosphoenolpyruvate Carboxykinase Gene Transcription by Protein Kinase A and Insulin. IDENTIFICATION OF A DISTINCT COMPLEX FORMED IN CELLS THAT MEDIATE INSULIN INHIBITION. J. Biol. Chem. 275, 17814-17820.
33. Long, Y.C. and Zierath, J.R. (2006) AMP-activated protein kinase signaling in metabolic regulation. J Clin Invest 116, 1776-83.
34. Zang, M., Xu, S., Maitland-Toolan, K.A., Zuccollo, A., Hou, X., Jiang, B., Wierzbicki, M., Verbeuren, T.J. and Cohen, R.A. (2006) Polyphenols stimulate AMP-activated protein kinase, lower lipids, and inhibit accelerated atherosclerosis in diabetic LDL receptor-deficient mice. Diabetes 55, 2180-91.
35. Fryer, L.G.D., Parbu-Patel, A., and Carling, D. (2002) The Anti-diabetic Drugs Rosiglitazone and Metformin Stimulate AMP-activated Protein Kinase through Distinct Signaling Pathways. J. Biol. Chem. 277, 25226-25232.
36. Dasgupta, B., and Milbrandt, J. (2007) Resveratrol stimulates AMP kinase activity in neurons. Proc. Natl. Acad. Sci. USA 104, 7217-7222.

37. Lochhead, P.A., Salt, I.P., Walker, K.S., Hardie, D.G. and Sutherland, C. (2000) 5-aminoimidazole-4-carboxamide riboside mimics the effects of insulin on the expression of the 2 key gluconeogenic genes PEPCK and glucose-6-phosphatase. Diabetes 49, 896-903.
38. Koo, S.H., Flechner, L., Qi, L., Zhang, X., Screaton, R.A., Jeffries, S., Hedrick, S., Xu, W., Boussouar, F., Brindle, P., Takemori, H. and Montminy, M. (2005) The CREB coactivator TORC2 is a key regulator of fasting glucose metabolism. Nature 437, 1109-11.

FATIGUE HARDENING IS THE KEY MECHANISM OF AGING AND DISEASES

[1]Valery P. Kisel and [2]Danik M. Martirosyan

[1]Institute of Solid State Physics, Department of Crystal Growth, 142 432 Chernogolovka, Moscow District, Russia
[2]Functional Foods Center, Dallas, TX, USA

I don't want to achieve immortality through my work. I want to achieve it through not dying. Woody Allen

Corresponding Author: Valery P. Kisel, E-mail: kisel@issp.ac.ru

Keywords: fatigue hardening, aging, cancer, cardiovascular diseases and hypertension

Running Title: Physical-chemical Softening of Biological Tissues is the only Way for Long Life Span without Diseases

ABSTRACT

This work evidences for physical base of arterial blood pressure (ABP) and its prognostic sense for aging and diseases. The new paradigm about the universal nature of phase-interface stresses and microscopic mechanisms of plastic deformation (MPD) in the origin and development of phase transitions in biological tissues (BT) provides a consideration of aging and diseases (local aging) as a result of mechanical fatigue durability and stress hardening of molecular and cell structures under body metabolism and external geomagnetic storms. This means that any practical guide to general medical treatment of aging and chronic diseases with drugs and herbs must be concerned with physical-chemical stress softening of BT.

INTRODUCTION

The aging of organisms is characterized by a gradual functional decline of all organ systems and progressive cellular damage. The idea that aging and illness are manifestations of the same biological processes, and that they can be understood only by working across various disciplines is very outdated. Mutations in a variety of many genes can extend the lifespan of *Caenorhabditis elegans* up to six fold (S. Alvarez, Gordon J. Lithgow). Knocking out certain genes endowed organisms such as yeast, worms and flies with unusually long life spans[1]. Oxidative stress increases the rate of telomere attrition, but the telomere dysfunction may be triggered by puzzling effects other than overt telomere shortening[2]. But oddly enough, scientists started to find that mutating these genes (atomic scale length) to slow aging could also stave off endogenous ailments such as cancer and diabetes[3], which do effect many more old people than young but isn't usually seen in the aging process itself. So the idea that growing old and growing ill are two sides of the same coin remains controversial at the first glance. Backers of this concept made headlines with findings that a chemical in red wine called resveratrol extends one's life span and might also prevent diabetes–like symptoms in mice[4]. Of course, aging is the deterioration of the body cells and a major cause of many diseases, but not the only one, aging has some effects that aren't considered diseased states. Therefore it is necessary to find the common parameters that characterize aging and diseases, and the pharmaceutical drugs and herbs that propose to extend the lifespan will also be an interesting class of drugs to treat diseases as well.

LITERATURE REVIEW

Aging and ailments

It is common knowledge that sometimes even healthy eaters (with a normal cholesterol level in blood) get heart diseases. They suffer from heart attacks even if they take precautionary measures such as avoiding high-cholesterol foods. Part of the solution to this age-old puzzle has recently been supplied by the data of work[5] in which the authors studied genetically engineered mice who have especially leaky mitochondria in their blood vessel cells. They found that although normal mice very rarely acquire arteriosclerosis, the mutant mice all developed the clogged arteries, even when fed on a low-fat, low-cholesterol diet[5].

It is commonly known that as the cells in blood-vessel walls grow old, their energy-generating tiny pockets of energy, shouldering the important task of converting food into usable energy forms – mitochondria – become less efficient and begin to leak extremely reactive forms of oxygen - reactive oxygen species (ROS) or reactive molecules - into the rest of the cell. The same behavior is found in other cells of the aging organism. The oxygen escapes after molecules called uncoupling proteins, found in the mitochondrion's membrane (due to cell metabolism), start to let charge flow across the membrane, triggering the plastic deformation and damaging (aging) of the blood vessel's walls[6]. Then the body mounts an immune response to repair this damage with local inflammations and a form of cholesterol sticks to arterial walls that ultimately clogs up arteries with plaques resulting in the increase of the risk of a heart attack. This means that cutting cholesterol from one's diet does not always prevent cardiovascular troubles, therefore suggesting that there is something more going on[5]. The authors of the work[7] reinforce the earlier suggestions that many diseases and disabilities of an older age have their roots in previous exposures to various infectious agents (deformation of BT) and other sources of inflammation in early human life.

Oxidation of the proteins by the action of ROS resulting from the mitochondria respiration causes collateral damage to the body's own tissues which might shorten the life span. The work[8] shows that the critical concentration (C_{crit}) of faults (mutations) in body cells for living doesn't depend on differentiated state, type or age of the cells. The rate of ROS production in cells, first of all depends on the intensity of respiration (cell metabolism) and correlates with the rate of their regeneration, that is to say that it correlates with the concentration of sites of protein synthesis – the ribosome's and their effective work. It is important to stress that during aging, at the age of higher than an estimated half of the life span, the concentration level (C) of faults in proteins begins to exceed the C_{crit} practically identically in various organs and tissues of men and rats, as well as in flies – in the species with different metabolic rates and life spans[8]. This means that the ROS concentration is the result of the important process and its puzzling mechanisms. The data of work[9] demonstrates that under γ-radiation dose, a variation of reactive species concentration in polymer depends on the time and temperature in the same manner as the production of point defects during ordinary deformation of solids and liquids[6].

It is well known that internal and external constraints govern the various processes of the growth and survival of living organisms. The mechanical properties of the cell membrane are fundamental in this respect: on the one hand, resistance to deformation allows the cell to control its shape and to resist high shear stresses, while, on the other hand, membrane tension (due to mechanical hardening) governs ion exchanges with the extra cellular environment. The dynamic inner cytoskeleton of cells is a three-dimensional network composed of three kinds of protein filaments. This cytoskeleton serves as an active mechanical support system for the cell body and contributes to cell shape maintenance. It controls important conditions for cell differentiation: chromosome migration during cell division, offers a guide network for intracellular trafficking, and contributes to the formation of membrane extrusions necessary for cell movement. Cellular functions and physical properties (membrane rigidity and tension, cytoskeleton elasticity, cytoplasmic viscoelasticity, etc.) thus all appear to be intricately related (cell scale length).

Recent experiments have shown that tiny living organism bacteria do appear to grow old: the bacteria *Escherichia coli* inheriting the older end reproduced 1% more slowly than their counterparts with each cell division[10]. The reproductive cloning efficiency for young (differentiated embryonic) stem cells is five to ten times higher than that for adult stem cells[11].

Checkpoint proteins (molecular scale length) control the survival of the post mitotic cells in nematode *Caenorhabditis elegans.* They promote the arrest of cell division, alter cellular stress resistance in response to DNA damage or stalled replication forks[12].

The nematode worm, *C. elegans,* has been used usually as a model of aging for decades, helping scientists to uncover details about what happens to human bodies as they grow old. It is worth stressing that when both genotype and environment are held constant, nevertheless 'chance' variation in the lifespan of individuals in a population is still quite large. Using isogenic population of the nematode the authors of work[13] showed that, on the first day of adult life, chance variation in the level of induction of a green fluorescent protein (GFP) reporter coupled to a promoter from the gene *hsp-16.2* predicts as much as a fourfold variation in subsequent survival. But the level of GFP induction is not heritable, and the expression of GFP levels in various reporter constructs is not associated with differences in longevity. It is worth noting that the gene *hsp-16.2* is probably not responsible for the observed differences in survival but instead probably reflects a certain, hidden and heterogeneous, but now quantifiable and puzzling physiological state that dictates the ability of an organism to deal with the rigors of living. There is no mere chance that the same reporter is also intimately coupled with the ability to withstand a subsequent lethal thermal stress[13.] First, these statements point to the fact that it is the initial molecular family structure (molecular scale length of intermolecular mechanical deformations and stresses) around the gene hsp-16.2 (the so-called hidden and heterogeneous but physiological state) that strictly coupled with the variation of longevity of the first generation of population of the isogenic nematode. Second, it is this deformable cell structures that are very sensitive to external physiological stress agitations – deformation hardening or softening of biological tissues (BT) in the frames of our microscopic plastic deformation (MPD) model[6]. Third, their differences in longevity for the next generations are determined with the variations of their individual genetic structures (deformability of new DNA, RNA structures, etc.) and new variable environmental states (deformation stress agitations in terms of the MPD model). The last ones correspond to the standard mechanical fatigue hardening or softening of the molecular and cell structures with their standard durability (mechanical rupture time or the life span for BT) up to their fracture (apoptosis of cells)[6].

Dozens of years of study have provided physicists a deep understanding of the same micro-mechanisms of plasticity on various scale lengths, and their key role in phase transitions for all structures of the world around us[6,14,15].

Numerous literature data irrefutably provides evidence for the deformation origin of physiological, physical and chemical stress effects on BT, cell growth and proliferation, differentiation, diseases and aging, which is irresistibly identical to the stiffening or softening deformation of solids with the appropriate production of lattice defects with various dimensions (ROS in mitochondria of BT stimulate the additional oxidative stress in cells, etc.)[1].

It is well known that mechanical stiffening of BT in living organisms (due to mechanical (trauma) and physiological stress, current flow, irradiation, physical and chemical effects, virus and bacteria invasions, etc.[6]) produces the cytotoxic ROS (defects of BT) accompanied with oxidative stress and inflammation, reduces their rate of growth and induces fast aging. The short-term inhibition of free radical production provokes the short-term acceleration of their growth (softening of BT) that may be seen in the process of tissue regeneration[6]. According to the free radical theory, aging is due to the accumulation of ROS, which inflict tissue and DNA damage including mitochondrial DNA (mtDNA) mutations which have been shown to accumulate with age in several species.

First of all, It has been confirmed by the conclusions of work[15] and means that any type of deformation (including particle and photon irradiation) results in the conventional deformation of material with its standard defects[15].

These results are in line with the data of work[12] where the authors engineered knock-in mice expressing a mutant of the mtDNA polymerase POLG, which is imparative for DNA proofreading. These mice exhibit a premature aging (progeria) phenotype beginning at 9 months, which includes greying, kyphosis, age-related hearing loss and reduction in skeletal muscle because they had accumulated mtDNA mutations at a frequency of three to eight times that of wild-type littermates. To test whether ROS accumulation contributes to the aging phenotype, Kujoth and colleagues did not observe any difference between POLG-defective and wild-type mice. Levels of lipid peroxidation or RNA and DNA oxidative damage were also similar between mutant and wild-type mice, suggesting that oxidative stress does not contribute with the premature aging phenotype seen in POLG-defective mice. Again these findings show that mtDNA mutations (faults) result in increased apoptosis in aging tissues and that aging precedes ROS accumulation and occurs independently of oxidative stress damage. This result and ROS model of aging are in line with the same strict sequence of deformation and defect accumulation in our mechanical-fatigue model of BT aging and the origin and progress of endogenous diseases[6.]

As a cell transitions from an undifferentiated state to a differentiated cell type, its gene expression profile changes, which in part reflects chemical and physical changes in chromatin structure. In a complementary approach, the authors of the work[16] have examined the macroscopic deformation properties of the nucleus during differentiation. Aspiration with a micropipette revealed that the nuclei of pluripotent human embryonic stem cells could be deformed relatively easily; however, as the cells differentiated, the nuclei became stiffer. Hematopoietic stem cells (from bone marrow) were able to differentiate into fewer cell types than embryonic stem cells and, similarly, presented an intermediate level of deformability.

Progression toward the differentiated state was accompanied by an increase in the filamentous protein lamin A/C and greater condensation of chromatin. When lamin A/C was knocked down in epithelial cells, their flow behavior resembled that of hematopoietic stem cells. Further analysis showed that the fluid character of the nucleus is determined primarily by chromatin but that the degree of nuclear deformability is set by the lamina. Variations in *the physical plasticity of the nucleus* may be important for allowing less or more (in metastasis) differentiated cells to move through blood and tissues[16].

Cancer

Extensive investigations of the ability of cells to grow *in vitro* demonstrate that this process is governed by the strong effect of cells and their phases on other ones due to mechanical stress mismatch at their interfaces first of all[6]. The controlling role of cell interactions (due to stress between old and new-born cells) in their growth is confirmed with the next observation: robust cells may become cancerous only after neighboring tissues, rather than the cells themselves, have been exposed to carcinogens. The researchers studied cultural tissues of the rat mammary gland, a model of breast cancer in which malignancies are thought to arise in the epithelium rather than the adjacent stroma. They transplanted epithelial cells that had been exposed to either saline or chemical mutagen into the stromal fat tissue that had also been exposed to either saline or a chemical mutagen. Maffini et al. found that healthy epithelial cells transplanted into the chemically treated fat tissue became cancerous, regardless of whether the cells had been mutated themselves. Conversely, chemically treated epithelial cells grew into apparently normal mammary gland ducts when injected into healthy stromal tissue[17].

The experiments[18] showed *in vitro* that the old (hard) human fibroblasts stimulated proliferation of pre-malignant and cancer epithelial cells, which are able to make tumors at their entwinement to bare mice. This increases the mismatch stresses at the interfaces of hard cells/pre-cancer cells thus intensifying the well-known high proliferation rate of the cancer cells. At young (soft) fibroblast grafting this ability is less expressed thus directly showing evidence for the key role of interface mechanical stresses on malignant cell growth[6].

Consequently tumors are generally stiffer than the surrounding healthy tissue, and this characteristic has been exploited in certain diagnostic procedures such as breast self-examination, etc.[19]. Tumor rigidity reflects not only intrinsic properties of the tumor cells but also an increased stiffness of the extracellular matrix (ECM). Recent work[20] showed that ECM hardening played an active role in tumor cell growth as well.

Again these results suggest that factors causing a sustained increase in matrix stiffness, for example a chronic inflammatory response due to ROS, may promote malignant transformation[21].

Evolution of species

Living organisms exhibit tremendous diversity in a large repertoire of forms and considerable size range. Scientists have discovered the conserved mechanisms control the development of all organisms. More recently, much has been learned about the control of tissue growth, tissue shape and their coordination at the cellular and tissue levels. *Drosophila* has proved to be a particularly powerful model signaling pathways that organize tissue patterns. New models integrate how specific signals and mechanical forces shape tissues and may also control their size[22].

The contemporary evolutionary theory of aging stipulates that aging is an inevitable consequence of low effectiveness of natural selection acting on traits expressed late in the life span of the organisms.

Uncovered softening/hardening effect of magnetic field on phase transitions (endogenous diseases[15], aging[23] (macroscopic scale length), the so-called magnetoplastic effect in plasticity[15], etc), a persistent universal trend of amino acid gain and loss in protein evolution (molecular scale length)[24] in the frames of gradual mechanical hardening at numerous successive phase transitions directly prove the main principles of evolution of species, aging and the universality of origin and evolution of inorganic and organic forms of matter in the Universe.

In terms of MPD mechanism of phase transitions[6] this confirms the key role of hardening mismatch at cell interfaces as a driving force in the cell growth and transformations as it is during various phase transitions. The softening of BT with physical, biochemical, physiological, etc. methods changes the mechanical fatigue limit of the materials, durability or lifespan of BT, and the rate of their hardening (aging).

Recent work shows that antidepressant drugs boost the lifespan by mimicking starvation[25].

It is interesting to note that the different metabolic rates and therefore the rate of body aging can be qualitatively estimated from the type of its immune response to deal with infections. Cutting back on calories is a sure way to extend the lifespan of any organism, from yeast, flies to mice and apes due to the lowering of body's metabolic rate. A new study on rodents reveals that closely related species deal with infection in many different ways: the fast-living mice are in such a hurry to get on with things that they prefer a fever over a full-fledged immune response to deal with infections and also show little or no illness behavior. But the slower-living mice had little or no fever at all. A third slow-living species, the California mouse, (*P. californicus*), actually lowered its body temperature. They adopt "sickness behavior" – that is, decreasing food intake, becoming lethargic, and a reduction of activities such as sex and aggression[26]. The explanation for this difference is simple: high metabolic rate in fast-living organisms rapidly develops a high fever more easily than in slower-living species which would prefer to rely on slow growing concentration of antibodies.

Dual role of deformation mechanisms in the origin and progress of endogenous diseases and their medical treatment.

It is important to note that the experimental forms of standard diseases in laboratory animals (diabetes, hypertension, heart diseases, cancer, etc.) are usually concerned with chemical or mechanical (trauma) influences on BT thus confirming the deformation nature of these diseases and their medicinal treatment.

It is worth stressing that the right vibrations transform salamander larva into a killing machine. The "predator" metamorphosis, with its larger head and aggressive attitude, is better adapted to grabbing larger prey. Visual, chemical or sound signals can trigger striking morphological changes in a range of aquatic animals and amphibians. For example, in the presence of the salamander *Hynobius retardatus,* tadpoles of the frog *Rana prica* transform into fatter forms that are too big for the predator to grab. Recently the authors of the work[27] have found a counter-adaptation in the predator that is triggered by the vibrations of the tadpoles' tails. The shape-shifting salamander larvae develop into either "standard" or "predator" morphs, with longer, more substantial bodies, and heads about a third broader. It was the first time vibrations have been shown to cause a morphological transformation. When this team reared the salamanders with tadpoles that had had their tales removed, they were half as likely to change into the predator morph. What's more, they found that artificial fins flapping like tadpole tails were about as effective as real tadpoles at inducing the larger morph, confirming that mechanical rather than chemical cues are the main trigger[27].

EXPERIMENTAL METHODS AND RESULTS

Physical base of oscillating arterial blood pressure (ABP) and its prognostic sense for aging and diseases

We found that the arterial blood pressure may be the simple indicator of vascular stiffening due to aging, cardio-vascular and other diseases. It is simple to show that the threshold stresses for vascular microdeformation are ($2ABP_{diastolic} - ABP_{systolic}$), where ABP is the arterial blood pressure, and the $ABP_{diastolic}$ is the threshold stress for vascular macrodeformation. It is very important to emphasize that in the whole range of ($2ABP_{diastolic} - ABP_{systolic}$), $ABP_{systolic}$ variations from the early childhood up to the old age these biological tissue parameters show the strict correlation ($2ABP_{diastolic} - ABP_{systolic}) \approx constant \bullet ABP_{systolic}$ in the same manner like it was observed earlier for all materials from solid helium up to the amorphous materials and diamond under various deformation conditions[6]. This agreement of the mechanisms of plastic deformation (MPD) in living BT under arterial blood pressure oscillations, metabolic transformations and physiological stresses, and in various states of different materials under load shows that MPD are strictly the same on atomic-to-cosmic scale lengths. This is the latest irrefutable argument in place in order to prove the new paradigm of the decisive role of MPD and phase mismatch-interface stresses between growing and differentiating cells, in each stage of their phase transitions from biochemical reactions up to kinetics of aging and growth of robust and cancerous cells, adaptation, origin of species and populations, etc. And this is in sync with all epidemiological, clinical and experimental investigations (tissue softening as a measure of a health)[6,19]. This also confirms the same mechanical properties of all materials (solids, glasses, polymers, biological tissues, liquids, plasma and gases[6]) and the same their micromechanisms of fatigue durability (the aging of biological tissues).

The oscillating character of $ABP_{systolic}$ and $ABP_{diastolic}$ values changes with time and reflects the dynamics of serrated hardening–softening of BT tissues during body metabolism under environmental conditions and may be the direct physical way to control the aging processes in BT under stress. Our numerous investigations show that the amplitudes and frequencies of ABP oscillations grow under stress, aging, different diseases, and geomagnetic storms in the same manner as the serrated plastic deformation in solids, liquids, glasses and ceramics, polymers and gases under mechanical hardening[6].

Author's experience in strictly dosed stress and herb therapy of aging, hypertony and endogenous diseases.

Recent experimental investigations show that gradual adaptation to moderate hypoxia results in activated synthesis of nucleic acids and proteins not only in the heart, lungs and blood system as well as other organs of oxygen transport and distribution but also in systems of neuroendocrinal regulation and immunity[6]. As we know from the above parts of this work and the work[6], the common deformation origin of resultant "systemic structural trace" (or GTA) forms on various scale lengths of phase transitions in the living organism is a material base of adaptation of BT to hypoxia. It increases the resistance of BT not only to ischemia and stress-induced damages of the heart and other organs but also to a much wider spectrum of factors, namely, to spontaneous hereditary hypertension, hallucinogens, audiogenic epilepsy, stress-induced atherogenic dyslipidemia, allergic and autoimmune diseases, tumor growth, etc.[6]. In the case of strictly calculated individual course of activation therapy[6] or interval hypoxy training[6], the BT is markedly softened and negative exacerbations of endogenous diseases are temporary weakened and grow to become rare. For example:

Patient P., age 62, suffered from chronic glomerulonephritis for 39 years with nephrotic syndrome (treated by a 35-year, long-term phytotherapy with a *Solidago virgaurea L.*, diet and homoeopathy[28]), breakdown, and frequent jumps of arterial blood pressure (ABP) up to 170-190/100-115 with sharp head-aches, myalgia, periodontitis, rare toenail, hemorrhoid and gingivitis hemorrhages, embarrassing temporary night urination, sleep disorders and easy an irritable state and depression during days of solar wind-magnetic storms (temporary aging due to hardening of BT with increasing concentration of ROS[6,14,15]). The standard hypertensive therapy didn't give a significant effect. The patient took weak intermittent stress stimulation – in small doses of any unknown for the organism physical, physiological, chemical or biological active factors (the normobaric interval hypoxic training, steeled in cold or hot water, dosed training and hydrogen peroxide H_2O_2 intake, activation therapy with 5-month specialized drop course of adaptive herb infusions (ginseng (*Panax*), *Paeonia anomala L.,* etc.)[6,14]. For example, there are the first three ranges of the number of *P. a. L.* drops of herb infusion on the basis of exponential dependence between the intensity of stimuli and the individual type of adaptation reactions[6,14] (50 g of dry herb + 50 g of dry roots and rhizome per 1 L of 40% alcohol) per ~30 mL H2O every morning just after the time for getting up and 20 min before food: 25,23,21,1 9,17, 12 (+ 23 drops after 3 hours), 11(+ 23 – 4h later), 10(+23 – 5 h later),9(+23 – 5h later);

26,24,22,20,18,16,11 (+23 –3h),10(+23 – 4h), 9(+23 – 4h), 6(+23 – 3h after and +18 drops in the evening), 5(+23 – 4h and +18 drops in the evening);

25,23,16 (+23 – 2.5h), 14(+23 – 3h 15 min later), 13 (+23 – 4h 30 min), 12 (+23 – 5h 30min), 8 (+23 – 4h after and +18 in the evening), 7(+23 –4h later +18 ev.), 6(+23 - 4h +18ev.), 6 (+23 – 5h +18 ev.), 5(+23 – 5h + 18ev.), and so on.

The above numerous exacerbations of aging were slowing down, the arterial pressure jumps decreased down to 165/100 during extremely hard magnetic stormy days only and a scarce amount of weakened head-aches remained; the strengths were increased up to the normal state of health. In calm days ABP was reduced to 95/65-145/95. The periods between the various exacerbations increased from ~6 to ~12/22/33 days and even more without any drugs. The rejuvenation of skinspots pigmentation (noticeably weakened) is in accordance with the rejuvenation of different organs under activation therapy: it increases the level of the hemoglobin in blood and immune strength, darkening of head hairs, 0.8°C lowering the body temperature in the morning, destroys the dandruff, etc. The patient lost 5% weight. The continuous four times per 10-days courses of *Huatuo Pills* (10 herbs, only one pill per day, Tsisin Pharmaceutical Comp., Guanjou, P. R. of China) with 1-day rests in between and then two-month periods free of pills prevent the ordinal head-aches and various hemorrhages. Now (2007) the patient uses these drop and herbs courses with additional hypertensive therapy (*Normodipine* (*amlodipine*), Gedeon Richter Joint-Stock Comp., Budapest, Hungary; 5 mg per day) in a continuous regimen.

1. Patient C., age 78, suffered from numerous ABP jumps per day even up to 210/140 occurred with head and heart aches, serious hemorrhoid with hemorrhage, severe arthritis in the knees, shortened sleep during days of solar wind-magnetic storms. The standard hypertensive therapy gives the permanent effect for ABP drop only down to the 120/80. The course of activation therapy was with 3% hydrogen peroxide H_2O_2 drops in 30-50 g (2 or 3 table spoons) of water: started with 1 drop three times a day 30 min before food intake or 1.5-2 hours after, then 2,3,4,5,6,7,8,9,10 drops 3 times a day, then stopped for 2-3 days and then again took the constant value of 10 drops 3 times a day during every 2-3 days with 1-day rest in between or in a continuous regime. Only after 7 months this H_2O_2 drop course gave a surprising extremely significant effect: ABP reduced down to 120/60-70 with the rare jumps down to 100/50, 90/45 before the heavy magnetic storms without other special drugs. The above exacerbations of aging disappeared.

The herbs for the additional medical treatment of aging and its concomitant diseases are:

a) For the slowing down of AP: peppermint, sweet clover, knotweed, dandelion, plantain, Leonorus, hawthorn, valerian, mountain ash, clover flowers, cudweed, sweet clover, Astragal, immortelle flowers, yarrow, birch leaf, etc.

b) For the treatment of atherosclerosis and vascular spasms : herb and roots of dandelion, sweet clover, St. John's wort, plantain, clover flowers, etc.

DISCUSSION

Thus literature and our data irrefutably proves the deformation origin of physiological, physical and chemical stress effects on biological tissues (BT), cells growth and proliferation, differentiation, aging and evolution of species, endogenous diseases, species, populations with various deformation modes of development from atomic to global scale lengths. The last statement means the phase transitions: DNA and RNA reorganization and replication, cell growth and proliferation, metastasis, reproduction of BT and body of the species, populations, etc. up to their fracture - apoptosis of cells and death, disappearance of populations, etc. These transformations are identical to the stiffening or softening deformation of solids, etc. with the appropriate production of lattice defects with various dimensions. The analogous defects in BT (reactive oxygen species, ROS, in mitochondria of BT) stimulates the additional oxidative stress in cells and their stiffening as a result of mechanical fatigue durability and stress hardening of molecular and cell structures under body metabolism, infections and external geomagnetic storms. Our comparison of the mechanisms of plastic deformation (MPD) in living BT under arterial blood pressure oscillations (this work), metabolic transformations and physiological stresses, various states of different materials under load[6] show that MPD are strictly the same on atomic-to-cosmic scale lengths. The BT softening processes under external low dosed oscillating stresses (weak variations of geomagnetic and electromagnetic fields, temperature, oxygen content in air, physical and intellectual exercises, activation therapy, etc.) are the natural sanitation of the sick organisms. They are absolutely necessary for the operation of living organisms and expand their growth and life span[6].

In the course of this study and the previous works[6,14,15,28] it is shown that up to now the puzzling common factor that determines the aging and sickliness of biological cells in numerous experiments is their mechanical stiffening. This work gives direct evidence for the physical base of oscillating arterial blood pressure (ABP): it is the APB levels that directly concerned with serrated mechanical properties of cardio-vascular system: the steady level of $ABP_{diastolic}$ points to their state of aging first of all. In the frames of this new paradigm, the aging and diseases are the gradual products of irregular local mechanical fatigue stiffening of BT under oscillating body metabolism and external stress influences where the mechanical durability of living cells in their life span. The versatility of micromechanisms of plasticity at phase transitions in inorganic and organic matter[6,14,15] is also confirmed by the versatility of fundamental mathematical laws and their variations, which describe the evolution of phase transformations in various forms of matter, for example, heat exchange, diffusion, zone melting, electro-diffusion and thermal diffusion, radioactive decay, extraction, chemical kinetics, human embryonic growth, change of body mass and tumor size in time, kinetics of digesting food, relationships between the basal metabolic rate and body mass, the genome length (number of base pairs) as a function of cell mass for a variety of unicellular organisms, the heart rate as a function of body mass for a variety of mammals, the number of trees of a given size as a trunk diameter, a dimensionless mass variable (mammals, birds, fish, and crustacea) against a dimensionless time variable, relative temperature and energy density – the time after the Big Bang (cosmic evolution) and so on[29].

CONCLUSION

It is worth stressing that this new approach shows the way to overcome aging and the body's chronic diseases to a considerable extent: we can really slow down these negative signs and cure the sick organism by means of oscillating BT hardening-softening with the help of weak stress stimuli (fractional activation therapy[6,14] and herb infusions) in accordance with the same well-developed methods of deformation softening of materials.

The programmed weak stimulation therapy again confirms the common nature of endogenous diseases, physical, chemical and biological processes, and is a good assistant for the traditional methods of healing and drugs for raising the level of life. We believe that this new approach may be a powerful additional resource for all scientific researchers and practitioners in the treatment of well-known incurable diseases.

REFERENCES

1. Kenyon C. Pinkston J.M., Garigan D., Hansen M. Cell. 2005, v. 120, pp 449-460.
2 Herbig U., Ferreira M., Condel L., Carey D., Sedivy J.M. Science. 2006, v.311, p. 1257.
3. Pinkston J.M., Garigan D., Hansen M., Kenyon C. Science. 2006, v. 313, pp 971-975.
4. Baur J., Sinclair D. Nature (London). 2006, v. 444, pp 337-342.
5. Bernal-Mizrachi C. Nature (London). 2005, v. 435, pp 502-506.
6. Kisel V. P., Kisel N.S. In: Proc. 2nd Int. Conf. "Functional foods for chronic diseases", Ed. by D.M. Martirosyan, Richardson, TX, USA, 2006, pp 213 – 234.
7. Finch C.E., Crimmins E.M. Science. 2004, v. 306 (17 Sept.), pp 1736-1739.
8. Ryazanov A.G. Molekuljarnaja biologia. 2001, v. 35, pp 727-730 (in Russian).
9. Pavlenko V.I.,Epifanovskii I.S. Perspektiv. materialy. 2003, No 5, pp 29-33 (in Russian)
10. Stewart E.J. PloS Biol. 2005, e45 doi: 10.1371/journal.pbio.0030045 (2005).
11. Sung L.-Y., Gao S., Shen H., Yu H., Song Y., Smith S.L. et al. Nature Genetics. 2006,

v. 38, pp 1323-1328.
12. Olsen A., Vantipalli M.C., Lithgow G.J. Science. 2006, v. 312, pp 1381-1385
13. Rea S.L., Wu D., Cypser J.R., Vaupel J.W., Johnson T.E. Nature Genetics. 2005, v. 37, pp 894-898.
14. Kisel V.P. and Kisel N.S. In: "Functional foods for Cardiovascular Diseases". Ed. by D.M. Martirosyan, Richardson, TX, USA, 2005, pp 235-239.
15. Kisel V.P.In:"Untraditional natural resources, innovation technologies and products". Coll.scientificworks.No 10.Moscow,RANS. 2003.pp 183-196. (in Russian)
16. Pajerowski N. Proc. Natl. Acad. Sci. U.S.A. 2007, v.104, p. 15619-15626.
17. Maffini M.V. J. Cell Sci. doi:10.1242/jcs.01000(2004); Nature, 2004,v.428, p.383.
18. Krtolica A., Campisi J. Uspekhi Gerontologii. 2003, v. 11, pp 109-116 (in Russian).
19. Physics News Update No 848, November 27, 2007: http://www.aip.org/pnu)
20. Paszek N. Cancer Cell. 2005; Science, 2005, v. 309, p. 1967.
21. Kujoth T. Science. 2005, v. 309, pp 481-484.
22. Lecuit T., Le Goff L. Nature. 2007, v. 450, pp 189-192.
23. Izmailov D.M., Obukhova L.M., Konradov A.A. Khimicheskaja. Fizika, 1995, v.14, pp 95-101 (in Russian).
24. Zuckerkandl E., Derancourt J., Vogel H. J. Mol. Biol. 1971, 59, p. 473-490.
25. Jordan I.K., Kondrashov F.A., Adzhubei I.A.,Wolf Y.I., Koonin E.V., Kondrashov A.S., Sunyaev S.R. Nature, 2005, v. 433, pp 633-638.
26. Bazykin G.A., Kondrashov F.A., Ogurtsov A.Y., Sunyaev S.R., Kondrashov A.S. Nature. 2004, v. 429, pp 558-562.
27. Petrascheck M. Nature (London). 2007, v. 450, pp 553-557.
28. http://sciencenow.sciencemag.org/cgi/content/full/2007/1030/3?etoc
29. New Scientist. 2005, v. 185, No 2489, p. 21.
Michimae H. Biology Letters, DOI: 10.1098/rsbi.2004.0242171.
30. Kisel V.P. Materials of the 1st Int. Conf.of Phytotherapeutists. Moscow. Overlei, 2006, pp 98-104.
31. Nikolaev D.A., Kholpanov L.P. The Law of Evolution. Moscow, "Nauka", 1999, 64 p.(in Russian).
32. West G.D., Brown J.H. Physics Today, 2004, v. 57, pp 36-42.

PART TWO

BOTANICALS AND FOOD SUPPLEMENTS AS A SOURCE OF FUNCTIONAL FOODS

POTENTIAL CHEMO-PREVENTATIVE COMPOUNDS ISOLATED FROM *IPOMOEA BATATAS* L. LEAVES

Shahidul Islam

Department of Agriculture, University of Arkansas at Pine Bluff, AR 71601, USA

Corresponding author: Shahidul Islam, E-mail: islams@uapb.edu

Keywords: sweetpotato leaves, polyphenolic components, oxidation

ABSTRACT

Sweetpotato leaves contain at least fifteen biologically active anthocyanins that have significant medicinal value for certain human diseases and may also be used as natural food colorants. The anthocyanins were acylated cyanidin and peonidin type. But the content of cyanidin in leaves is much higher that that of peonidin, suggesting that the anthocyanin composition of sweetpotato leaf is cyanidin type. The cyanidin type anthocyanins are superior to the peonidin type in antimutagenicity and antioxidative activity. Structure-activity studies show that the number of sugar units and hydroxyl groups on aglycons is associated with biological activities of anthocyanins. The activities appear to increase with a decreasing number of sugar units, and with an increasing number of hydroxyl groups on aglycons. The six different polyphenolic compounds were identified and quantified by NMR, FAB-MS spectra and RP-HPLC analysis procedures. The result suggests that the main phenolic compound in Ipomoea batatas leaves is 3, 5-di-O-caffeoylquinic acid followed by 4, 5-di-O-caffeoylquinic acid. The results revealed that total polyphenol was positively correlated with the different caffeoylquinic acid derivatives of sweetpotato leaves. The result also suggests that all the different bioactive caffeoylquinic acid derivatives and total polyphenol were positively correlated with the RSA of sweetpotato leaves. These caffeoylquinic acid derivatives (CQAD) showed very high 1, 1-diphenyl-2-picrylhydrazyl (DPPH) free radical scavenging activity (RSA), and effectively inhibited the reverse mutation induced by Trp-1 on Salmonella typhimurium TA 98 and TA 100. The physiological function of CQAD with the plural caffeoyl group is more effective than with a monocaffeoyl one. The radical scavenging activity and the anti-mutagenicity of these derivatives in order of efficacy is triCQA > diCQAs > monoCQA, suggesting that the number of caffeoyl groups bound to QA plays a role in the radical scavenging activity of the CQA derivatives. In other words, additional caffeoyl groups bound to QA are necessary for higher function. Thus, the Ipomioea batatas leaves contained distinctive polyphenolic components with high content of mono-, di-, and tricaffeoylquinic acid derivatives and could be a source for bioactive compounds, which may have the significant medicinal values for certain human diseases linked to oxidation, such as cancer, hepatotoxicity, allergies, aging, human immunodeficiency virus, and cardiovascular problems.

INTRODUCTION

The sweetpotato (*Ipomoea batatas* (L.) Lam.) has become a component of an ever increasing range of products. Internationally, the plant has diverse uses, including ornamental, livestock feed, starch and alcohol manufacture, human consumption, biofuel and bioplastic production. The sweetpotato is the seventh most important food crop in the world [1], and is among the crops selected by the U. S. National Aeronautics and Space Administration to be grown in a controlled ecological life support system as a primary food source [2]. While human consumption of the swollen storage roots is well established, eating of shoot tips, leaves and petioles is becoming more widespread. Although USA is experiencing escalating imports of sweetpotato products, the level of domestic production has progressively increased. Sweetpotato roots are used for a beverage, a paste, a powder, an alcohol drink and a natural colorant have been developed in this decade [3, 4]. While human consumption of the swollen storage roots is well established, eating of shoot tips, leaves and petioles is becoming more widespread. Although USA is experiencing escalating imports of sweetpotato products, the level of domestic production has progressively increased. The consumption of sweetpotato leaves as a fresh vegetable in some parts of the world[5]. Food scientists are becoming increasingly interested in sweetpotato leaves from the points of view of a hungry world and of good use of natural resources. As sweetpotato leaves can be harvested several times in a year, their yield is ultimately higher than many other leafy vegetables.

The anthocyanins are the predominant group of visible polyphenols and comprise the red, purple and blue pigmentation of many plants. They represent a diverse group of secondary metabolites[19-20] including important natural antioxidants and food colorants. Anthocyanins have been shown to have some positive therapeutic effects such as in the treatment of diabetic retinopathy[21], fibrocystic disease of the breast[22], and on vision[23]. Anthocyanins may also have other potential physiologic effects as antineoplastic agents[24], radiation-protective agents[25], vasotonic agents[26], vasoprotective and anti-inflammatory agents[27], chemoprotective agents against platinum toxicity in anticancer therapy[28], hepatoprotective agents against carbon tetrachloride damage[29], and possible other beneficial effects due to their interaction with various enzymes and metabolic processes[30, 31]. No adverse effects were observed in animals fed a grape color extract containing principally anthocyanins[32], nor were there any adverse effects on humans consuming a grape skin extract that has been approved by the Food and Drug Administration as a food colorant[23]. Phenolic compounds exist universally in vegetables, which are also rich sources of natural antioxidants [6-9] and in fruits [10, 11]. Dietary antioxidants have attracted special attention because they can protect the human body from oxidative stress, which may cause many diseases including cancer, aging and cardiovascular diseases [11-16]. Therefore, sweetpotato leaves with high nutritive value[17] and antioxidants, namely phenolics [18] may become an excellent source material for physiological functional foods. This paper describes the potential chemo-preventative properties of sweetpotato leaves.

MATERIALS AND METHODS

The sweetpotato leaves were harvested, washed gently, put into pre-labeled vinyl bags, and immediately frozen at –85 ^{O}C. The following day the frozen samples were freeze-dried for 48 hours in a Vacuum Freeze Dryer (Model TR-PK-3-80, Trio Sciences Co, Ltd., Tokyo, Japan) with a plate temperature of 27- 30°C. The freeze-dried samples were powdered using a blender prior to analysis.

Extraction and measurement of total phenolics

Total phenolics were measured by the procedure described by Coseteng and Lee [10] with a slight modification. The lyophilized powdered sweetpotato leaf tissue (10 mg) was vigorously mixed with 1:10 (mg/mL) 80% aqueous ethanol solution. The mixture was boiled for 5 min under a hood, centrifuged at 5000 x g for 10 min, and the supernatant was collected. The residue was mixed with 5 mL of 80% aqueous ethanol, boiled for 5 min to re-extract the phenolics and centrifuged under the same conditions. The extracts were combined, made up to 10 mL, and used for the measurement of total phenolics. The alcohol extract was diluted to obtain an absorbance reading within the range of the standards (40-800 μg chlorogenic acid/mL). The absorbance was measured at 600 nm with a dual wavelength flying spot scanning densitometer (Shimadzu Co., Kyoto, Japan), with a microplate system. The results were expressed as g/100g dry mass (DM).

Identification of Anthocyanin compositions

The crude extract was vacuum dried. The residue was redissolved in 15% acetic acid and filtered through a 0.2 μm filter membrane (Advantec, Tokyo, Japan) and used for anthocyanin identification and quality analysis. The HPLC analysis was performed according to the method of Odake et al33 with slight modification using LC-10ADvp (Shimadzu Co., Tokyo, Japan) liquid chromatograph with SPD-M10Avp detector, CTO-10Avp column oven and SIL-10Advp auto sampler. Analytical HPLC was run on a Luna C18 (2) column (100 mm X 4.60 mm, 3μm, Phenomenex, Torrance, California, USA) at 35 OC with 120 kg/cm2 pump pressure and monitored at 520 nm. The following solvents in water with a flow rate of 1 ml min-1 were used: (A) 1.5% phosphoric acid, and (B) 1.5% phosphoric acid, 20% acetic acid, 25% acetonitrile. The elution profile was a linear gradient elution for B of 15% to 50% during 90 min solvent A. The chromatograms were recorded and the relative concentration of pigments was calculated from the peak areas. In all samples, 20 μL of the leaf extract was injected. Identification of anthocyanins as carried out by comparing the peaks with authentic standard peaks of APL cell line and purple-fleshed sweetpotato YGM. Anthocyanins were YGM–0a, [cyanidin 3–sophoroside–5–glucoside]; YGM–0b, [Peonidin 3–sophoroside–5–glucoside]; YGM-1a, [cyanidin 3–(6,6'–caffeoyl-p–hydroxybenzoylsophoroside)–5–glucoside]; YGM-1b, [cyanidin 3–(6,6'– dicaffeoylsophoroside) –5–glucoside]; YGM–2, [cyanidin 3–(6– caffeoylsophoroside)–5–glucoside]; YGM–3, [cyanidin 3–(6,6'–caffeoylferuloylsophoroside)–5–glucoside]; YGM-4b, [peonidin 3–(6,6'–dicaffeoylsophoroside)–5–glucoside]; YGM-5a, [peonidin 3–(6,6'–caffeoyl-p–hydroxybenzoylsophoroside)–5–glucoside]; YGM-5b, [cyanidin 3–(6– caffeoylsophoroside)–5–glucoside] and YGM-6, [peonidin 3–(6,6'– caffeoylferuloylsophoroside) –5–glucoside] YGM–2, [cyanidin 3–(6–caffeoylsophoroside)–5–glucoside]; YGM–3, [cyanidin 3–(6,6'–caffeoylferuloylsophoroside)–5–glucoside]; YGM-4b, [peonidin 3–(6,6'– dicaffeoylsophoroside) –5–glucoside]; YGM-5a, [peonidin 3–(6,6'–caffeoyl-p– hydroxybenzoylsophoroside)–5–glucoside]; YGM-5b, [cyanidin 3–(6– caffeoylsophoroside)–5–glucoside] and YGM-6, [peonidin 3–(6,6'– caffeoylferuloylsophoroside)–5–glucoside] 33-35.

Identification of isolated sweetpotato leaf phenolics

The dried leaves of sweetpotato (150 g) were extracted twice by shaking with 2 L of 100% methanol at room temperature. The dried extract (17 g) was partitioned between benzene and water (50:50). The water layer (8 g) was fractionated on a adsorption chromatography using MCI gel CHP20P column (50 mm x 350 mm i.d., 75-150μm, styrene polymer, Mitsubishi Chemical Ind. Ltd., Tokyo, Japan) equilibrated with deionized water and adsorbed components were eluted with 20%, 40%, 60%, 80% and 100% methanol successively. The 40% methanol eluate contained mainly caffeic acid (CA) and chlorogenic acid (ChA), while the 60% methanol eluate mainly di-O-caffeoylquinic acids and 3, 4, 5-tri-O-caffeoylquinic acid. These eluates were each further fractionated on a reversed-phase chromatography using an ODS column (25 mm x 140 mm i.d., 30-50μm, Fuji Silisia Ltd., Nagoya, Japan) using 20-70% methanol give CA (15 mg/150 g DM), ChA (400mg/150 g DM), 3,4-di-O-caffeoylquinic acid (3,4-diCQA) (2 mg/150 g DM), 3,5-di-O-caffeoylquinic acid (3,5-diCQA) (60 mg/150 g DM), 4,5-di-O-caffeoylquinic acid (4,5-diCQA) (21 mg/150 g DM) and 3,4,5-tri-O-caffeoylquinic acid (3,4,5-triCQA) (2 mg/150 g DM). The above phenolics were identified as described in a previous paper [18].

Quantification of phenolic acids by RP-HPLC.

The lyophilized, powdered sweetpotato leaf tissue (50 mg) was vigorously mixed with 4 mL of 80% ethanol in a capped centrifuge tube. The mixture was boiled for 5 min in a hood and centrifuged at 3000 x g for 10 min. The supernatant was filtered through a cellulose acetate membrane filter (0.20 μm, Advantec, Tokyo, Japan). A 5-μL portion of the filtrate was injected into the HPLC system and eluted as described below. The HPLC system consisted of two Model LC-10AT pumps, a Model SIL-10AXL autoinjector, a Model CTO-10AC column oven and a Model SPD-M10AVP photodiode array UV-VIS detector (Shimadzu, Co., Kyoto, Japan). The column was a YMC-Pack ODS-AM AM-302 (150 mm x 4.6 mm i.d., 5 μm particles; YMC, Kyoto, Japan). The column oven temperature was set at 40 ^{0}C. The mobile phase consisted of water containing 0.2% (v/v) formic acid (A) and methanol (B). Elution was performed with a linear gradient as follows: 2% B from 0 to 15 min, 2% to 45% B from 15 to 50 min, and 45% B from 50 to 65 min. The flow rate was 1 mL/min. The phenolics were detected at 326 nm. The retention times (t_R) of the phenolics compounds were compared with those of purified phenolics from sweetpotato leaves used as authentic standard (purity > 97% estimated by the HPLC analysis) [18]

Measurement of radical scavenging activity

The radical scavenging activity (RSA) as measured according to Brand-Williams et al. [36] with slight modifications[37]. 1,1-Diphenyl-2-picrylhydrazyl (DPPH) was used as a stable radical. All reactions were run in a 96 well microplate with a total volume of 300 μL. A 75 μL sample (80% ethanol extract) was combined with 150 μL 0.1 M 2-(N-morpholino) ethanesulfonic acid (MES) (pH 6.0) and 75 μL DPPH in 50% ethanol in the microplate well and mixed. For the control, 75 μL of 80% ethanol was used in place of the sample. The reaction mixtures were shaken and held for 2 min at room temperature in the dark. Trolox (6-Hydroxy-2, 5, 7, 8-tetramethyl-chroman-2-carboxylic acid) was used as the reference antioxidant compound and 80% ethanol was used as the blank solution (without DPPH). The decrease in absorbance of DPPH at 520 nm was measured within 2 min. All samples were analyzed in triplicate. The RSA of samples (antioxidants) was expressed in terms of IC_{50} (concentration in μmole Trolox/mg DM required for a 50% decrease in absorbance of the DPPH radical). A plot of absorbance vs. concentration was made to calculate IC_{50}. The results were expressed as μmole Trolox/mg DM.

Assay of antimutagenicity

The antimutagenicity assay was performed by a slight modification of the method of Yahagi and others[38]. The antimutagenic activity was evaluated for *Salmonella typhimurium* TA 98 using a mutagen, Trp-P-1. These mutagens require metabolic activation to induce mutation in TA 98. S-9 mix contained 50 μmol of sodium phosphate buffer (pH 7.4), 4 μmol of MgCl2, 16.5

μmol of KCl, 2.5 μmol of glucose-6-phosphate, 2 μmole of NADH, 2 μmol of NADPH, and 50 μL of S-9 fraction in a total volume of 0.5 mL. For the inhibition test, 0.1 mL of mutagen, 0.1 mL DMSO-dissolved polyphenolics solution, and 0.5 mL of S-9 mix or phosphate buffer were simultaneously incubated with 0.1 mL of bacterial suspension at 37 ^{O}C for 20 min, and then poured on minimal-glucose-agar plates with 2 mL of soft agar. The colony number of each plate was accounted after 48 h cultivation at 37 ^{O}C.

Statistics

A completely randomized design with five replications was used. Data for the different parameters were analyzed by analysis of variance (ANOVA) procedure, and the level of significance was calculated from the F value of ANOVA. The relation between total polyphenols and RSA were described with correlation analysis using the above statistical program.

RESULTS AND DISCUSSION

Nutritional value of sweetpotato leaves

Sweetpotato leaves posses a variety of chemical compounds which are relevant to human health. Depending on genotypes and growing conditions, sweetpotato leaves are comparable with spinach in nutrient contents[17, 39, 40]. The average contents of minerals and vitamins in recently developed cultivars 'Suioh' were 117 mg calcium, 1.8 mg iron, 3.5 mg carotene, 7.2 mg vitamin C, 1.6 mg vitamin E, and 0.56 mg vitamin K/100 g fresh weight of leaves. Levels of iron, calcium and carotene rank among the top, as compared with other major vegetables[40]. Sweetpotato leaf is also rich in vitamin B, β-carotene, iron, calcium, zinc and protein, and as a crop is more tolerant of diseases, pests, and high moisture than many other leafy vegetables grown in the tropics[39, 41, 42]. We previously reported that sweetpotato leaves were an excellent source of antioxidative polyphenolics, among them anthocyanins [3, 35, 43, 44] and phenolic acids such as caffeic, monocafeoyl quinic (chlorogenic), dicaffeoylquinic, and tricaffeoylquinic acids[3, 18, 37, 43], superior in this regard to other commercial vegetables [4, 17, 40]. Sweetpotato leaves and other tropical leafy vegetables have been shown to contain higher levels of oxalic acid than common temperate climate vegetables, with the exception of spinach [45]. Oxalate concentrations in food crops have long been a concern in human diet. Because of the negative health effects associated with high intake of oxalate levels can cause acute poisoning, resulting in hypocalcaemia, or chronic poisoning in which calcium oxalate is deposited as crystals in the kidneys, causing renal damage. Furthermore, oxalic acid and soluble oxalates can bind calcium, reducing its bioavailability and calcium oxalate itself is poorly utilized by humans. The average content of oxalic acid of sweetpotato variety 'Suioh' leaves was 280 mg/100 g fresh weight. This content was not high compared to the 930 mg/100 g fresh weight in spinach [40]. Oxalic acid contents of other sweetpotato varieties tested were also several times less than that of spinach [4].

Anthocyanin compositions

We reported that sweetpotato leaves represent one of the richest and most commonly available sources of anthocyanin among fruits and vegetables 3, 35, 43, 44. Fifteen anthocyanin compounds were identified and characterized in sweetpotato leaves (Table 1). The HPLC chromatogram of the sweetpotato leaves anthocyanin is presented in Figure 1. The anthocyanins were acylated

Table 1. Chemical Names of the Fifteen Anthocyanins in *Ipomoea batatas* L. Leaves

Common name	Chemical name
YGM[Z]–0*a*	Cyanidin 3–sophoroside–5–glucoside
YGM–0*b*	Peonidin 3–sophoroside–5–glucoside
YGM–0*c*	*p*-hydroxybenzoylated (cyanidin 3-sophoroside–5–glucoside)
YGM–0*d*	Caffeoylated (cyanidin 3–sophoroside–5–glucoside)
YGM–0*e*	*p*-hydroxybenzoylated (peonidin 3–sophoroside–5–glucoside)
YGM–0*f*	Caffeoylated (peonidin 3–sophoroside–5–glucoside)
YGM–0*g*	Feruloylated (cyanidin 3–sophoroside–5–glucoside)
YGM-1*a*	Cyanidin 3–(6,6′–caffeoyl-*p*–hydroxybenzoylsophoroside)–5–glucoside
YGM-1*b*	Cyanidin 3–(6,6′–dicaffeoylsophoroside)–5–glucoside
YGM–2	Cyanidin 3–(6–caffeoylsophoroside)–5–glucoside
YGM–3	Cyanidin 3–(6,6′–caffeoylferuloylsophoroside)–5–glucoside
YGM-4*b*	Peonidin 3–(6,6′–dicaffeoylsophoroside)–5–glucoside
YGM-5*a*	Peonidin 3–(6,6′–caffeoyl-*p*–hydroxybenzoylsophoroside)–5–glucoside
YGM-5*b*	Cyanidin 3–(6–caffeoylsophoroside)–5–glucoside
YGM-6	Peonidin 3–(6,6′–caffeoylferuloylsophoroside)–5–glucoside

[Z]YGM=Yamagawamurasaki

Cyaniding and peonidin type. But the content of cyanidin in leaves are much higher that that of peonidin, suggesting that the anthocyanin composition of sweetpotato leaf is cyanidin type[3, 35, 43]. The cyanidin type anthocyanins are superior to the peonidin type in antimutagenicity[17, 46] and antioxidative activity [47]. The sweetpotato anthocyanins are ubiquitous bioactive compounds because of its potential antioxidant activities that may exert cardioprotective effects [48, 49] and anticancer effects [50, 51]. As indicated above, sweetpotato anthocyanins possess multifaceted action, including antioxidation, antimutagenicity, anti-inflammation and anticarinogenesis. Extensive structure-activity studies show that the number of sugar units and hydroxyl groups on aglycons is associated with biological activities of anthocyanins. The activities appear to increase with a decreasing number of sugar units, and with an increasing number of hydroxyl groups on aglycons [3, 17, 51].

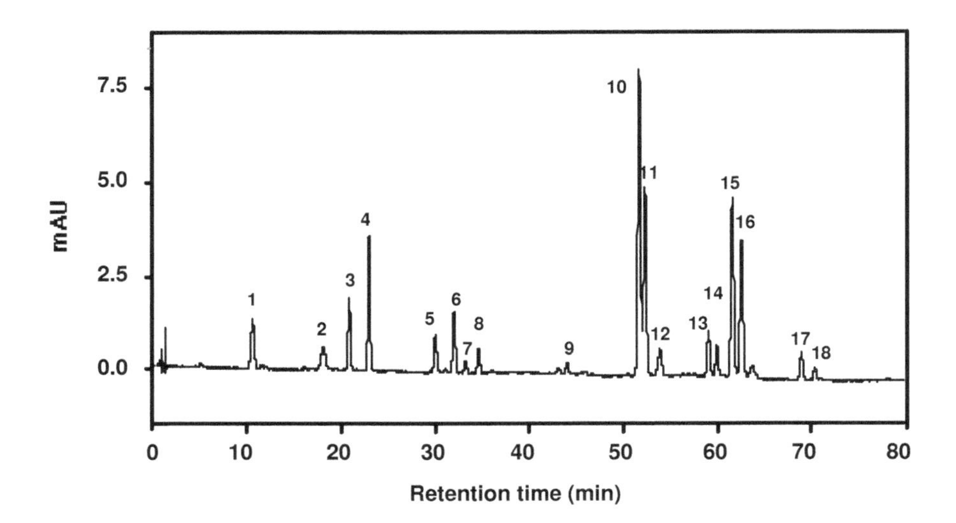

Figure 1. Reverse-phase HPLC separation chromatograms for anthocyanin compositions in sweetpotato leaf extracts. The peak no. 1 (t_R=10.8 min) is YGM-0*a*, peak no. 2 (t_R=18.3 min) is YGM-0*b*, peak no. 3 (t_R= 21.0 min) is YGM-0*c*, peak no. 4 (t_R=23.1 min) is YGM-0*d*, peak no. 5 (t_R= 30.2 min) is YGM-0*e*, peak no. 6 (t_R= 33.4 min) is YGM-0*f*, peak no. 8 (t_R= 34.8 min) is YGM-0*g*, peak no. 10 (t_R= 51.9 min) is YGM-1*a*, peak no. 11 (t_R= 52.5 min) is YGM-1*b*, peak no. 12 (t_R= 54.1 min) is YGM-2, peak no. 14 (t_R=61.1 min) is YGM-3, peak no. 15 (t_R=61.8 min) is YGM-4*b*, peak no. 16 (t_R=62.8 min) is YGM-5*a*, peak no. 17 (t_R=63.9 min) is YGM-5*b* and peak no. 18 (t_R=69.1 min) is YGM-6.

Polyphenolics content

Sweetpotato leaves contain a much higher content of total polyphenols than any other commercial vegetables, including sweetpotato roots and potato tubers. A preliminary study (n=10) revealed that the highest polyphenol concentration was in leaves (6.19 ± 0.14 g/100 g dry weight), followed by petioles (2.97 ± 0.26 g/100 g dry weight), stems (1.88 ± 0.19 g/100 g dry weight), and finally roots (< 1.00 g/100 g dry weight), indicating that polyphenolic concentrations are organ-dependent [18]. The distribution of total leaf polyphenolic content of all 1,389 genotypes in the germplasm collection is shown in Figure 2. The highest content found was 17.1 g/100 g dry weight and the lowest was 1.42 g/100 g dry weight. Most of the genotypes (75%) contained >6.00 g/100 g dry leaf powder of total polyphenolics, which was a very high concentration compared to any other commercial vegetables.

Polyphenolics composition

Caffeic acid (CA) and five caffeoylquinic acid (CQA) derivatives: 3-mono-*O*-CQA (chlorogenic acid; ChA), 3,4-di-*O*-caffeoylquinic acid (3,4-diCQA), 3,5-di-*O*-caffeoylquinic acid (3,5-diCQA), 4,5-di-*O*-caffeoylquinic acid (4,5-diCQA), and 3,4,5-tri-*O*-caffeoylquinic acid (3,4,5-triCQA) are found in sweetpotato leaves (Islam and others 2002b). Recently we isolated, identified and characterized the above compounds in sweetpotato leaves for the first time (Figure 3). ChA, di- and triCQA are esters of quinic acid (QA) and bear one-, two-, and three-caffeoyl groups. Chlorogenic acid and diCQA derivatives have been isolated from various plants including sweetpotato (Shimozono and others 1996; Walter and others 1979), as mentioned above, but there are very few reports on 3, 4, 5-triCQA. Isolation of 3, 4, 5-triCQA was reported in *Securidaka longipedunculata* (*Polygalaceae*)[52] and *Tessaria integrifolia* (*Asteraceae*)[53]. Several varieties of sweetpotato

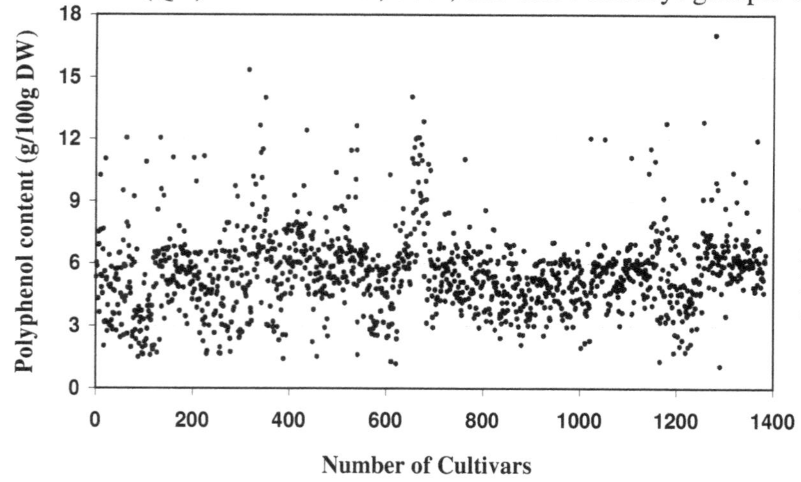

Figure 2. Distribution of total Polyphenol contents in the leaves of 1389 Sweetpotato Cultivars

contain a high content (>0.2%) of 3, 4, 5-triCQA[18, 54], suggesting that the sweetpotato leaf is a source of not only mono- and diCQA derivatives but also 3, 4, 5-triCQA.

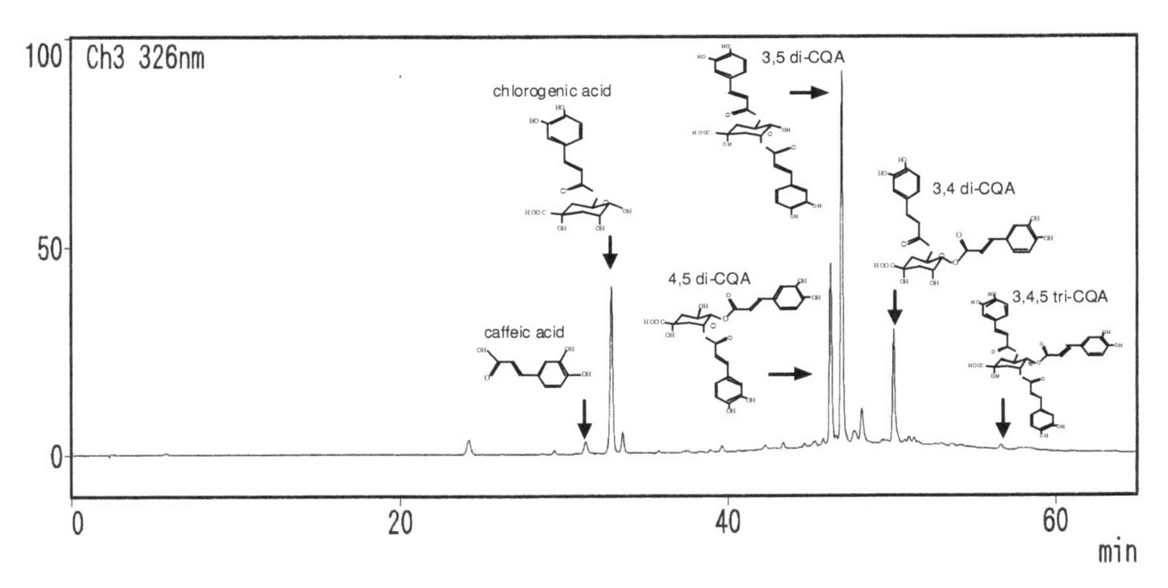

Figure 3. Reverse-phase HPLC (C_{18}) separation chromatograms for phenolic acids in sweetpotato leaf extracts.

Physiological function of sweetpotatoes

In the recent past, active research has been conducted to determine health-promoting functions of sweetpotato leaves. The following aspects (Table 2) of these functions are important to look into when considering new uses of sweetpotato leaves:

Table 2. Physiological functions *of Ipomoea batatas* L. and their related components

Physiological Function	Related components	References
Antioxidative activity/Radical scavenging activity	Polyphenol, anthocyanin	6, 37, 43, 44
Antimutagenicity	Polyphenol, anthocyanin	3, 16, 46, 54
Anticarcinogenesis	Polyphenol, anthocyanin	14, 50, 51, 56
Antihypertension	Polyphenolics, anthocyanin	46, 49
Antimicrobial activity	Fiber, pectin-like polysaccharide	3, 46
Antiinflammation	Polyphenol	53
Antidiabetic effect	Anthocyanin, polyphenol	3
Anti HIV	Polyphenolics	52
Promotion of Bibidobacterium growth	Dietary fiber	3, 46
Reduction of liver injury	Polyphenol	3, 49
Relief from constipation	Dietary fiber, jalapin	3,46
Ultraviolet protection effect	Polyphenolics	3, 46

Antioxidative and radical scavenging activity

Although lipids are essential for human health, certain polyunsaturated fatty acids have many double bonds and cause radical chain reactions with oxygen, and thereby produce various lipid peroxide and oxidized resolvents. Lipid peroxides cause deterioration in cell functions, arterial sclerosis, liver disorders, retinopathy and are also involved in carcinogenesis and aging.

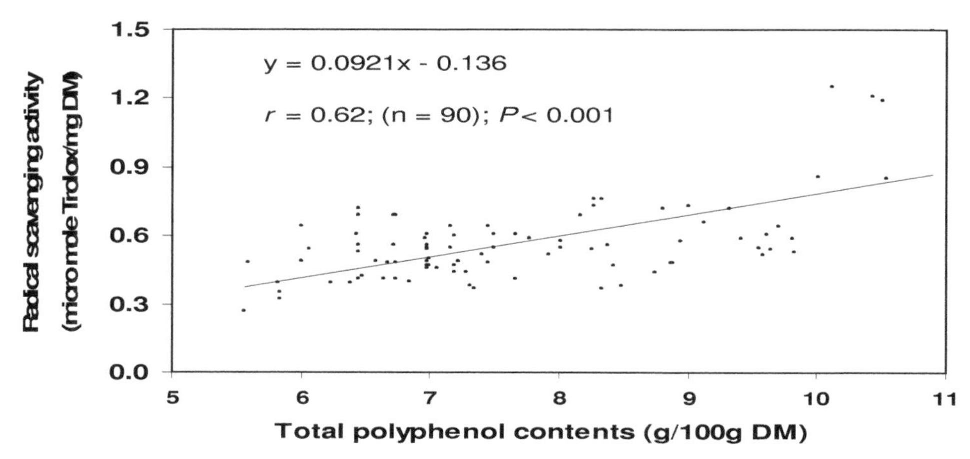

Figure 4. Correlation between the total polyphenol content and radical scavenging activity of sweetpotato leaves.

From this viewpoint, antioxidative substances contained in plants have attracted much attention recently. Several authors have reported the antioxidative and radical scavenging activities of sweetpotato leaves [18, 37, 54]. Polyphenolic content and antioxidative activity in vegetables show a good correlation, with the antioxidative activity of edible chrysanthemum (*Chrysanthemum morifolium*), which has the highest polyphenolic content being the most effective among 43 commercial vegetables[6]. Sweetpotato leaves also revealed relatively much higher activity in DPPH (1, 1-Diphenyl-2-picrylhydrazyl) radical scavenging activity than the vegetables described above[37, 54]. There were significant positive correlations between radical scavenging activity and polyphenol contents of sweetpotato leaves (Figure 4).

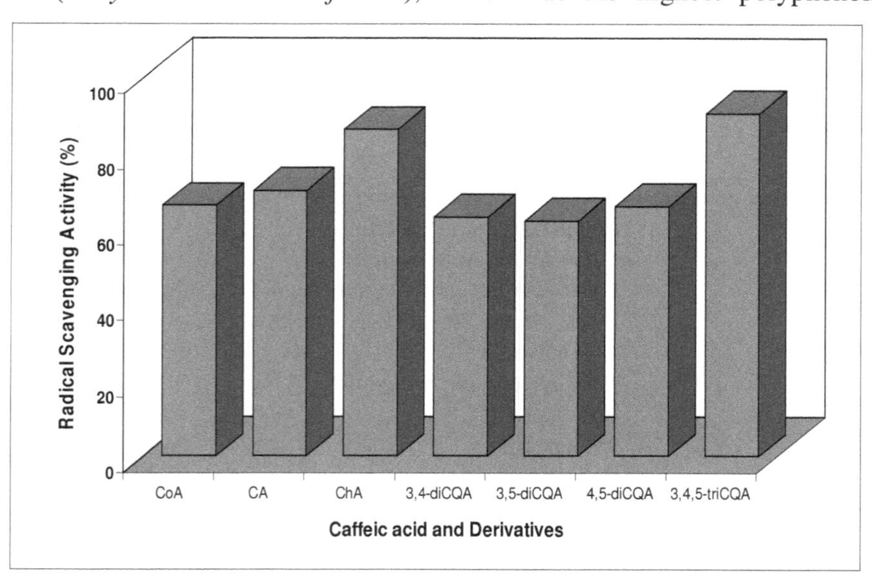

Figure 5. Radical scavenging activity (%) of caffeoylquinic acid derivatives isolated from sweetpotato leaves. Values are mean of five separate experiments. CoA = p-coumaric acid, CA = Caffeic acid; ChA = Chlorogenic acid; 3,4-diCQA = 3,4-di-o-caffeoyl quinic acid; 3,5-diCQA = 3,5-di-o-caffeoyl quinic acid; 4-5-diCQA = 4,5-di-o-caffeoyl quinic acid; 3,4,5-triCQA = 3,4,5-tri-o-caffeoyl quinic acid.

The radical scavenging activity of CQA derivatives in order of effectiveness was 3, 4, 5-triCQA ≥ ChA > 3, 4-diCQA = 3, 5-diCQA = 4, 5-diCQA > CA = CoA (Figure 5). These data indicate that sweetpotato leaves are a good supplementary resource of antioxidants.

Antimutagenicity and anticarcinogenicity

Cancers occur through such processes as initiation, promotion and progression in body cells. Initiation is a kind of mutation that occurs in cancer and anticancer genes. Thus, controlling the gene mutation, brought about by the carcinogens, leads to cancer prevention. The mutagens contained in food may include ingredients of vegetables, mold toxins and other pollutants in food. These mutagens are considered as factors involved in occurrence of human cancers. The recent development of screening methods for environmental carcinogens by determining their mutagenicity has enabled various types of mutagens and carcinogens to be detected and identified in daily foods 55. It is now known that various types of inhibitors that act against mutagens and carcinogens are present in our daily food, and that they play an important role in reducing the risks of mutagenesis and carcinogenesis 17. CQA derivatives effectively inhibited the reverse mutations induced by Trp-P-1 on Salmonella typhimurium TA and the anti-mutagenicity of these derivatives in order of effectiveness was 3, 4, 5-triCQA > 3, 4-diCQA = 3, 5-diCQA = 4, 5-diCQA > ChA> CA > C*0*A [54] (Figure 6).

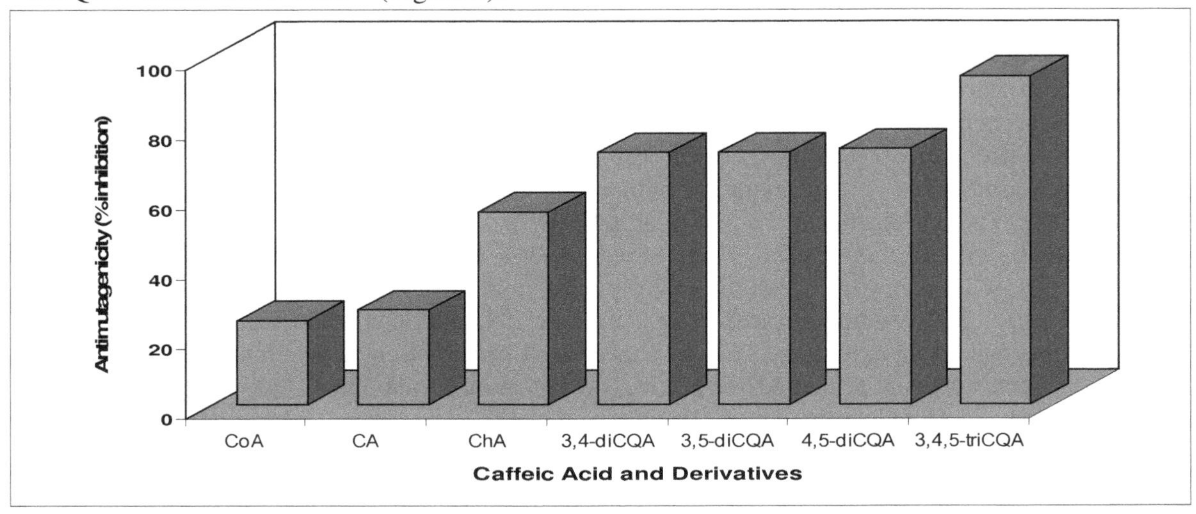

Figure 6. Effect of caffeoylquinic acid derivatives isolated from sweetpoatto leaves on antimutagenicity (% inhibition) of Trp-P-1 against Salmonella typhimurium TA 98. Trp-P-1 was added at a dose of 0.075 μg per plate. The mutagenicity was tested with S-9 mix. The values shown have had the spontaneous mutation frequency subtracted. CoA= p-coumaric acid, CA= Caffeic acid; ChA= Chlorogenic acid; 3,4-diCQA= 3,4-di-o-caffeoyl quinic acid; 3,5-diCQA= 3,5-di-o-caffeoyl quinic acid; 4-5-diCQA= 4,5-di-o-caffeoyl quinic acid; 3,4,5-triCQA= 3,4,5-tri-o-caffeoyl quinic acid.

Growth suppression of three kinds of cancer cells–stomach cancer (Kato-III), colon cancer (DLD-1), and promyelocytic leukemia (HL-60) cells–by QA, CA, and CQA derivatives is shown in Fig. 4. QA had no effect on the growth of each kind of cancer cells. 3, 4, 5-triCQA remarkably suppressed the growth of each kind of cancer cells at a concentration of 500 μM. HL-60 was especially sensitive to the CQA derivatives compared with the others (Fig. 4C). CA and the three kinds of di-CQA derivatives (3, 4-diCQA, 3, 5-diCQA, and 4, 5-diCQA) suppressed the growth of HL-60 cells at concentrations of 100 μM and 500 μM, respectively. CA exceptionally suppressed the cell multiplication of HL-60. These results show the necessity of the caffeoyl group bound to QA as well as the differential sensitivity of tumor cells to these compounds. In order to clarify the mechanism of the cell growth suppression, we examined morphological changes in HL-60 cells treated with 3,4,5-triCQA at a concentration of 500 μM. Nuclear granulation and DNA fragmentation [56] suggested that the cellular death was due to apoptosis induction.

SUMMARY AND CONCLUSION

The six different polyphenolic and anthocyanins were identified and quantified from sweetpotato leaves. The phenolic acids which were composed of caffeic acid (CA) and five kinds of caffeoylquinic acid derivatives, 3-mono-O-caffeoylquinic acid (chlorogenic acid, ChA), 3,4-di-O-caffeoylquinic acid (3,4-diCQA), 3,5-di-Ocaffeoylquinic acid (3,5-diCQA), 4,5-di-O-caffeoylquinic acid (4,5-diCQA), and 3,4,5-tri-O-caffeoylquinic acid (3,4,5-triCQA). Sweetpotato leaf also contain at least fifteen biologically active anthocyanins that have significant medicinal value for certain human diseases and may also be used as natural food colorants. The anthocyanins were acylated cyanidin and peonidin type. The result suggests that the main phenolic compound in *Ipomoea batatas* leaves is 3, 5-di-*O*-caffeoylquinic acid followed by 4, 5-di-*O*-caffeoylquinic acid. The average levels in sweetpotato leaf of 60 genotypes were in the following order: 3, 5-di-*O*-caffeoylquinic acid> 4, 5-di-*O*-caffeoylquinic acid> 3-*O*-caffeoylquinic acid> 3, 4-di-*O*-caffeoylquinic acid> 3, 4, 5-tri-*O*-caffeoylquinic acid> caffeic acid. These caffeoylquinic acid derivatives (CQAD) showed very high antioxidant activity, and effectively inhibited the reverse mutation induced by Trp-1 on *Salmonella typhimurium* TA 98 and TA 100. The physiological function of CQAD with the plural caffeoyl group is more effective than with a monocaffeoyl one. Sweetpotato leaf contains high concentrations of polyphenolics, when compared with the major commercial vegetables such as spinach, broccoli, cabbage, lettuce, etc. Sweetpotato leaf is a physiologically functional food that offers protection from diseases linked to oxidation, such as cancer, and cardiovascular problems. Therefore, the *Ipomioea batatas* leaves contained distinctive polyphenolic components with high content of mono-,di-, and tricaffeoylquinic acid derivatives and could be a source for bioactive compounds, which may have the significant medicinal values for certain human conditions. Thus, sweetpotato leaves may be a valuable resource for physiological functional foods that protect the human body from diseases linked to oxidation.

REFERENCES

1. Food and Agriculture Organization (FAO). 1997. FAO production yearbook of 1997. Vol.51. FAO (United Nations), Rome, Italy. Hoff JE, Howe JM, Mitchell CA. 1982. Nutritional and cultural aspects of plant species selection for a controlled ecological life support system. NASA Contractor Rep. 166324, Moffett Field, CA. USA. Islam, S. 2006. Sweetpotato (Ipomoea batatas

L.) Leaf: Its Potential Effect on Human Health and Nutrition. Journal of Food Science, 71: R13-R21.

2. Yoshimoto M. 2001a. New trends of processing and use of sweetpotato in Japan, Farming Japan 35: 22-28.
3. Villareal RL, Tsou SC, Lo, HF, Chiu SC. 1982. Sweetpotato tips as vegetables, In: R L Villareal and TD Griggs (Eds.). SweetPotato. Proceedings of the first international symposium, AVRDC, Shanhua, Taiwan, pp. 313-320.
4. Tsushida T, Suzuki M, Kurogi M. 1994. Evaluation of antioxidant activity of vegetable extracts and determination of some active compounds. Nippon Shokuhin Kogyo Gakkaishi 41:611-618.
5. Murata M, Noda I, Homma S. 1995. Enzymatic Browning of apples on the market: relation between browning, polyphenol content, and polyphenol oxidase. Nippon Shokuhin Kagaku Kogaku Kaishi 42: 820-826.
6. Chuda Y, Ono H, Kameyama M, Nagata T, Tsushida, T. 1996. Structural identification of two antioxidant quinic acid derivatives from garland (*Chrysanthemum curonarium* L.). J Agric Food Chem 44:2037-2039.
7. Murayama T, Yada H, Kobori M, Shinmoto H, Tsushida T. 2002. Evaluation of three antioxidants and their identification and radical scavenging activities in edible chrysanthemums. J Japan Soc Hort Sci 71:236-242.
8. Coseteng MY, Lee CY. 1987. Changes in apple polyphenoloxidase and polyphenol concentrations in relation to degree of browning. J Food Sci 52: 985-989.
9. Robards K, Prenzler PD, Tucker G, Swatsitang P, Glover W. 1999. Phenolic compounds and their role in oxidative processes in fruits. Food Chem 66: 401-436.
10. Huang MT, Ferraro T. 1991. Phenolic compounds in food and cancer prevention, p. 220-238. In: MT Hung, CT Ho, CY LEE (Eds.). Phenolic compounds in food and their effects on health: Antioxidant cancer prevention. ACS Symp. Ser. 5078. Amer. Chem. Soc., Washington, D.C.
11. Shahrzed S, Bitsch I. 1996. Determination of some pharmacologically active phenolic acids in juices by high-performance liquid chromatography. J Chromatography A 741: 223-231.
12. Shimozono H, Kobori M, Shinmoto H, Tsushida T. 1996. Suppression of the melanogenesis of mouse melanoma B 16 cells by sweetpotato extract. Nippon Shokuhin Kagaku Kogaku Kaishi. 43: 313-317.
13. Prior RL, Cao G, Martin A, Sofic E, McEwen J, Brien CO, Lischner N, Ehlenfeldt M, Kalt W, Krewer G, Mainland CM. 1998. Antioxidant capacity as influenced by total phenolic and anthocyanin content, maturity and variety of Vaccinium species. J Agric Food Chem 46: 2686-2693.
14. Yoshimoto M, Okuno S, Kumagai T, Yoshinaga M, Yamakawa O, Yamaguchi M, Yamada J. 1999. Antimutagenicity of sweetpotato (*Ipomoea batatas*) roots. Biosci Biotechnol Biochem 63: 537-541.
15. Yoshimoto, M., Okuno, S., Suwa, K., Sugawara, T. and Yamakawa, O. 2002a. Effect of harvest time on nutrient content of sweetpotato leaves. Proc.12th Symp. of the International Society for Tropical Root Crops, p. 319-323.
16. Islam S, Yoshimoto M, Yahara S, Okuno S, Ishiguro K, Yamakawa O. 2002b. Identification and characterization of foliar polyphenolic compositions in swetpotato (*Ipomoea batatas* L.) genotypes. J Agric Food Chem 50: 3718-3722
17. Harborne JB. 1994. Flavonoids: Advances in research since 1986. Chapman & Hall, London, U.K.
18. Meyer A. S, Heinonen M, Frankel EN. 1998. Antioxidant interactions of catechin, cyanidin, caffeic acid, quercetin, and ellagic acid on human LDL oxidation. Food Chem 61:71–75.
19. Scharrer A, Ober M. 1981. Anthocyanosides in the treatment of retinopathies. Klin Monatsbl

Augenheikd 178:386-389.
20. Leonardi M. 1993. Treatment of fibrocystic disease of the breast with *Vaccinium myrtillus* anthocyanins. Our experience. Minerva Ginecol 45: 617-621.
21. Timberlake CF, Henry BS. 1988. Anthocyanins as natural food colorants. Prog Clin Biol Res 280:107–121.
22. Kamei H, Kojima T, Hasegawa M, Koide T, Umeda T, Yukawa T, Terade K. 1995. Suppression of tumor cell growth by anthocyanins *in vitro*. Cancer Invest 13:590–594.
23. Akhmadieva AK, Zaichkina SI, Ruzieva RK, Ganassi, EE. 1993. The protective action of a natural preparation of anthocyanin (Pelargonidin-3,5-diglucoside). Radiobiologiia 33:433–435.
24. Colantuoni AS, Bertugli M, Magistretti J, Donato, L. 1991. Effects of *Vaccinium myrtillus* anthocyanosides on arterial vasomotion. Arzneim-Forsch 41:905-909.
25. Lietti A, Cristoni A, Picci M. 1976. Studies on *Vaccinium myrtillus* anthocyanosides. I. Vasoprotective and anti-inflammatory activity. Aezneim-Forsch 26:829-832.
26. Karaivanova M, Drenska D, Ovcharov RA. 1990. Modification of the toxic effects of platinum complexes with anthocyanins. Eksp Med Morfol 29:19–24.
27. Mitcheva M, Astroug H, Drenska D, Popov A, Kassarova M. 1993. Biochemical and morphological studies on anthocyans and vitamin E on carbon tetrachloride induced liver injury. Cell Mol Biol 39:443–448.
28. Saija A, Princi P, D'Amico N, Pasquale R, Costa G. 1990. Effects of *Vaccinium myrtillus* anthocyanins on triiodothyronine transport in the rat. Pharmacol. Res. 22 (Suppl.3):59-60.
29. Costantino L, Rastelli G, Albasini A. 1995. Anthocyanidines as inhibitors of xanthine oxide. Pharmazie 50:573–574.
30. Becci PJ, Hess FG, Gallo MA, Johnson WD, Babish JG. 1983. Subchronic feeding study of color extract in beagle dogs. Food Chem Toxicol 21:75–77.
31. Odake, K., Terahara, N., Saito, N., Toki, K., and Honda, T., 1992. Chemical structures of two anthocyanins from purple sweetpotato, *Ipomoea batatas*. Phytochemistry, 31: 2127-2130
32. Islam S, Yoshimoto M, Terahara N, Yamakawa O. 2002a. Anthocyanin compositions in sweetpotato leaves. Biosci Biotech Biochem 66: 2483-2486.
33. Islam MS, Jalaluddin M, Garner J., Yoshimoto M, Yamakawa O. 2005. Artificial shading and temperature influenced on anthocyanin composition of *Ipomoea batatas* leaves. HortScience 40: 176-180
34. Brand-Williams, W., M. E. Cuvelier, and C. Berset. 1995. Use of a free radical method to evaluate antioxidant activity. Lebensm. Wiss. Technol. 28:25-30.
35. Islam S, Yoshimoto M, Ishiguro K, Okuno S, Yamakawa O. 2003c. Effect of artificial shading and temperature on radical scavenging activity and polyphenolic compositions in sweetpotato leaves. J Amer Soc Hortic Sci 128: 182-187
36. Yahagi T, Nagao M, Seino Y, Matsushima T, Sugimura T, Odaka M. 1977. Mutagenicities of N-nitrosoamines on *Salmonella*. Mutat Res 48:121-129.
37. Woolfe, JA. 1992. Sweetpotato. An untapped food resource, Cambridge University Press, Cambridge, U. K. pp. 118-187.
38. Ishiguro K, Toyama J, Islam MS, Yoshimoto M, Kumagai T, Kai Y, Yamakawa O. 2004. Suioh, a new sweetpotato cultivar for utilization in vegetable greens. Acta Hortic 637: 339-345
39. Asian vegetable research and development center (AVRDC). 1985. Composition of edible fiber in sweetpotato tips. AVRDC Progress Report 1985:310–313.
40. Yoshimoto, M, Okuno S, Islam MS., Kurata R, Yamakawa O. 2003. Polyphenol content and antimutagenicity of sweetpotato leaves in relation to commercial vegetables. Acta Hortic 628:

677-685
41. Islam S, Yoshimoto M, Yamakawa O, Ishiguro K, Yoshinaga M. 2002c. Antioxidative compounds in the leaves of different sweetpotato cultivars. Sweetpotato Res Front. 13: 4
42. Islam S, Yoshimoto M, Ishiguro K, Yamakawa O. 2003b. Bioactive and Functional properties of *Ipomoea Batatas* L. Leaves. Acta Hortic 628: 693-699
43. Evenson SK, Standal BR. 1984. Use of tropical vegetables to improve diets in the Pacific region. HITAHR Res. Ser. 028, HITAHR, University of Hawaii, U.S.A.
44. Yoshimoto M, Okuno S, Yamaguchi M, Yamakawa O. 2001b. Antimutagenicity of deacylated anthocyanins in purple-fleshed sweetpotato. Biosci Biotechnol Biochem 65: 1652-1655.
45. Rice-Evans, CA, Miller NJ, Bolwell PG, Bramley PM, Pridham, JB. 1995. The relative antioxidant activities of plant-derived polyphenolic flavonoids. Free Radical Res 22: 375-383.
46. Furuta S, Suda I, Nishiba Y, Yamakawa O. 1998. High *tert*-butylperoxyl radical scavenging activities of sweetpotato cultivars with purple flesh. Food Sci Technol Int 4:33–35.
47. Suda I, Yamakawa O, Matsugano K, Sugita K, Asuma K, Irisa K, Tokumaru F. 1998. Changes of serum γ–GTP, GOT and GPT levels in hepatic function-weakling subjects by ingestion of high anthocyanin sweetpotato juice. Nippon Shokuhin Kagaku Kogaku Kaishi 45:611–617.
48. K-Islam I, Yoshinaga M, Hou DX, Terahara N, Yamakawa.O. 2003. Potential chemopreventive properties of anthocyanin rich aqueous extracts *in vitro* produced tissue of sweetpotato (*Ipomoea batatas* L.). J Agric Food Chem 51:5916–5922.
49. Hou, DX. 2003. Potential mechanisms of cancer chemoprevention by anthocyanins. Current Mol Med 3: 149-159.
50. Mahmood, N., Moore, P.S., De Tommasi, N., De Simone, F., Colman, S., Hay, A.J. and Pizza, C. 1993. Inhibition of HIV infection by caffeoylquinic acid derivatives. Antiviral Chem. Chemother. 4:235-240.
51. Peluso G, Feo VD, Simone FD, Bresciano E, Vuotto ML. 1995. Studies on the inhibitory effects of caffeoylquinic acids on monocyte migration and superoxide anion production. J Natu Prods 58: 639-646.
52. Islam S, Yoshimoto Y, Yamakawa O. 2003a. Distribution and physiological function of caffeoylquinic acid derivatives in sweetpotato genotypes. J Food Sci 68: 111-116
53. Ames BN, McCann J, Yamasaki E. 1975. Methods for detecting carcinogens and mutagens with *Salmonella*/mammalian microsome mutagenicity test. Mutat Res 31:347-364.
54. Kurata, R., M. Adachi, O. Yamakawa and M. Yoshimoto. 2007. Growth suppression of human cancer cells by polyphenolics from sweetpotato (*Ipomoes batatas* L.) leaves. J. Agric. Food Chem., 55: 185-190.

EGYPTIAN SUGAR CANE WAX: THERAPEUTIC SIGNIFICANCE OF A NEW LIPID-LOWERING AGENT IN RAT

Mahmoud A Mohammad, Kadry Z Ghanem, Mohamed H Mahmoud, Sahar S Abdel-aziz and Magda S Mohamed

Department of Food Science Nutrition, National Research Center, Dokki, Giza, Egypt.

Correspondind author: Mahmoud A Mohammad, E-mail: mohammad@bcm.tmc.edu

Key words: Policosanol, Lipid Peroxidation, Management of Atherosclerosis, Coronary Heart Disease

ABSTRACT

Data from the epidemiological studies indicate that increases in serum cholesterol levels are associated with an increased risk of death from coronary heart disease. Policosanol is a mixture of high-molecular-mass aliphatic alchohols isolated and purified from sugar cane (*Saccharum Officinarum* L.). Policosanol is a drug currently in use to reduce elevated LDL-C and total cholesterol levels in combination with dietary therapy in patients with hypercholesterolemia. In the present study, Egyptian sugar cane wax was extracted, its polycosanols were determined qualitatively and quantitatively with GC-MS and GC and evaluated for its potency to reduce hypercholesterolemia in rats. The results showed that polycosanol constitutes 3.25% of the core wax. Octacosal (C28OH) is the major component, amounting to 85% of Policosanols in wax. An intervention study was adopted to evaluate the effect of polycosanols on the cholesterol levels. 30 male albino rats were divided into five groups of equal mean body weights. The first group of rats was fed on basal diet (control). The second group was fed on basal diet containing 1% cholesterol. Groups 3, 4 and 5 were fed on basal diet containing 1% cholesterol and sugar cane wax extract (15, 45 and 100 mg per kg diet, respectively). Total cholesterol and low density lipoprotein cholesterol showed a significant increase ($P<0.05$) in hypercholesterolemic group compared to control. Supplementation of sugar cane wax extract in diets of both groups 4 and 5 significantly ($P<0.05$) decreased total cholesterol level compared to the second group. A significant increase was observed in HDL-C in group 5 compared to the second group. The data obtained revealed that triacylglycerol was significantly decreased due to supplementation with sugar cane wax extract of groups 4 and 5 compared to hypercholesterolemia in rats from the second group. Insignificant differences were noticed in the mean value of body weight, organ weight and relative organ weights between different experimental groups. We concluded that sugar cane wax seems to be a very promising photochemical alternative to classic lipid-lowering agents.

INTRODUCTION

Epidemiological studies have shown a continuous relationship between total serum cholesterol and the risk of coronary heart disease (CHD) (1, 2); National Cholesterol Education

Program (NCEP) 1988 has identified elevated low-density lipoprotein cholesterol (LDL-C) as a primary risk factor for CHD (3). NCEP Adult Treatment Panel II Report 1993 has recommended aggressive dietary and drug therapy for patients with known CHD (4). The most potent drugs that are currently used to lower elevated (LDL-C) levels are the 3-hydroxy – 3-methylglutaryl-coenzyme A (HMG-CoA) reductase inhibitors (statins) (5, 6). Due to patient reluctance to be treated with chemically derived drugs, especially for primary prevention which may contribute to the above discrepancy, there is a need for effective, safe and ideally naturally derived cholesterol-lowering drugs. Policosanol is a drug currently in use to reduce elevated LDL-C and total cholesterol levels in combination with dietary therapy in patients with hypercholesterolemia (7). Moreover, policosanol generally increases high-density lipoprotein (HDL-C) levels (8). Policosanol is a natural mixture of high-molecular-mass aliphatic alcohols isolated and purified from sugar cane (Saccharum Officinarum L.) wax. Octacosanol with molecular weight (M.W.) equals 410.7 is its major component, followed by triacosanol (M.W. 438.5) and hexacosanol (M.W. 382.4) while other alcohols (tetracosanol, heptacosanol, nonacosanol, dotriacontanol and tetratriacosanol) are minor components (9). Most pharmacological effects of policosanol, including cholesterol reduction, have been proven for octacosanol, not for other constituents (8).

Previous studies have demonstrated that serum cholesterol levels of pigs are significantly decreased by supplementation of the diet with policosanol (10). Also, oral administration of policosanol decreases the serum cholesterol level in normo- and hypercholesterolaemic rabbits (11-13), healthy human voluntreers (14, 15) and patients with hypercholesterolaemia (16-18). The present study was designed to compare the effects of a higher dose of Egyptian sugar cane wax polycosanols, a cholesterol –lowering drug 0.05 mg / day (100 mg wax /kg diet) with the effects of 0.01 (15 mg wax /kg diet) and 0.02 (45 mg wax /kg diet) mg /day on hypercholesterolaemic rats corresponds to, 45 and 15 mg/kg diet of wax respectively.

MATERIALS AND METHODS

The standards tetracosanol (C24), hexacosanol (C26) and octacosanol (C28) were purchased from Sigma, 99% GC purity. Hexane and cholorform, were obtained from Merck (Germany), silylation reagent, N-methyl-N- (trimethylsilyl) trifuoracetamide (MSTFA) was supplied by Sigma, 99% GC purity.

Extraction of sugar wax

Sugar cane was hand-peeled. The peels were grounded and hydrolyzed by refluxing with 1.0 N NaOH in methanol for 30 min. The mixture was cooled and filtered through glass wool using a glass funnel. Millipore water was added to the filtrate. Then the solution was extracted with diethyl ether. The extraction was repeated three times using equal volumes of diethyl ether. The diethyl ether phases collected from three extractions were combined and washed with Millipore water until reaching neutrality. The ether extract was evaporated using rotary evaporator and then evaporated dryness under nitrogen after drying over anhydrous sodium sulfate (19).

Gas chromatography- Mass spectrometer:

Finnigan Instrument GC- 2000 (THERMO) model SSQ 7000 series, coupled to a computerized-data processor, equipped with a capillary column DB5, 0.25mm id X 30m.

Operating condition: the temperature of column kept start at 50°C for 3 min, then column programmed from 50°C to 300°C at 10°C/min and subsequently kept at 300°C for 5 min; helium carrier gas flow 1mL/min. Ion source and interface temperatures were 250 and 200 °C, respectively. Ionization energy was 70 eV (20).

Capillary Gas chromatography

Hewlett Packard HP 6890 series GC equipped with flame ionization detection (FID) system with an SPB-5 fused –silica capillary column (30mx0.25mm I.D., 0.5 µm film thickness) from Supelco. The conditions for the gas chromatography were: injector temperature, 300°C; detector temperature, 300°C and oven temperature gradient from 100 to 200°C at 40°C/min, then increased by 10°C/min from 200 to 320°C and subsequently kept at 320°C for 30 min. Helium was used as a carrier gas at a flow-rate of 1 ml/min (20).

Qualitative determinations with GC-MS

Identification of tetracosanol (C24OH), hexacosanol (C26OH) and octacosanol (C28OH) in the chromatographic profile of sugar cane wax extract, after derivatization to the trimethylsilyl (TMS) derivative, by direct comparison with pure polycosanol standards (Tetracosanol, Hexacosanol and Octacosanol) to the sample using its relative retention and the characteristic mass fragmentation. The sample was subjected to GC-MS analysis, monitoring the most abundant M-15 ion at m/z of 411, 439 and 467, respectively and other characteristic fragments for the trimethylsilyl -TMS alcohols derivatives.

Quantitative determinations with GC:

Stock solutions were prepared by dissolving 10 mg of each of Tetracosanol, hexacosanol and octacosanol in 20 ml acetone. Portions of these stock solutions were taken and dried under nitrogen. For derivatization, 20 µl of MSTFA were added; the tubes were heated for 15 min at 60°C. The tubes were dried under nitrogen, then 500µL of chloroform were added. Calibration curves were developed from these solutions in the range of 0.05 to 5µg/ml through injection of 1µ into the GC (20). The sugar cane wax sample was prepared using the same procedure.

Animals:

The experiment was done on 30 male albino rats weighing 80g ± 1.0. The rats were housed in individual stainless steel screen bottom cages. The rats were fed on basal diet for one week for adaptation, water was available ad-libitum. The rats were equally divided into 5 groups.

Diet :

Basal diet was provided in accordance with (21) and (22), as shown in Table 1.

Animal Experiment:

The first group was fed on basal diet (control).The second group was fed on basal diet containing 1 % cholesterol for six weeks. The third, fourth and fifth groups were fed on basal diet containing 1 % cholesterol plus sugar cane wax extract (15, 45 and 100 mg/kg diet, respectively). After 6 weeks the animals were fasted overweight, blood samples were with drawn by a fine capillary glass tube from the orbital plexus vein. The blood was collected on heparin and centrifuged at 3000 rpm for 15 min. and stored at –20°C until analysis.

METHODS

Plasma samples were analyzed for enzyme activity of AST and ALT (23), urea (24), total cholesterol (25), triglycerides (26), HDL, LDL cholesterol (27), creatinine (28), and malondialdhyde (29).

Statistical analysis.

The results were expressed as mean±SD. Statistical significance was calculated using student`s t test according to (30). Differences were considered statistically significant if the p value < 0.05.

RESULTS

Wax extract was subjected to GC-MS analysis, monitoring the most abundant M-15 ion at m/z 411, 439 and 467 in order to confirm the presence of tetracosanol-, hexacosanol- and octacosanol-TMS derivative respectively (Figure 1). Figure (2) shows the GC chromatograms of polycosanol standards: 0.5μg/ml tetracosanol at 13.6 min, 0.5μg/ml hexacosanol at 15.25 and 2.5 μg/ml octacosanol at 16.67 minutes, respectively (A) and chromatogram of sugar cane wax (B). The quantitative determination of (C24 OH), (C26OH) and (C28OH) for calibration curve of the method was carried out according to the standard technique (Figure 3). The results showed that octacosanol is the major component of polycosanol (Figure 4). Body weight (initial, final and weight gain), organs weight (liver, spleen, kidney, heart and lung) and relative organs weight (organ weight / final body weight) are shown in Table 2. Non significant differences were noted in the mean value of body weight, organs weight and relative organs weight between different experimental groups. Table 3 illustrates the hemoglobin and activities of transaminase (AST & ALT) in plasma of male albino rats in addition of plasma creatinine and urea of the different experimental groups. The results showed no significant changes in AST & ALT activities between different treated groups. No significant effect was observed in plasma urea and creatinine or hemoglobin in the different treated groups compared to control group.

Table 4 illustrates the total cholesterol (TC), high density lipoprotein cholesterol (HDL-C), low density lipoprotein cholesterol (LDL-C), triacylglycerols and risk ratio TC /HDL-C. The Total cholesterol and low density lipoprotein cholesterol showed significant increase in hypercholesterolemic rats in group 2 (52 % and 169 %) compared with group 1 (control). Supplementation of sugar cane wax extract in both groups 4 and 5 decreased its level compared with cholesterol group 2. Thus, administration of cholesterol to rats fed diet containing sugar cane wax extract (45 mg/kg and 100 mg/kg diet), led to significant reduction (P<0.05) of TC by (31.5 %and 27.9 %) and LDL-C (by 36.7 %and 54.7 %), respectively in plasma, compared with the non-supplemented hypercholesterolaemic rats of group 2 .

Significant increase was observed in HDL-C of group 5 compared to control. The data obtained revealed that triacylglycrol was significantly decreased due to supplementation with cane sugar wax extract of groups 3 and 5 compared to hypercholesterolemia in rats of group 2 (Table 4). Also, the risk ratio TC /HDL-C was lowered from 3.04 in group 2 to 1.70 in group 5 as compared to control. It also improves the risk ratio by decreasing its value; such ratio is commonly used as an index of coronary heart disease. The mean values of malondialdehyde

content in different experimental groups are presented in Figure 5. Significant decrease (P<0.05) in MDA level was observed in group 5 compared to control.

Table 1: The Composition of Diet (g/kg).

Groups	Group (1)	Group (2)	Group (3)	Group (4)	Group (5)
Casein	120	120	120	120	120
Sucrose	100	100	100	100	100
*Salt mixture	50	50	50	50	50
**Vitamin mix	10	10	10	10	10
Corn oil	60	60	60	60	60
Cellulose	40	40	40	40	40
Cholesterol	0	10	10	10	10
Sugar cane wax	0	0	0.015	0.045	0.100
Starch	620	610	610	610	610

* Ref, (21) , ** Ref, (22)

Table 2: Initial, terminal body weights, weight gain and relative organs weight among different treatment groups.

Group	Group (1)	Group (2)	Group (3)	Group (4)	Group (5)
	Mean ±SE				
Initial weight (g)	81.89± 3.26	81.34±4.75	81.30±3.85	81.28±4.40	81.40±5.11
Final weight(g)	183.40±6.94	169.00±7.53	174.38±13.79	175.23±6.46	175.00±8.00
Wight Gain(g)	101.50±7.27	87.60±9.86	89.92±6.91	92.73±2.69	93.60±8.96
Liver weight (g)	6.26±0.30	5.68±0.22	6.47±0.48	6.43±0.18	6.11±0.37
%	3.42±0.15	3.37±0.10	3.72±0.04	3.71±0.12	3.3.1±0.16
Heart weight (g)	0.60±0.03	0.55±0.01	0.57±0.03	0.59±0.01	0.57±0.03
%*	0.33±0.02	0.33±0.02	0.33±0.01	0.32±0.01	0.32±0.03
Spleen	0.85±0.10	0.70±0.07	0.75±0.02	0.74±0.03	0.68±0.06

weight (g)					
%*	0.46±0.04	0.42±0.05	0.42±0.03	0.40±0.04	0.38±0.03
Kidney weight (g)	1.61±0.12	1.31±0.50	1.49±0.14	1.46±0.04	1.44±0.07
%*	0.88±0.04	0.78±0.04	0.85±0.05	0.83±0.03	0.82±0.03
Lung weight (g)	1.27±0.14	1.17±0.06	1.12±0.08	1.19±0.06	1.08±0.06
%*	0.690.06	0.70±.04	0.64±0.01	0.68±0.20	0.61±0.04

* (organ weight / final body weight)x 100

Table 3: Plasma liver and kidney functions and hemoglobin among different treatment groups.

Group	Group (1)	Group (2)	Group (3)	Group (4)	Group (5)
	Mean ±SE				
Hemoglobin (g/dl)	15.20±1.31	14.90±2.11	14.70±1.70	14.89±2.01	15.01±2.11
GOT (Unit/L)	22.63±5.80	36.20±3.15	35.47±4.58	31.20±5.11	33.59±5.38
GPT (Unit/L)	44.02±5.81	43.75±6.54	49.59±5.27	50.62±4.31	48.61±8.10
Urea (mg/dl)	40.33±5.22	41.11±4.25	42.55±3.99	41.45±4.29	39.49±5.21
Creatinine (g/dl)	2.97±.0.45	3.73±0.74	2.89±0.46	3.23±0.51	4.69±0.44

Table 4: Plasma lipid profile among different treatment groups (mg/dl).

Group	Group (1)	Group (2)	Group (3)	Group (4)	Group (5)
	Mean ±SE				
Total cholesterol	46.53±1.42^{a}	70.81±1.93^{b}	67.52±1.84^{ab}	48.50±3.77^{ac}	51.06±3.21^{ac}
HDL	29.62±2.62	23.35±1.90^{a}	17.60±3.37	18.69±3. 4	30.06±2.36^{b}
LDL	17.66±1.34^{a}	47.45±3.21^{b}	49.98±4.57^{ab}	29.91±5.55^{c}	21.50±2.70^{ac}
TG	124.0±4.17^{a}	170.12.49^{b}	134.00±13.00^{ab}	131.00±4.17^{ac}	120.00±13.68^{ac}
TC/HDL	1.57±0.13^{a}	3.03±0.14^{b}	3.83±0.15^{ab}	2.59±0.17^{ab}	1.70±0.15^{ac}

a, b , c same scripts in the same row indicate no significant differences ($p<0.05$)

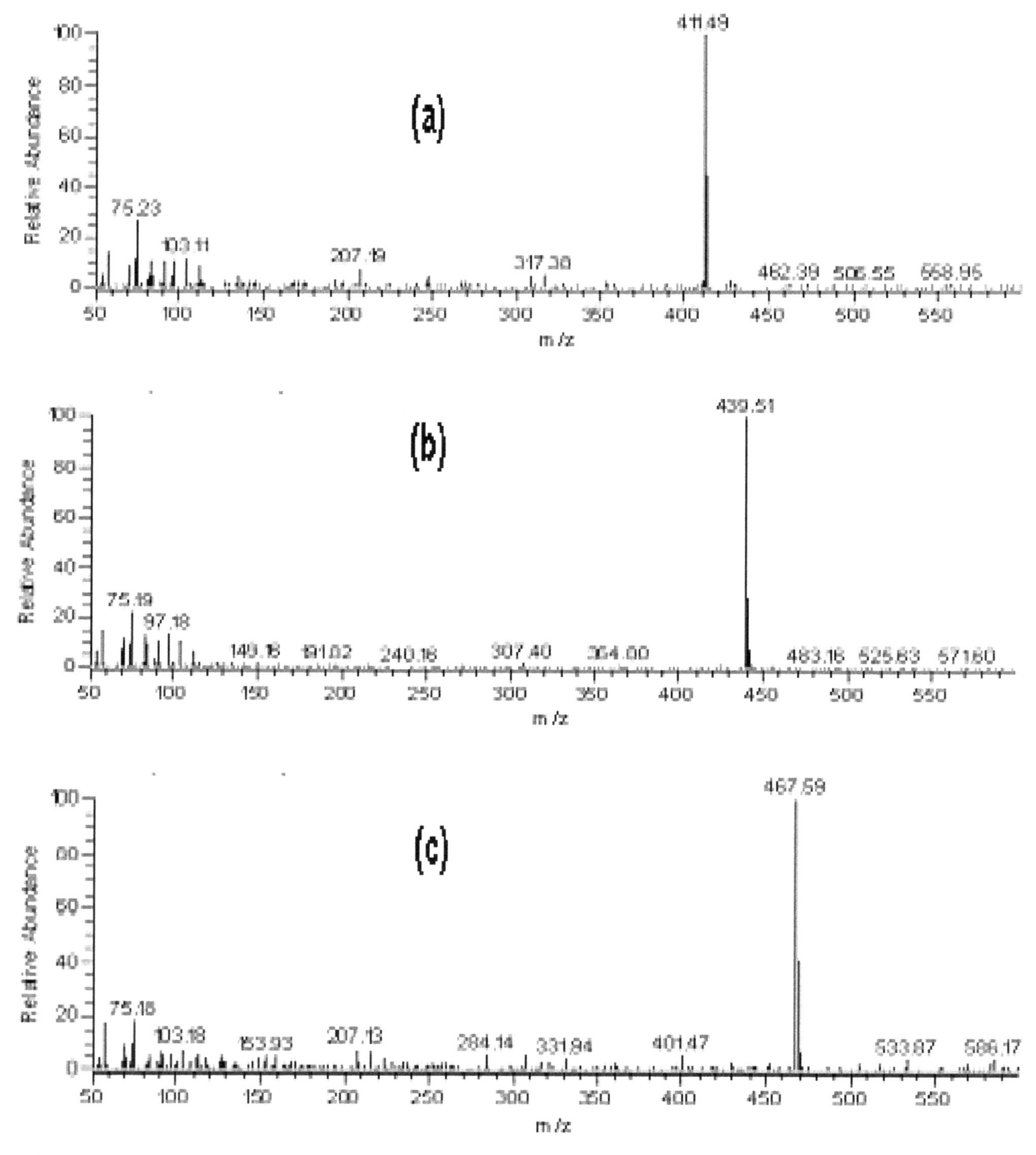

Figure 1: Mass ion fragments of (a) tetracosanol, (b)hexacosanol and (c) octacosanol-MSTFA derivative.

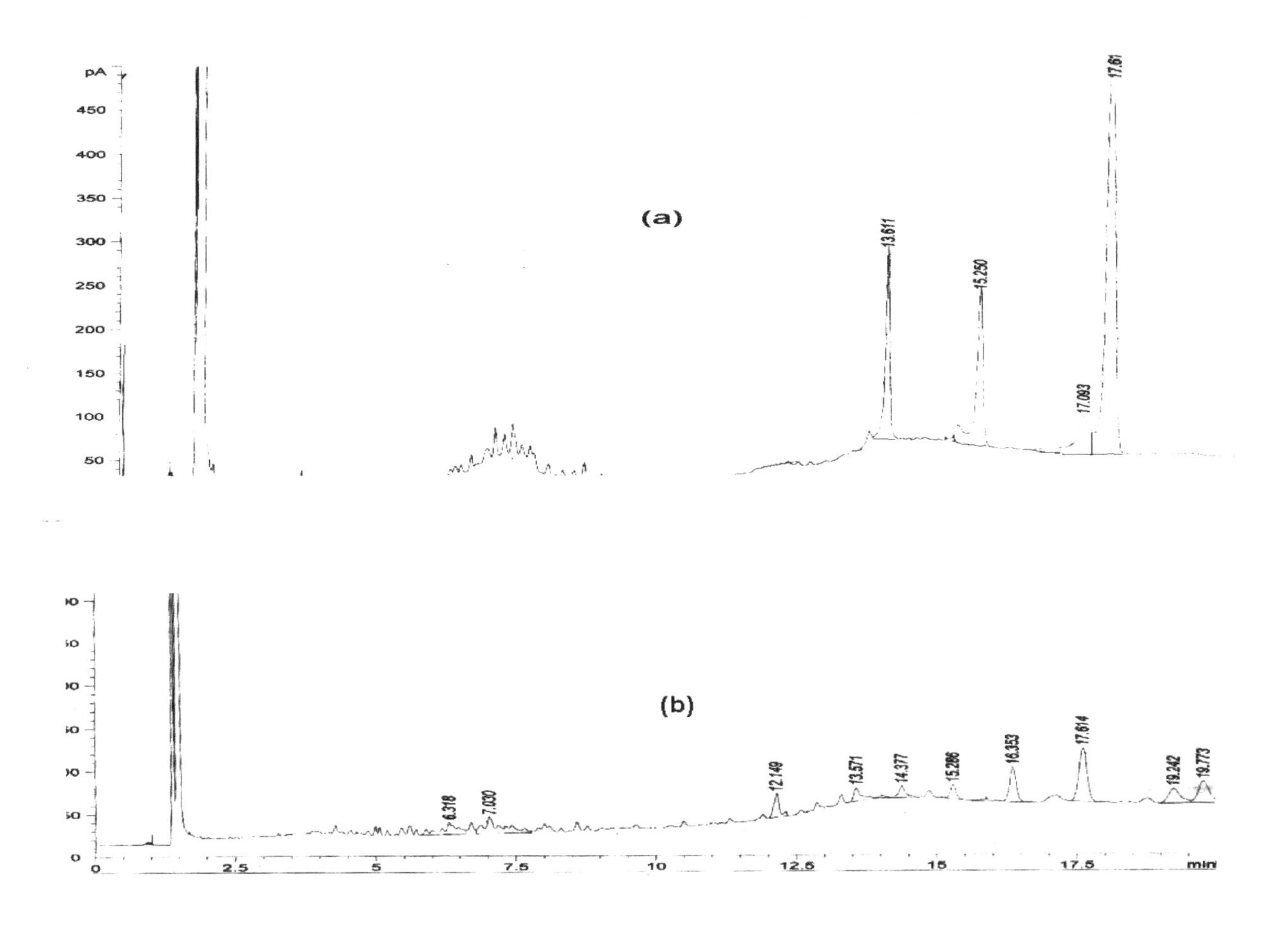

Figure 2: GC Chromatograms of polycosanol-TMS derivative's standards: 0.5μg/ml tetracosanol at 13.6 min, 0.5/ml μg hexacosanol at 15.25 and 2.5/ml μg octacosanol at 16.61 min, respectively (a) and chromatogram of sugar cane wax (b).

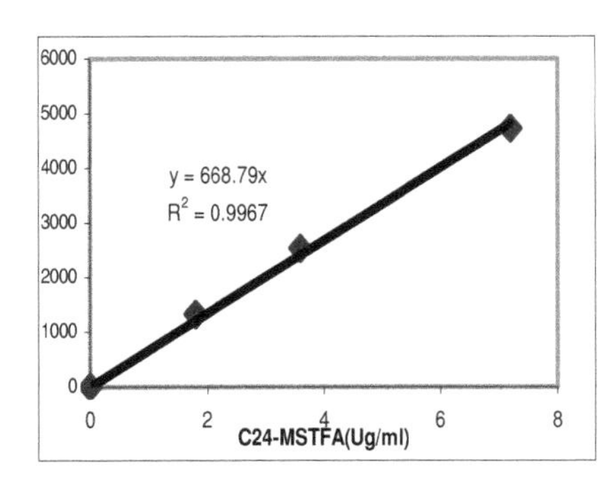

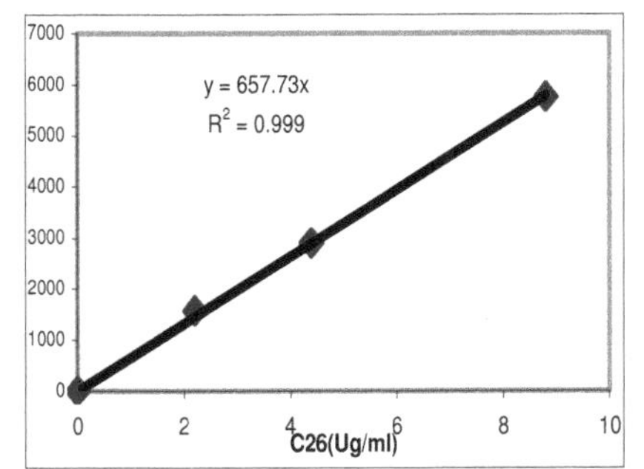

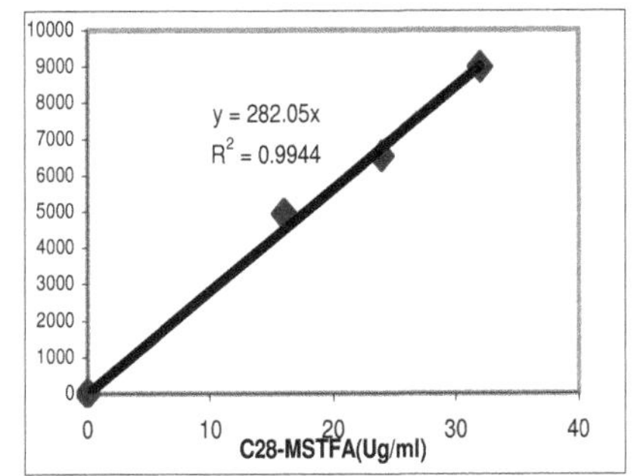

- a - - b- - c-

Figure 3: Standard curves for polycosanol- MSTFA derivatives (a, tetra-: b, Hexa- and c, octa-cosanol) as developed by GC by plotting concentration against peak area.

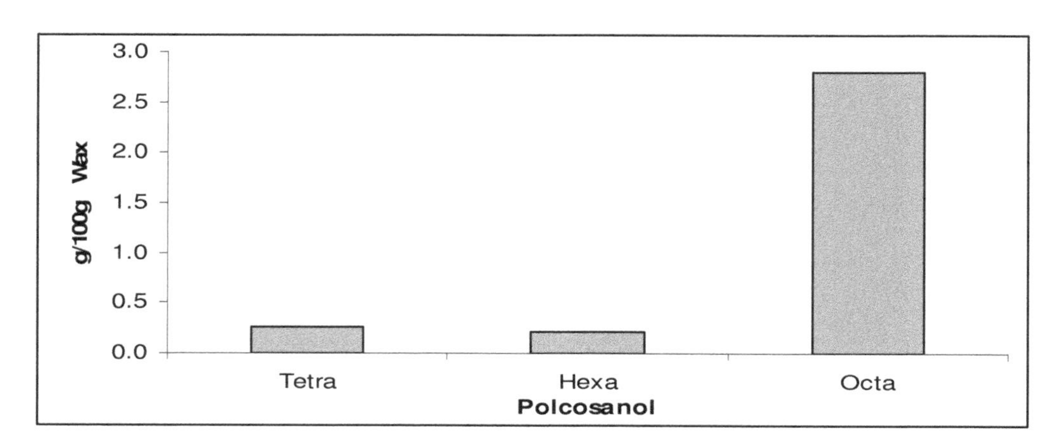

Figure 4: Content of polycosanols in sugar cane wax extract.

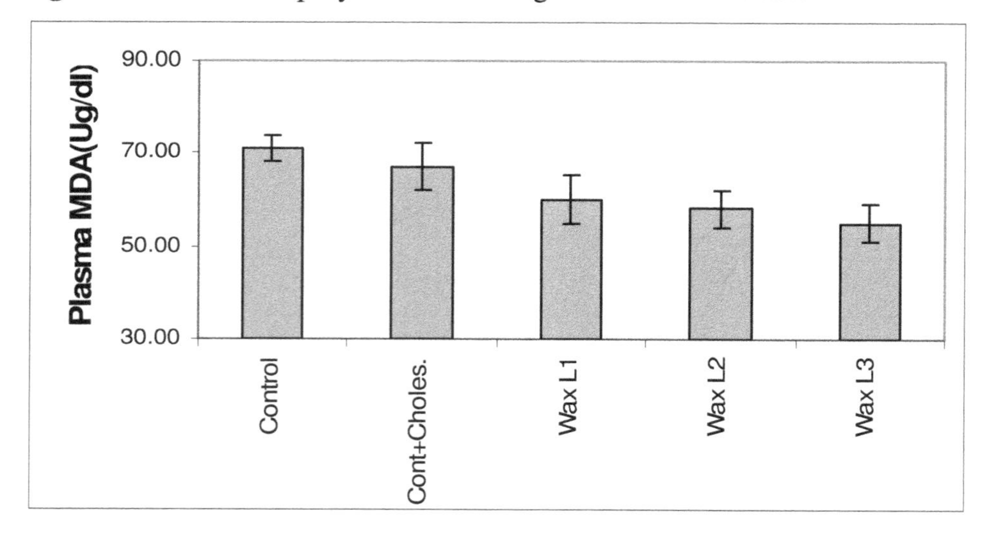

Figure 5: Effect of sugar cane wax on the level of malondialdehyde.

DISCUSSION

Octacosanol is the major compound of policosanol, a mixture of high-molecular-mass primary fatty alcohols obtained from sugar cane wax. These results are in accordance with

(11,31,19). The results concerning relative organ weights indicated non significant difference between supplementation with sugar cane wax extract and control group.

The present results demonstrated that supplemented animals with sugar cane wax extract showed insignificant difference in activities of transaminase (ALT and AST). The obtained results revealed that neither urea nor creatinine levels were affected by sugar can wax extract consumptions compared to control. Oral toxicity of policosanol was evaluated in a 12-months study in which daily doses from 0.5 to 500 mg / kg were given orally to Sprague Dawley rats (32).

The effects of sugar can wax extracts on body weight gain, organ weight (%) and blood parameters (ALT, AST hemoglobin, creatinine and urea) were similar in control and treated groups. Toxicity of policosanol obtained from sugar cane wax in beagle dogs was studied by (33). This study has shown that no drug –related toxicity was induced by policosanol administered up to 180 mg /kg/day for 52 weeks to beagel dogs. On the other hand, (16) indicated that single oral doses of 1000 mg of policosanol administered to healthy volunteers were tolerated without adverse drug reaction. The results showed that supplementation with sugar cane wax extract (45mg/kg and 100 mg/kg) for 6 weeks significantly reduced hypercholesrolaemia induce rats fed on diets containing 1.0 %_cholesterol diet. Policosanol diminished the increase in plasma TC, LDL-cholesterol and triacylglycerols, moreover level of HDL-C was significantly increased. These results are in accordance with (11, 31). Lipid lowering effects of policosanol in a variety of animal species, including rabbits (11-13) and rats (32, 34-37), and other rodents (38), beagle dogs (33), monkeys (39), swine and chickens was studied. Since policosanol exerts its main effect on LDL-C, a mechanism of action through cholesterol synthesis inhibition or enhanced LDL-C catablism would be conceivable (7). In cultured human fibroblasts policosanol decreased carbon 14-labeled acetate incorporation into cholesterol, whereas incorporation of ^{14}C-labeled mevalonate was not affected suggesting inhibition of cholesterol synthesis in vitro at a step before mevalonate formation (40). LDL-C binding, uptake, and degradation were enhanced at concentrations that did not significantly decrease cholesterol synthesis (40). Moreover, a decrease in cholesterol synthesis in vivo in rats uising tritiated water could not be demonstrated (41). There was also no competitive or noncompetitive inhibition of hydroxymethylglutaryl (HMG) CoA reductase in liver homogenates (41). In rabbits with hypercholesterolemia induced by wheat starch-casein diet, polycosanol decreased tritiated water incorporation into hepatic sterols (11). In addition, the rate of removal of iodine 125-labeled LDL-C from blood and the hepatic LDL-C binding capacity were enhanced, suggesting that, at least in part, increased receptor mediated uptake by the liver is involved in the policosanol-induced LDL-C decrease. On the basis of the studies mentioned, the precise mechanism leading to LDL-C reduction remains unclear, but inhibition of cholesterol synthesis, increased hepatic LDL uptake, and increased serum LDL catabolic rates may play a role (7). Supplementation of sugar cane wax extract (level 3) reduces malondialdhyde content in plasma. These results are in accordance with (37, 31). This beneficial effect of policosanol on membrane lipid peroxidation may be useful in protecting to some extent against free radical-associated diseases (34).

CONCLUTION

Policosanol, in addition to its cholesterol-lowering effect, has other property that enables it to reduce the potential of lipid peroxidation. Such effect can be considered as a promising value in the management of atherosclerosis.

REFERENCES:

1. Kannel, W.B.; Castelli, W.P.; Gordon, T. et al.: Lipoprotein cholesterol in the prediction of atherosclerotic disease: new perspective based on the Framingham Heart study. Ann. Intern. Med. 90: 85-91, 1995.
2. National Heart Foundation of Australia and The Cardiac Society of Australia and New Zealand. Lipid management guidelines - 2001. Med. J. Aust. 175 (9 Suppl): S57-S88, 2001.
3. Expert Panel. Report of the National Cholesterol Education Program Expert Panel on detection, evaluation, and treatment of high blood cholesterol in Adults. Arch. Intern. Med. 148: 36-69, 1988.
4. Expert Panel on Detection, Evaluation, and Treatment of High Blood Cholesterol in Adults. Summary of the second report of the National Cholesterol Education Program (NCEP) expert panel on detection, evaluation, and treatment of high blood cholesterol in adults (Adult Treatment Panel II). JAMA 269: 3015-3023, 1993.
5. Laufs, U. and Liao, J.K.: Direct vascular effects of HMG-CoA reductase inhibitors. Trends Cardiovasc. Med. 10: 143–148, 2000.
6. Law, M.R.; Wald, N.J. and Rudnicka, A.R.: Quantifying effect of statins on low density lipoprotein cholesterol, ischaemic heart disease, and stroke: systematic review and meta-analysis. BMJ 326(7404): 1423-1429, 2003.
7. Gouni-Berthold, I.; Berthold, H.K. and Fulda, R.: Policosanol: Clinical pharmacology and therapeutic significance of a new lipid-lowering agent. Am. Heart J. 143(2): 356-365, 2002.
8. Más, R.: Policosanol: Drugs of the Future. 25:569–586, 2000.
9. Castano, G.; Mas, R.; Fernandez, J.C.; Illnait, J.; Fernandez, L. and Alvarez, E.: Effects of policosanol in older patients with type II hypercholesterolemia and high coronary risk. J. Gerontol. A: Biol. Sci. Med. Sci. 56(3): M186–M192, 2001.
10. Cruz-Bustillo, D.; Mederos, D.; Mas, R.; Arruzazabala, L.; Laguna, A.; Barreto, D.; Martinez, O.: Cholesterol-lowering effect of Ateromixol (PPG) on fattening hogs. Review CENIC Cinecias. Biologicas. 22: 62-63, 1991.
11. Menendez, R.; Arruzazabala, L.; Mas, R.; Del Rio, A.; Amor, A.M.; Gonzalez, R.M.; et al.: Cholesterol-lowering effect of Policosanol on rabbits with hypercholesterolaemia induced by a wheat starch-casein diet. Br. J. Nutr. 77(6): 923-932, 1997.
12. Mendoza, S.; Gdmez, R.; Noa, M.; Más, R.; Castaiio, G.; Mesa, R.; et al.: Comparison of the effects of D-003 and policosanol on lipid profile and endothelial cells in normocholesterolemic rabbits. Curr. Ther. Res. 62(3): 209-220, 2001.
13. Gamez, R.; Maz, R.; Arruzazabala, M.L.; Mendoza, S. and Castano, G.: Effects of concurrent therapy with policosanol and omega-3 fatty acids on lipid profile and platelet aggregation in rabbits. Drugs R D 6(1):11-9, 2005.
14. Hernandez, G.; Illnait, J.; Más, R.; Castano, G.; Fernandez, L.; Gonzalez, M.; et al.: Effect of Policosanol on serum lipids and lipoproteins in healthy volunteers. Curr. Ther. Res. 51: 568-575, 1992.
15. Menendez, R.; Mas, R.; Amor, A.M.; Gonzalez, R.M.; Fernandez, J.C.; Rodeiro, I.; et al.: Effects of policosanol treatment on the susceptibility of low density lipoprotein (LDL) isolated from healthy volunteers to oxidative modification in vitro. Br. J. Clin. Pharmacol. 50(3):255-262, 2000.
16. Pons, P.; Rodriguez, M.; Robaina, C.; Illnait, J.; Más, R.; Fernandez, L. and Fernandez, J.C.: Effect of successive dose increases of policosanol on the lipid profile of patients with type II

hypercholesterolaemia and tolerability to treatment. Int. J. Clin. Pharmacol. Res. 14(1): 27-33, 1994.

17. Más, R.; Castano, G.; Fernández, L.; Illnait, J.; Fernández, J.C. and Alvarez, E.: Effects of policosanol in older hypercholesterolemic patients with coronary disease. Clin. Drug Invest. 21:485–497, 2001.
18. Castano, G.; Más, R.; Fernández, L.; Illnait, J.; Mesa, M.; Alvarez, E. and Lescay, M.: Comparison of the efficacy and tolerability of policosanol and atorvastatin in elderly patients with type II hypercholesterolemia. Drug Aging 20(2):153–163, 2003.
19. Irmak, S.; Dunford, N.T. and Milligan, J.E.: Policosanol contents of beeswax, sugar cane and wheat extracts. Food Chem. 95: 312–318, 2006.
20. Delange, D.M. and Bravo, L.G.: Trace determination of 1-octacosanol in rat plasma by solid–phase extraction with tenax GC and capillary gas chromatography. J. Chromatogr. B 762: 43-49, 2001.
21. Müller, R.: Vorschrift zur proteinbewertung in Versuchen an wachsenden ratten. Z.Tierphysiol. Tierernahr. Futtermittelkd. 19,305, 1964.
22. Reeves, P.G.; Nielsen, F.H. and Fahey, G.C.: AIN-93 purified diets for laboratory rodents: final report of the American Institute of Nutrition ad hoc writing committee on the formulation of the AIN-76A rodent diet. J. Nutr. 123, 1939 – 1951, 1993.
23. Reitman, S. and Frankel, S.: A colorimetric method for the determination of serum glutamic-oxaloacetic and glutamic pyruvic transamination. Am. J. Clin. Path. 28 (56): 56-63, 1957.
24. Patton, C.J. and Grouch, S.R.: Spectrophotometric and kinetics investigation of the Berthelot reaction for the determination of ammonia. Anal. Chem. 49: 464-469, 1977.
25. Roeschlau, P.; Berndt, E. and Gruber, W.: Clin. Chem. Biochem. 12,403, 1974.
26. Schettler, G. and Nusse, l.E.: Arb. Med. Soz. Med. Prav. Med. 10: 25, 1975.
27. Wieland, H. and Seidel, D.: A simple specific method for precipitation of low density lipoproteins. J. Lipid Res. 24(7): 904-909, 1983.
28. Bartles, H.; Bohmer, M. and Heierli, C.: Serum kreatinin bestimmung ohne enteiweissen. Clin. Chim. Acta. 37: 193, 1972.
29. Uchiyama, M. and Mihara, M.: Determination of malonaldehyde precursor in tissues by thiobarbituric acid test. Anal. Biochem. 86(1): 271–278, 1978.
30. Statgraphics. Statistical Graphics System Version 2.6. Statistical Graphic Corporation, Graphic Software System Inc.U.S. STSC.Inc; 1987.
31. Arruzazabala, M.L.; Molina, V.; Más, R.; Fernández, L.; Carbajal, D.; Valdes, S. and Castano, G.: Antiplatelet effect of Policosanol (20 and 40 mg/day) in healthy volunteers and dyslipidaemic patients. Clin. Exp. Pharmacol. Physiol. 29(10): 891-897, 2002.
32. Aleman, C.L.; Más, R.; Hernandez, C.; Rodeiro, I.; Cerjido, E.; Noa, M.; et al.: A 12-month study of Policosanol oral toxicity in Sprague Dawley rats. Toxicol. Lett. 70(1): 77-87, 1994.
33. Mesa, A.R.; Más, R.; Noa, M.; Hernandez, C.; Rodeiro, I.; Gamez, R.; et al.: Toxicity of Policosanol in beagle dogs: one-year study. Toxicol. Lett. 73 (2): 81-90, 1994.
34. Fraga, V.; Menendez, R.; Amor, A.M.; Gonzalez, R.M.; Jimenez, S. and Más, R.: Effect of Policosanol on in vitro and in vivo rat liver microsomal lipid peroxidation. Arch. Med. Res. 28(3):355-360, 1997;.
35. Menendez, R.; Mas, R.; Amor, A.M.; Ledon, N.; Perez, J.; Gonzalez, R.M.; et al.: Inhibition of rat lipoprotein lipid peroxidation by the oral administration of D003, a mixture of very long-chain saturated fatty acids. Can. J. Physiol. Pharmacol. 80(1):13-21, 2002.
36. Noa, M.; Más, R.; de la Rosa, M.C. and Magraner, J.: Effect of policosanol on lipofundin-induced atherosclerotic lesions in rats. J. Pharm. Pharmacol. 47(4): 289-291, 1995.

37. Menendez, R.; Fraga, V.; Amor, A.M.; Gonzalez, R.M. and Mas, R.: Oral administration of Policosanol inhibits in vitro copper ion-induced rat lipoprotein peroxidation. Physiol. Behav. 67(1):1-7, 1999.
38. Wang, Y.; Ebine, N.; Jia, X.; Jones, P.J.H.; Fairow, C. and Jaeger, R.: Very long chain fatty acids (policosanols) and phytosterols affect plasma lipid levels and cholesterol biosynthesis in hamsters. Metab. Clin. Exp. 54: 508– 514, 2005.
39. Noa, M. and Más, R.: Protective effect of policosanol on atherosclerotic plaque on aortas in monkeys. Arch. Med. Res. 36(5): 441–447, 2005.
40. Menedez, R.; Fernandez, S.I.; Del Rio, A.; Gonzalez, R.M.; Fraga, V.; Amor, A.M. and Mas, R.M.: Policosanol inhibits cholesterol biosynthesis and enhances low density lipoprotein processing in cultured human fibroblasts. Biol. Res. 27(3-4): 199-203, 1994.
41. Menendez, R.; Amor, A.M.; Gonzalez, R.M.; Fraga, V. and Más, R.: Effect of policosanol on the hepatic cholesterol biosynthesis of normocholesterolemic rats. Biol. Res. 29(2):253-257, 1996.

ANTIDIABETIC POTENTIAL OF NEPALESE HERBAL AND FOOD PLANTS

Megh Raj Bhandari, Nilubon Jong-Anurakkun and Jun Kawabata

Laboratory of Food Biochemistry, Division of Applied Bioscience, Graduate School of Agriculture, Hokkaido University, Sapporo- 8589, Japan

Corresponding author: Megh Raj Bhandari, E-mail: mrjbhandari@yahoo.com

Key words: Herbal plants, Diabetes, Traditional medicine, Nepalese plants

ABSTRACT

Diabetes mellitus is caused due to deficiency in production of insulin by the pancreas or by the ineffectiveness of the insulin produced. The rapidly increasing diabetes mellitus is becoming a serious threat to human health in all over the world. Medicinal plants are well-known natural sources for the treatment of various diseases since antiquities. To treat diabetes, the rural Nepalese people use herbal treatments either alone or in combination with other form of treatments, but therapeutic importance of many such herbs have not been investigated yet. The aim of this study was to investigate *in vitro* antidiabetic potential of selected food and herbal plants of Nepal. The 50% methnolic extracts of 39 plant species, prevalently using in Nepalese folk medicine system, were studied. Antidiabetic activity was evaluated by three separate methods: Rat intestinal sucrase and maltase inhibitory activities, and Porcine pancreatic α-amylase (PPA) inhibitory activity. Studied plants showed a wide variation in these enzymes inhibitory activities ranging from moderate (≤ 50%), high (≤ 75%) to very high (≥ 75%). The results indicated that a number of Nepalese herbs have remarkable antidiabetic potential. Among them, *Acacia catechu, Bergenia ciliata, Cassia fistula, Cedrus deodara, Centella asiatica, Holarrhena pubescens, Inula recemosa, Justicia adhatoda, Myrica esculenta, Neopicrorhiza scrophulariiflora, Nymphaea stellata, Rheum australe, Rhododendrom arboreum, Rubia manjith, Sapindus mukorossi, Swertia angustifolia, Taxus baccata, Terminalia bellirica, Terminalia chebula, Valeriana villosa,* and *Woodfordia fruticosa* showed significant inhibitory activities for these targeted digestive enzymes and could be considered as potential sources of antidiabetic agents. Dietary α-amylase and α-glucosidase inhibitors from natural sources are potentially safer, and therefore, studied herbs may be a preferred alternative for the use in the formulation of new therapeutic drugs, nutraceutical and functional foods that help to prevent or treatment of diabetes. Furthermore, such comprehensive research will justify the potential efficacy of these traditional foods and herbal medicines and will open avenue for further scientific study.

Outline of content

- Introduction
- Materials and methods
    - ►Materials
    - ►Preparation of the methanol extracts
    - ►Rat intestinal sucrase inhibitory activity assay

INTRODUCTION

Diabetes mellitus is a metabolic disorder in the endocrine system and is found in all over the world. People suffering from diabetes are not able to produce or properly use insulin in the body, so they have a high content of blood glucose. As a very common chronic disease, diabetes is becoming the third "killer" of the human health along with cancer, and cardio- and cerebrovascular diseases because of its high prevalence, morbidity and mortality (1).

Diabetes mellitus is currently one of the most costly and burdensome chronic diseases, and is a condition that is increasing in epidemic proportions throughout the world (2). Diabetes affects about 5% of the global population and management of diabetes without any side effects is still a challenge to the medical system (3, 4). To treat diabetes, a number of synthetic drugs are in use but unfortunately, the risk of severe hypoglycemia associated with the use of these medications can be contradicted because of congestive heart failure, renal insufficiency, or liver disease, or may not be tolerated because of gastrointestinal adverse effects (5). Consequently, the alternative therapeutic agents, such as α-glucosidase inhibitors, have been investigated.

α-amylase and α-glucosidases are exoenzymes hydrolyzing terminal glycosidic bonds and releasing α-glucose from the non-reducing end of the substrate chain; starch, oligosaccharides and disaccharides (Fig.1). Thus, the inhibition of these digestive enzymes could delay the degradation of starch, oligosaccharides and disaccharides that brings about a decrease in the intestinal absorption rate of glucose and the postprandial blood glucose rise in non-insulin dependent diabetes mellitus (6, 7, 8, 9, 10).

To this end, research has begun to embrace traditional medicines and foods from various cultures as scientists search for clues to discover new therapeutic drugs for diabetes and associated hypertension. Traditional Indian and Chinese medicines have long been using plant and herbal extracts as antidiabetic agents (11, 12). Indeed, many medicinal plants and herbal extracts have been found to inhibit the enzymatic activity of α-glucosidase and α-amylase (13, 14, 15, 16, 17, 18, 19, 20, 21). Herbal drugs are prescribed widely because of their effectiveness, less side effects and relatively low cost. Therefore researches are mounting on natural herbal plants to search for clues to discover new therapeutic drugs for diabetes (1).

To treat diabetes, the rural Nepalese communities were found to use herbal treatments either alone or in combination with other form of treatments (22). Based on the richness of the local flora and interest of the population in medicinal plants, the Nepalese flora should be subjected to comprehensive scientific evaluation of their chemical, pharmacological, and biological properties. Such comprehensive research will justify the potential efficacy of these traditional herbal medicines. However, Nepalese herbal plants remain uninvestigated in this respect. Therefore, in this study, we investigated the *in vitro* antidiabetic potential (anti-amylase and anti-glucosidase activities) of selected herbal and food plants of Nepal which have not been studied yet.

EXPERIMENTAL MATERIALS

We collected the plants which have a good reputation in popular traditional medicine. The plants were either collected in different parts of Nepal or purchased from the local herbal shops in Kathmandu and were identified and authenticated using descriptive literature and in comparison with the herbarium collection in the National Herbarium Center, Godawari, Kathmandu, Nepal. The family, botanical names, local names and the parts used for the study are shown in Table 1. Rat intestinal acetone powder, porcine pancreatic α-amylase and starch azure (RBB-starch) were supplied by Sigma Aldrich Japan Co., Tokyo, Japan. The ICN Alumina B, Akt.I was purchased from ICN Biomedical GmbH, Germany. Other chemicals were of analytical grade and were purchased from Wako Pure Chem. Co. Osaka, Japan.

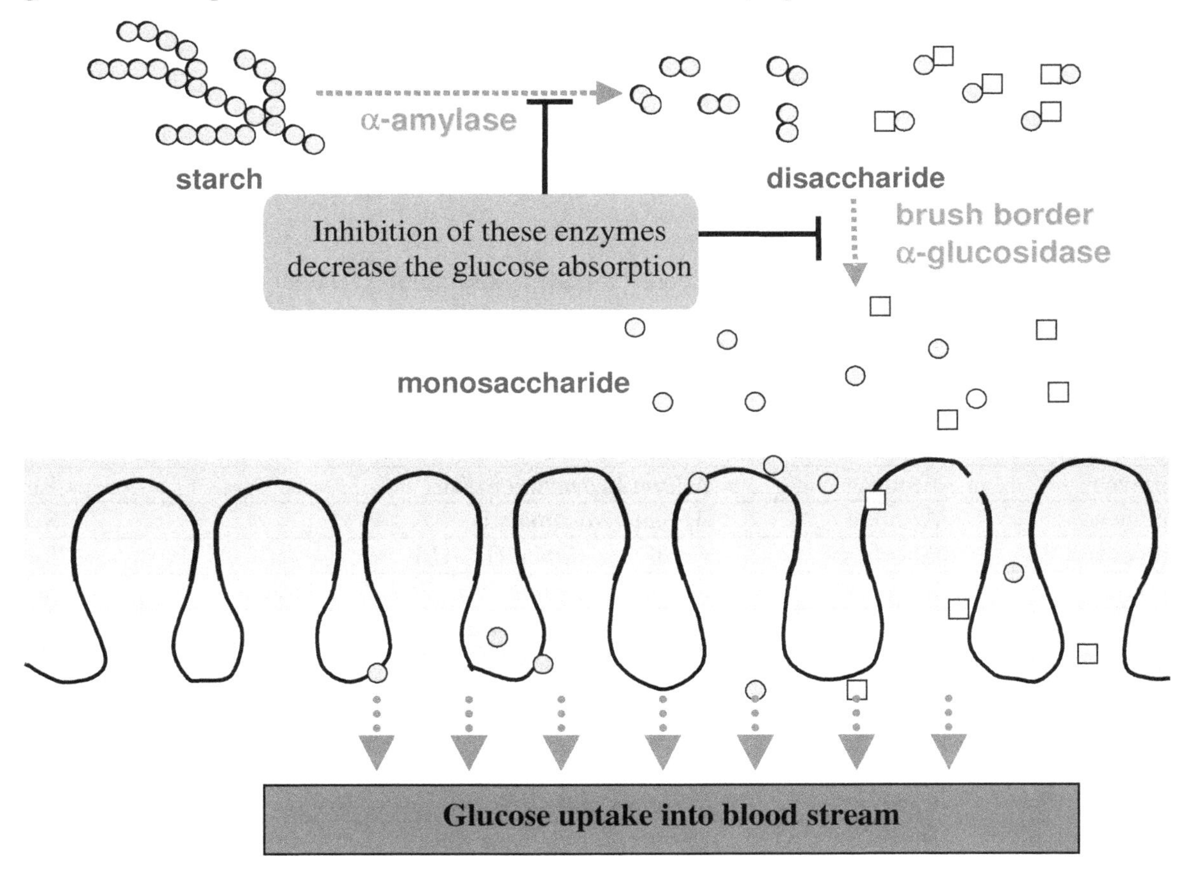

Figure 1: Schematic diagram showing the role of digestive enzyme in carbohydrate digestion.

METHODS

Preparation of the methanol extracts

Five grams of dried herbal plants were extracted with 50 ml of 50 % aqueous methanol for 24 hr at room temperature. The methanolic extracts were filtered using Whatman filter paper and then concentrated in vacuum at 40oC using Rotary Evaporator. The dried residue was redissolved in 50 % dimethyl sulfoxide (DMSO) with a final concentration of 15mg/ml and subjected to porcine pancreatic α-amylase (PPA), sucrase, and maltase inhibitory activity assay.

Rat intestinal sucrase inhibitory activity assay

Rat intestinal sucrase inhibitory activity was determined using the method described previously (15) with a slight modification. Sucrose solution (56 mM) and 0.1 M potassium phosphate buffer (pH 7) were pre-incubated at 37°C for 5 min. The test samples in 50% DMSO were added into each assay followed by the crude enzyme solution. After the reaction was carried out at 37°C for 15 min, 750 μl of 2 M Tris-HCl solution (pH 6.9) was added to stop the reaction. Then, the reaction mixture was passed through a basic alumina column (ϕ 6 mm x 3.5 mm h) to remove phenolic or acidic compounds and the glucose concentration was measured by a glucose oxidase method (Glucose-B Test Kit, Wako, Osaka, Japan). The mixture of 50 μl filtrate and 200 μl glucose kit solution was incubated in a 96-wells microplate at 37°C for 30 min. The optical density (OD) of the wells was measured at 490 nm. The inhibitory activity was calculated by the following equation.

$$\text{Inhibitory activity (\%)} = (\text{OD control} - \text{OD test sample}) / \text{OD control} \times 100$$

Table 1: Description of plants used in this study

Plant family	Local name	Scientific name	Plant part
Acanthaceae	Asuro	*Justicia adhatoda* L.	Leaf
Alliaceae	Jimbu	*Allium hypsistum* Stearn	Leaf
Apiaceae	Ajmoda	*Apium graveolens* L.	Seed
Apiaceae	Ghortapre	*Centella asiatica (*L.) Urb.	Leaf
Apocynaceae	Indrajau	*Holarrhena pubescens* Wall.& G.Don	Seed
Apocynaceae	Sarpagandha	*Rauvolfia serpentina* Benth.ex Kurz	Root
Asteraceae	Titepati	*Artemisia indica* Willd.	Leaf
Asteraceae	Bhringaraj	*Eclipta prostrata (*L.) L.	Leaf
Asteraceae	Rasana	*Inula cappa* (Buch.-Ham ex D.Don) DC.	Leaf
Bignoniaceae	Tatelo	*Oroxylum indicum* (L.) Kurz.	Seed
Combretaceae	Barro	*Terminalia bellirica* (Gaertn.) Roxb.	Fruit
Combretaceae	Harro	*Terminalia chebula* Retz.	Fruit
Convolvulaceae	Nishoth	*Operculina turpethum* (L.) Silva Manso	Root
Ericaceae	Laligurans	*Rhododendron arboreum* Sm.	Flower
Gentianaceae	Chiraito	*Swertia angustifolia* (Buch.-Ham ex D.Don)	Leaf
Lamiaceae	Tulsai	*Ocimum tenuiflorum* L.	Leaf
Lauraceae	Siltimur	*Lindera neesiana* (Wall. Ex Nees) Kurz	seed
Leguminosae	Khayar	*Acacia catechu* (L.f.) Willd.	Bark

Leguminosae	Rajbrikksha	*Cassia fistula* L.	Fruit
Leguminosae	Salparni	*Desmodium gangeticum* DC.	Leaf
Lythraceae	Dhaniyaro	*Woodfordia fruticosa (*L.) Kurz	Flower
Meliaceae	Neem	*Azadirachta indica* A.Juss.	Leaf
Myricaceae	Kaphal	*Myrica esculenta* Buch.-Ham ex D.Don	Bark
Myrsinaceae	Vayuvidanga	*Embelia ribes* Burm. f.	Fruit
Nyctaginaceae	Punarnawa	*Boerhavia diffusa* L.	Leaf
Nymphaeaceae	Nilakamal	*Nymphaea stellata* Willd	Flower
Pinaceae	Debdar	*Cedrus deodara* (Roxb. ex Lambert)G.Don	Bark
Plumbaginaceae	Seto kuro	*Plumbago zeylanica* L.	Leaf
Polygonaceae	Padamchal	*Rheum australe* D.Don	Bark
Rubiaceae	Majhitho	*Rubia manjith* Roxb. ex Flem.	Root
Rutaceae	Bel	*Aegle marmelos* (L.) Correa	Fruit
Rutaceae	Timur	*Zanthoxylum armatum* DC.	Fruit
Sapindaceae	Ritha	*Sapindus mukorossi* Gaertn.	Fruit
Saxifragaceae	Pakhanbed	*Bergenia ciliata* (Haw.) Sternb.	Root
Scrophulariaceae	Kutki	*Neopicrorhiza scrophulariiflora* (Pennell) D.Y. Hong	Root
Taxaceae	Talsipatra	*Taxus baccata* L.	Leaf
Valerianaceae	Jatamasi	*Nardostachys grandiflora* DC.	Root
Valerianaceae	Sugandhawala	*Valeriana villosa* Wall.	Root
Zingiberaceae	Aduwa	*Zingiber officinale* (Willd.) Roscoe	Rhizome

Rat intestinal maltase inhibitory activity assay

Rat intestinal maltase inhibitory activity was determined using the method described previously (17) with a slight modification. Rat maltase inhibitory activity assay was carried out in the same manner as the sucrase inhibitory activity assay described in earlier section except for using 35 mM maltose solution as a substrate. The inhibitory activity was calculated by the same manner as discussed earlier.

Porcine pancreatic α-amylase (PPA) inhibitory activity assay

PPA inhibitory activity was determined using the method described previously (16) with a slight modification. Starch azure (2 mg) used as a substrate was suspended in 0.5 M Tris-HCl buffer (pH 6.9) containing 0.01 M CaCl2 and soaked in boiling water for 5 min. Then, the starch azure solution was pre-incubated at 37°C for 5 min. The test samples in 50% DMSO and PPA were added into each assay and the reaction was carried out at 37°C for 10 min and stopped by adding 0.1 ml of 50% acetic acid. The reaction mixture was then centrifuged (3000 rpm, 4°C) for 5 min. The absorbance of the supernatant at 595 nm was measured. The inhibitory activity was calculated by the same manner as discussed earlier.

Statistical analysis

The values are expressed as percentage of inhibition compared to the control. The results represent the mean of three independent analyses and all the data are expressed as mean±SEM.

RESULTS AND DISCUSSION

Antidiabetic potential was evaluated by three separate methods: Rat intestinal sucrase and maltase inhibitory activities, and Porcine pancreatic α-amylase (PPA) inhibitory activity and the results are summarized in Table 2. The plant extracts were individually investigated and values are expressed as percentage of inhibition.

In our *in vitro* system, among the plant species studied, 11 species showed very high (50-100% inhibition) rat intestinal sucrase inhibitory activity, nine plant species showed moderate inhibitory activity (25-50% inhibition) and the rest were found low (≤ 25%) to this enzyme inhibitory activity. Plant species *Sapindus mukorossi* (96.9%), *Rubia manjith* (89.9%), *Justicia adhatoda* (86.3%), *Neopicrorhiza scrophulariiflora* (80.2%), *Valeriana villosa* (77.5%), *Inula recemosa* (67.4%), *Acacia catechu* (66.8%), and *Centella asiatica* (64.3%) showed comparatively high sucrase inhibitory activity. The plant species showed stronger inhibition to maltase compared to that of sucrase. A total of 15 species showed very high (50-100% inhibition) rat intestinal maltase inhibitory activity, 15 plant species showed moderate inhibitory activity (25-50% inhibition) and the rest were found low (≤ 25%) to this enzyme inhibitory activity. Plant species *Rhododendrom arboreum* (95.7%), *Sapindus mukorossi* (91.1%), *Terminalia chebula* (91.1%), Acacia *catechu* (84.5%), *Terminalia bellirica* (82.9%), *Woodfordia fruticosa* (77.1%), *Nymphaea stellata* (74.2%), *Cedrus deodara* (71.5%), *Bergenia ciliata* (68.6%), and *Rubia manjith* (66.6%) showed comparatively high maltase inhibitory activity. Although the extracts of Nepalese herbal plant showed moderate to strong anti-sucrase and anti-maltase activities, the same extracts did not perform as well against α-amylase.

Table 2: Rat intestinal sucrase, maltase and porcine pancreatic α-amylase (PPA) Inhibitory activities of plant extracts at 15mg/ml (mean±SEM; n=3).

Plant species	Enzyme inhibitory activity (%)		
	Sucrase	**Maltase**	**PPA**
Acacia catechu	66.8±1.6	84.5±1.6	19.9±0.1
Aegle marmelos	11.4±0.7	9.5±0.5	25.5±1.3
Allium hypsistum	38.2±1.2	46.5±2.4	36.7±2.0
Apium graveolens	35.6±3.8	40.7±1.2	3.6±0.2
Artemisia indica	24.3±1.8	40.2±0.6	25.5±1.4
Azadirachta indica	7.9±2.2	53.0±1.7	13.9±1.2
Bergenia ciliata	30.2±4.1	68.6±1.4	93.5±2.1
Boerhavia diffusa	25.9±2.7	27.4±4.1	29.2±1.0
Cassia fistula	47.8±4.8	52.5±4.3	2.7±0.1
Cedrus deodara	29.8±3.8	71.5±2.2	75.7±1.6
Centella asiatica	64.3±8.2	48.9±6.1	12.1±1.7
Desmodium gangeticum	19.7±1.7	27.4±1.0	1.5±0.0

Eclipta prostrate	16.0±3.2	19.8±0.5	11.9±0.7
Embelia ribes	35.4±2.7	8.1±0.0	4.6±0.1
Holarrhena pubescens	20.7±5.7	52.5±2.7	11.8±0.9
Inula recmosa	67.4±2.7	40.3±2.0	0.3±0.0
Justicia adhatoda	86.3±3.0	38.5±1.8	12.1±0.8
Lindera neesiana	13.3±3.5	7.4±2.3	17.5±2.0
Myrica esculenta	7.3±0.8	37.5±2.3	84.9±1.6
Nardostachys grandiflora	13.2±2.7	27.2±9.3	27.1±2.8
Neopicrorhiza scrophulariiflora	80.2±3.3	57.6±1.9	Nil
Nymphaea stellata	50.6±3.2	74.2±0.5	13.3±1.2
Ocimum tenuiflorum	7.6±1.1	34.1±3.6	4.9±0.1
Operculina turpethum	15.3±2.1	8.8±3.1	10.3±0.6
Oroxylum indicum	13.9±1.4	8.5±0.2	14.9±0.7
Plumbago zeylanica	12.6±1.3	13.1±1.7	22.4±1.1
Rauvolfia serpentina	8.1±1.1	4.9±0.4	24.3±2.1
Rheum australe	7.9±0.4	35.4±1.5	44.3±1.3
Rhododendrom arboreum	3.9±0.2	95.7±2.6	1.8±0.1
Rubia manjith	89.9±1.3	66.6±0.9	0.5±0
Sapindus mukorossi	96.9±1.2	91.1±0.9	1.7±0
Swertia angustifolia	20.7±2.9	44.0±3.0	23.6±1.8
Taxus baccata	17.3±1.1	35.4±1.4	25.8±1.7
Termialia bellirica	59.3±3.6	82.9±15	Nil
Terminalia chebula	39.8±3.5	91.1±0.6	3.8±0.1
Valeriana jatamansii	77.5±4.7	56.4±10	10.2±2.4
Woodfordia fruticosa	57.7±5.3	77.1±0.7	33.4±3.1
Zanthoxylum armatum	8.1±0.5	19.5±1.1	17.2±0.2
Zingiber officinale	28.3±2.3	37.9±3.2	21.3±0.9

Among the plant species studied, only three species showed very high (50-100% inhibition) porcine pancreatic α-amylase inhibitory activity, seven plant species showed moderate inhibitory activity (25-50% inhibition) and the rest were found low ($\leq$ 25%) to this enzyme inhibitory activity. Plant species *Bergenia ciliata* (93.5%), *Myrica esculenta* (84.9%), and *Cedrus deodara* (75.7%) showed significantly high PPA inhibitory activity. The results of present investigation clearly indicated that large a number of non-traditional Nepalese herbal plants possess from moderate to strong antidiabetic potential, and also suggest that studied herbs could be used as an ingredient in health food and in food that helps to prevent diabetes.

In general this study revealed that out of 39 plant species nearly half of them showed moderate to strong antidiabetic potential exhibiting higher inhibitory activity for either one, any two or all three enzymes (Sucrase, Maltase and PPA) investigated. The α-glucosidase and α-amylase are the most important enzymes responsible for the breakdown of carbohydrates, and its inhibition may provide an effective mean of controlling abnormal levels of blood glucose (23). Therefore inhibition of these digestive enzymes by studied plant species indicates their applicability as antidiabetic agents. In the present investigation, the plant species appeared to possess potential antidiabetic activity includes *A. catechu, B. ciliata, C. fistula, C. deodara, C. asiatica, H. pubescens, I. recemosa, J. adhatoda, M. esculenta, N. scrophulariiflora, N. stellata, R. australe, R. arboreum, R. manjith, S. mukorossi, S. angustifolia, T. baccata, T. bellirica, T. chebula, V. villosa,* and *W. fruticosa*. Among these potential antidaibetic plants only few species have previously reported to have hypoglycemic activity (24), while rest of them are reported first time here for their antidiabetic activity.

A large number of synthetic drugs are available to treat diabetes mellitus. However, the use of many of these drugs has been associated with negative side effects, leading researchers to seek safer and/or natural sources of new therapeutics (25, 26). Currently, much focus has been on the discovery of dietary sources of mild α-amylase and/or α-glucosidase inhibitors to delay the intestinal absorption of digested carbohydrates (21, 27). Here, we report the ability of several plant species from Nepal to inhibit α-amylase and α-glucosidase enzyme activities indicating their strong potential as an antidiabetic agent.

While reports in the literature are building evidence on medicinal plant and herbal extracts as potential diabetic therapeutic agents, information concerning Nepalese herbal plants is lacking. Our study results showed that some non-traditional Nepalese herbal plants contained significant antidiabetic activity *in vitro*, and suggest that the use of such herbs may represent a promising strategy to help control postprandial hyperglycemia through modulation of carbohydrate absorption. Dietary α-amylase and α-glucosidase inhibitors from natural sources are potentially safer, and therefore, studied herbs may be a preferred alternative for the use in formulation of new therapeutic drugs, nutraceutical and functional foods. Furthermore, such comprehensive research will justify the potential efficacy of these traditional herbal medicines and will open avenue for further scientific study. However, further studies are necessary for chemical characterization of the active principles and more extensive biological evaluations in order to elucidate putative component and plausible mechanisms of these inhibitions, which are under progress in our research group.

CONCLUSIONS

Our study results showed that some non-traditional Nepalese herbal plants contained significant antidiabetic activity and suggest that these herbs and food plants could have their applicability as important antidiabetic agents. The comprehensive results of current research will be highly useful for food scientist, biochemist and pharmaceutics. The medicinal herbs found with high antidiabetic potential can be used in formulation of functional foods and other important medicine useful in diabetes mellitus, the chronic human health problem. Nepalese medicinal herbs are presently limited only in traditional ayurvedic treatment. Therefore, the outcome of this study would highlight their functional values, which will promote the production, processing and utilization of plants for medicine and functional foods, and offer a significant means of income generation for the people, as well as offer a tremendous potential for increasing these socioeconomic benefits to developing country like Nepal.

Acknowledgements

This work was financially supported by Japan Society for the Promotion of Science (JSPS). The authors would like to thanks the staffs of the National Herbarium Center, Kathmandu, Nepal for their participation and assistance in the identification of the plant specimens. Lastly authors extend their gratitude to the local people and traditional healers who were willing to share their treasured knowledge with them.

REFERENCES

1. Li, W. L., Zheng, H.C., Bukuru, J., and De Kimpe, N., *Natural medicines used in the tradicinal Chinese medical system for therapy of diabetes mellitus.* J. Ethnopharmacol., 2004; 92: 1-21.
2. King, H., Aubert, R.E., and Herman, W.H., *Global burden of diabetes, 1995-2025; prevalence, numerical estimats, and projections.* Diabetes Care, 1998; 21: 1414-1431.
3. Chakraborty, R., and Rajagopalan, R., *Diabetes and insulin resistance associated disorders: disease and therapy.* Curr. Sci., 2002; 83: 1533-1538.
4. Kameswararao, B., Kesavulu, M.M., and Apparao, C., *Evaluation of antidiabetic effect of Momordica cymbalaria fruit inalloxan-diabetic rats.* Fitoterapia, 2003; 74: 7-13.
5. Josse, R. G., Chiasson, J. L., Ryan, E. A., and Lau, D. C. W., Acarbose *in the treatment of elderly patients with type 2 diabetes.* Diabetes Res. Clin. Pract., 2003; 59: 37-42.
6. Clissold, S. P., Edwards, C., and Acarbose. A., *Preliminary review of its pharmacodynamic and pharmacokinetic properties, and therapeutic potential.* Drugs, 1988; 35: 214-243.
7. Toeller, M., *α-Glucosidase inhibitors in diabetes: efficacy in NIDDM subjects.* Eur. J. Clin. Invest., 1994; 24: 31-35.
8. Saito, N., Sakai, H., Suzuki, S., Sekihara, H., and Yajima, Y., *Effect of and α-glucosidase inhibitor (voglibose), in combination with sulphonylureas, on glycaemic control in type 2 diabetes patients.* J. Int. Med. Res., 1998; 26:219-232.
9. Mooradian, A. D., and Thurman, J. E., *Drug Therapy of Postprandial Hyperglycaemia.* Drugs, 1999:57, 19-29.
10. Gavin, J. R., *Pathophysiologic mechanisms of postprandial hyperglycemia.* J Am Coll Cardiol., 2001; 88: 4-8.
11. Chen, H., Feng, R., Guo, Y., Sun, L., and Jiang, J., *Hypoglycemic effects of aqueous extract of Rhizoma polygonati odorati in mice and rats.* J. Ethnopharmacol., 2001; 74: 225-229.
12. Grover, J. K., Yadav, S., and Vats, V. *Medicinal plants of India with antidiabetic potential.* J. Ethnopharmacol., 2002; 81 (1): 81-100.
13. Nishioka, T. Watanabe, J., Kawabata, J., and Niki, R., *Isolation and activity of N-p-Coumaroyltyramine, an α-Glucosidase inhibitor in Welsh onion (Allium fistulosum).* Biosci. Biotechnol. Biochem., 1997; 61: 1138-1141.
14. Watanabe, J., Kawabata, J., Kurihara, H., and Niki, R., *Isolation and identification of α-Glucosidase inhibitors from Tochu-cha (Eucommia ulmoides).* Biosci. Biotechnol. Biochem., 1997; 61: 177-178.
15. Nishioka, T., Kawabata, J., and Aoyama, Y., *Baicalein, and α-Glucosidase inhibitor from Scutellaria baicalensis.* J. Nat. Prod., 1998; 61: 1413-1415.
16. Hansawasdi, C., Kawabata, J. and Kasai, T., *α-Amylase inhibitor from Roselle (Hibiscus sabdariffa Linn.) tea.* Biosci. Biotechnol. Biochem., 2000; 64: 1041-1043.
17. Toda, M., Kawabata, J., and Kasai, T., *α-Glucosidase inhibitor from Clove (Syzgium aromaticum).* Biosci. Biotechnol. Biochem., 2000; 64: 294-296.
18. Vats, V. Grover, J. K., and Rathi, S. S., *Evaluation of antihyperglycemic and hypoglycemic effect of Trigonella foenum-graecum Linn, Ocimum sanctum Linn, and Pterocarpus*

marsupium Linn in normal and alloxanized diabetic rats. J. Ethnopharmacol., 2002; 79: 95-100.

19. Kawabata, J., Mizuhata. K., Sato, E., Nishioka, T., Aoyama, Y., and Kasai, T., *6-Hydroxyflavonoids as α- Glucosidase inhibitors from Marjoram (Origanum majorana) leaves.* Biosci. Biotechnol. Biochem., 2003; 67: 445-447.
20. Kim, Y., Wang, M., and Rhee, H., *A novel glucosidase inhibitor from pine bark* Carbohydr. Res. 2004; 339: 715-717.
21. Maccue, P., and Shetty, K., *Inhibitory effects of rosmarinic acid extracts on porcine pancreatic amylase in vitro.* Asia Pac. J. Clin. Nutr., 2004; 13(1): 101-106.
22. Ghimire, U., *Screening of some Nepalese medicinal plants for the hypoglycemic activity and the potential cytotoxicity.* Central Deparment of Chemistry, Institute of Science and Technology. Kathmandu, Tribhuwan University, 2000.
23. Kubo, I., Muria, Y., Soediro, I., and Soetarno, S., *Efficient isolation of glycosidase inhibitory stilbene glycosides from Rheum palmatum.* J. Nat. Prod., 1991; 54: 115-1118.
24. Rahaman, A., and Zaman, K., *Medicinal plants with hypoglycemic activity.* J. Ethnopharmacol., 1989; 26: 1-55.
25. Dawara, R. K., Makkar, H.P., and Singh, B., *Protein-binding capacity of microquantities of tannins.* Anal. Biochem, 1988; 170 (1): 50-53.
26. Maccarty, M. F., *Towards a wholly nutritional therapy for type 2 diabetes.* Med. Hypotheses, 2000; 54(3): 483-487.
27. Scheen, A. J., *Is there a role for alpha-glucosidase inhibitors in the prevention of type 2 diabetes mellitus.* Durgs, 2003; 63(10): 933-951.

THE EFFECT OF MEADOWSWEET (FILIPENDULA ULMARIA) FLOWER EXTRACT AND HYPOTHIAZIDE ON RENAL PHYSIOLOGICAL FUNCTION IN RATS

J. Bernatoniene, A. Savickas, R. Bernatoniene, Z. Kalvėnienė, R. Klimas

Kaunas University of Medicine, A. Mickevičiaus 9, LT-44307 Kaunas, Lithuania

Corresponding author: J. Bernatoniene E-mail: jurgabernatoniene@yahoo.com

Keywords: Renal function, kinins, prostaglandins

SUMMARY

The aim of our study was to determine how meadowsweet flower extract and hydrochlorothiazide affect certain physiological renal functions – excretion of urine and electrolytes, and changes in plasma levels of prostaglandins and kinins following water and salt load. The study was performed on male Wistar rats (weight - 145-170 g, age - 4-6 months), using meadowsweet flower extract and hydrochlorothiazide. We determined that the diuretic effect of meadowsweet flower extract after 4 hours was similar to that of hydrochlorothiazide. This indicates that meadowsweet flower extract affects physiological renal function. Natriuretic and kaluretic effect of meadowsweet flower extract was weaker than that of hypothiazide. We found that meadowsweet flower extract increased blood plasma levels of prostaglandins and kinins following water and salt load. This shows that meadowsweet flower extract improves renal blood circulation, and reduces the osmotic gradient and reabsorption of water in renal collecting canals.

INTRODUCTION

Kidneys are vitally important organs that maintain the stability of the internal medium [1]. The influence of the antidiuretic hormone and angiotensin II on renal medullar cells or arterial walls results in the release of prostaglandins (PG) into the bloodstream (PG) [2]. Prostaglandins increase plasma levels of kinins and are strong vasodilators [3]. For experimental studies we chose diuretic preparations – hydrochlorothiazide (HT) and meadowsweet flower extract (MFE). When investigating physiological renal function in rats, induced by glycerol injection, Lebedev A. A. et al. found significant correlation between the cumulative excretion of HT with urine and natriuresis encountered in intact animals, which was weaker in rats with acute renal insufficiency [4]. Toussaint et al. determined that HT induced a prolonged diuresis which was not changed by either captopril or ramipril [5]. It has been found that vegetal preparations used for stimulation of urine excretion reduce the amount of extracellular fluid in tissues and in serous cavities of a diseased person [6, 7]. Meadowsweet contains polyphenol tannins, phenolic glycosides, essential oil, etc. The *British Herbal Compendium* describes that flowers of meadowsweet have anti-inflammatory, diuretic, stomachic, and astringent actions [5]. We did not find any data in literature on the effect of MFE on the renal physiological function, electrolyte excretion, and changes in plasma levels of kinins and prostaglandins. Therefore, the aim of our study was to determine how meadowsweet flower extract and HT affect certain physiological renal functions – excretion of urine and electrolytes, and changes in plasma levels of prostaglandins and kinins following water and salt load.

MATERIAL AND METHODS

Approval of the Lithuanian Ethics Committee for Laboratory Animal Use was obtained before the commencement of the experiments. For the study, we used male Wistar rats (weight - 145-170 g, age - 4-6 months), HT (manufactured by Merck Sharp & Dohme, INC., 1996, Whitehouse Station, NJ, U.S.A.) and meadowsweet flowers (supplied by "Acorus Calamus", Lithuania). Meadowsweet flower extract was produced according to the technique described in the European Pharmacopoeia 1997:0765.

The amounts of these preparations were selected with respect to their therapeutic doses and their pharmacological and toxicological effect - HT dose was 50 ml/kg, and MFE dose – 4.8 ml/kg.

Meadowsweet flower extract, HT, and isotonic sodium chloride solution were injected into the stomach of conscious animals via a special metal probe. Diuretic activity was investigated according to E. B. Berchin's technique [8]. Electrolyte levels were determined using flame photometry (Ph. Eur. 2002, 2.2.23), prostaglandin levels – by radioimmunology technique (Ph. Eur. 01/2002:0125) using H3 isotopes, and kinin levels – using the enzymatic method (Ph. Eur. 2002, 2.6.15).

The data are presented as means ± S.E.M. nonparametric methods were applied for making the inferences about the data. The differences between the mean values in dependent groups were tested using Wilcoxon matched pairs test. The differences between the mean values in independent groups were tested using the nonparametric Kruskal-Wallis test with Dunn's post-hoc evaluation. $P<0.05$ was taken as the level of significance. Statistical analysis was performed using the software Statistica 1999, 5.5 StatSoft Inc., USA.

RESULTS AND DISCUSSION

The main function of kidneys is to excrete the final products of metabolism from the organism [8]. According to literature data, the majority of the ingested fluid is excreted from the organism within the first two hours, and within 4 hours the body excretes all ingested fluid [10].

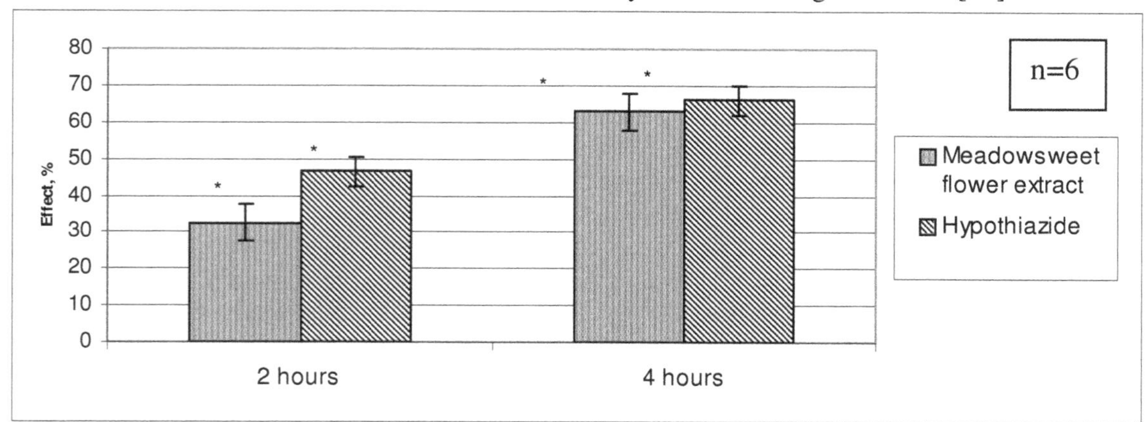

Fig. 1. Changes in the physiological renal function under the influence of meadowsweet flower extract and hypothiazide $p< 0.05$ *vs.* controls

The obtained experimental findings (Fig. 1) showed that MFE and HT statistically reliably increased the amounts of the excreted urine after 2 and 4 hours following administration, compared to controls. The diuretic effect of HT was by 12±0.4% (from 2.13±0.04 ml to 2.40±0.1 ml) and by 2±0.2% (from 4.30±0.08 to 4.40±0.07 ml) (after, accordingly, 2 and 4 hours following administration) greater, compared to MFE. It can be stated that the diuretic effect of MFE after 4 hours following the administration was similar to than of HT.

Electrolytes that are important for the organism – sodium and potassium salts - are excreted with urine [11]. Potassium ions are important for cardiac activity and carbohydrate and protein metabolism, and sodium ions are important for water metabolism, osmotic blood plasma pressure, acid-alkali balance, as well as for muscle and nerve irritability, conduction, and contraction processes. It has been found that the stability of cell volume is determined by the osmotic concentration of cell environment [12]. Disturbances in the regulatory mechanisms of fluids may result in the disturbance in the water and electrolyte balance. For this reason, it was relevant to determine how salt and water load influences diuresis and Na + and K+ excretion under the influence of HT and MFE (Table 1).

Table 1. Changes in diuresis and electrolytes under the influence of meadowsweet flower extract (MFE) and hypothiazide after water and salt (n=5) load

Experimental conditions	**Investigated markers**					
	Diuresis after 4 hours, %		*Na+ excretion, %*		*K+ excretion,%*	
	Hypothiazide	MFE	Hypothiazide	MFE	Hypothiazide	MFE
Water load	64.3±6*	62.3±5*	30.6±4*	25.3±5*	22±4*	10±2
Salt load	68.3±7*	71.2±4*	45.4±10*	14.2±2	18±3*	6±1

* $p < 0.05$ *vs.* controls

Statistically MFE and HT significantly increased diuresis after 4 hours following water and salt load, compared to the respective findings in controls. The amount of the excreted urine after water load did not differ statistically significantly from to the amount of excreted urine after salt load. After loading rat kidneys with water, MFE and HT statistically significantly increased the excretion of sodium ions, compared to the respective findings in controls. Han S. et al. proved that increasing levels of Na+ in tissues entails an increase in the osmotic pressure in tissues and retention of water, which conditions formation of edema [13]. Statistically non-significantly greater amounts of Na+ ions were excreted under the influence of hypothiazide, compared to meadowsweet flower extract. After salt load, Na+ excretion under the influence of HT increased by 45±10% (from 112.3±3.8 μ mol/ml to 162.8±4.4 μ mol/ml) ($p<0.05$), and under the influence of MFE – by 14±2 % (from 112.3±3.8 μ mol/ml to 128.0±2.9 μ mol/ml) ($p<0.06$). Na+ excretion after salt load was statistically reliably greater than that after water load: under the influence of HT it was 15.1±2.2%, and under the influence of MFE – 11.1±0.3%. Potassium is the most important cation of extracellular fluid [13]; its metabolism in the organism is regulated by adrenal cortical hormones and kidneys [12]. In further studies we determined that both after water and salt load, the excretion of potassium ions was statistically reliably greater under the influence of

HT that under the influence of MFE. In general, it can be stated that increased excretion of Na+ and K+ when using HT (compared to MFE) disturbs the balance of these electrolytes in the body. This results in disturbed water balance, since it is directly dependent on the concentration of Na+ and K+ ions.

It is known that prostaglandins (PG) have a diuretic and natriuretic (stimulate Na+ excretion) effect, which is useful in the presence of cardiovascular failure and hypertension [7]. It has been found that PG improves renal blood circulation, reduce the osmotic gradient and reabsorption of water in renal collection canals, and stimulate diuresis [8]. For this reason it was relevant to investigate the effect of MFE and HT on prostaglandin levels in blood plasma of rats (Fig. 2).

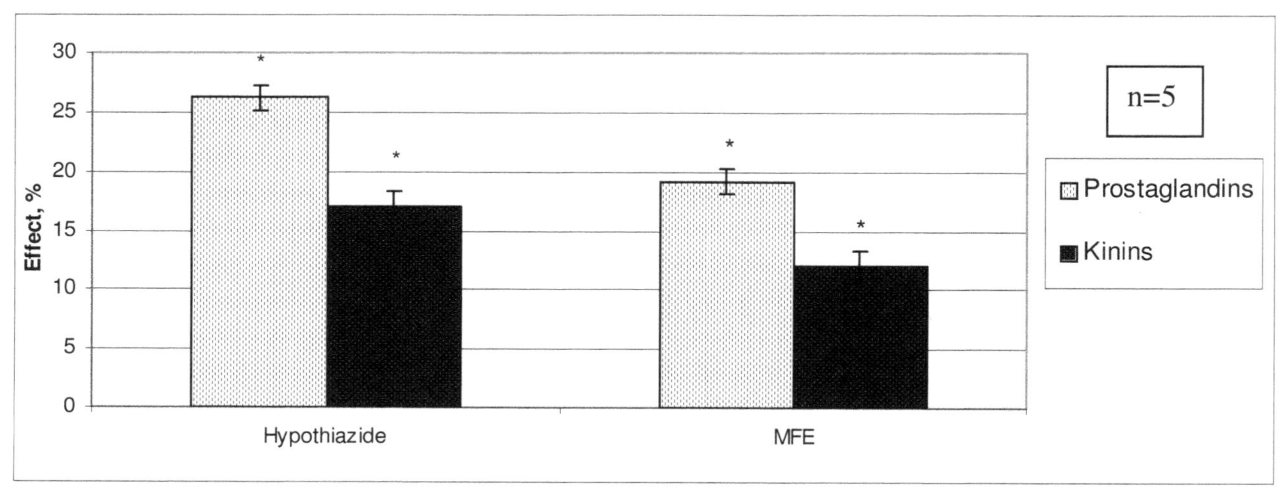

Fig. A.

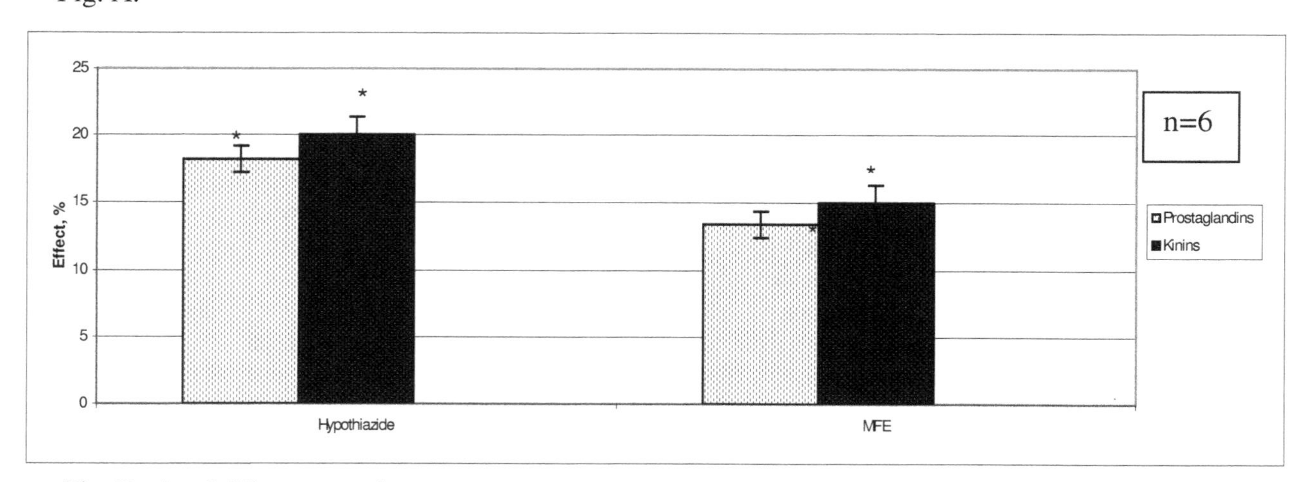

Fig. B. *p<0.05 vs. controls

Fig. 2. Changes in blood plasma levels of prostaglandins and kinins under the influence of hypothiazide and meadowsweet flower extract: Fig. A – after water load, Fig. B – after salt load

The results of the studies showed that after water and salt load, PG levels in rats following the intragastric injection of HT were by, respectively, 8.3±0.5% and by 5.3±0.2% greater than those following the injection of MFE. This can be explained by the fact increased excretion of urine under the influence of HT and MFE occurs due to the improvement of renal blood circulation, which is related to increased prostaglandin levels [9, 11, and 12].

Further on in the study we investigated the specific mechanism of the diuretic effect of MFE and HT in order to clarify the influence of these preparations on kinin system activity. The diuretic effect of kinins is related to their influence of glomerular filtration and inhibition of Na+ reabsorption in proximal renal canals. In addition to that, kinins activate PG synthesis, and prostaglandins improve renal blood circulation [10, 13]. After water load, kinin levels under the influence of HT were by 4±0.2% greater, compared to those under the influence of MFE. After salt load, kinin levels under the influence of HT were by 5±0.1% statistically non-reliably greater, compared to those under the influence of MFE. However, both HT and MFE statistically significantly increased kinin levels, compared to the respective levels in controls (by 20±0.4% and 15±0.2%, accordingly). This indicates that HT and MFE increased PG and kinin levels, and these substances rank among the regulators of the urine excretion function in kidneys.

CONCLUSIONS

In conclusion, we found that the diuretic effect of meadowsweet flower extract after 4 hours following the administration was similar to that of HT. This shows that meadowsweet flower extract influences the physiological renal function. The natriuretic and kaluretic effect of meadowsweet flower extract was lower than that of hypothiazide. We found that meadowsweet flower extract after water and salt load increased blood plasma levels of prostaglandins and kinins. This indicates that meadowsweet flower extract improves renal blood circulation, decreases the osmotic gradient and water reabsorption in renal collecting canals, and stimulates diuresis.

REFERENCES

1. Acierno MJ, Labato MA. Clin Tech Small Anim Pract 2005; 20: 23-30.
2. Vitela M, Herrera-Rosales M, Haywood JR, Mifflin SW. Am J Physiol Regul Integr Comp Physiol 2005; 288: 56-62.
3. Lohmeier TE, Hildebrandt DA, Warren S et al. Am J Physiol Regul Integr Comp Physiol 2005; 288: 828-36.
4. Lebedev AA, Grebenova SV. Eksp Klin Farmakol 1996; 59: 26-8.
5. Toussaint C, Masselink A. Klin Wochenshr 1989; 67:1138-46.
6. Nekrasova AA, Klembovskii AA, Levitskaia LuV et al. Kardiologiia 1986; 26:13-20.
7. Markov KhM, Kucherenko AG, Aralov MD et al. Kardiologiia 1988; 28: 37-41.
8. Berxin E.B. Chem. Pharm. Magazine 1977; 5:3-11
9. Al-Awwadi NA, Araiz C, Bornet A, et al. J Agric Food Chem 2005; 12: 151-7.
10. Bardai SE, Lyoussi B, Wibo M et al. Clin Exp Hypertens 2004; 26: 465-74.
11. Yetiser S, Kertmen M, Yildirim A. Acta Otorhinolaryngol Belg 2004; 58: 119-23.
12. Parfenov VA. Ter Arkh 2005; 77:56-9.
13. Han S, Zheng Z, Ren D. J Huazhong Univ Sci Technolog Med Sci 2002; 22: 302-4.

ANTIDIABETIC EFFECT OF CINNAMON (*CINNAMON ZEYLANICUM*) BARK EXTRACTS ON STREPTOZOTOCIN-INDUCED DIABETIC RATS

M. A. Hossain, M. B. Zaman, M. A. Hannan and M. T. Hossain

Department of Biochemistry, Bangladesh Agricultural University, Mymensingh-2202, Bangladesh

Corresponding author: M. A. Hannan, Email: mah_biochemist@yahoo.com

Keywords: *Cinnamon zeylanicum,* type II diabetes, serum glucose, cholesterol, triacylglycerol

ABSTRACT

Effects of alcoholic and aqueous extracts of cinnamon (*Cinnamon zeylanicum)* on the concentration of serum glucose, total serum cholesterol (TCh) and serum triacylglycerol (TAG), and activities of serum glutamate pyruvate transaminase (SGPT) and serum glutamate oxaloacetate transaminase (SGOT) were studied on type II diabetes in rats induced by streptozotocin. Compared with diabetic control (T_1), aqueous extract (T_3) and glibenclamide (T_6) showed the greatest reduction in plasma glucose concentration and T_3 and alcoholic extract (T_5) also reduced serum TCh concentrations toward normal. Serum TAG and SGPT activities were significantly decreased by the treatments of T_3, alcoholic extract (T_4), T_5 and T_6, and all the treatments increased SGOT concentration over control.

INTRODUCTION

Human diabetes is very common all over the world. Bangladesh ranked tenth among the diabetic populations of 40 countries in 2000 (Manning, 2004). Prevalence of impaired glucose tolerance (IGT) was 11.8% and 4.8% in rural and urban population, respectively, and the prevalence of diabetes among adults is estimated to be 4% of total population in Bangladesh (Sayeed *et al.* 1997). Recently, spices are becoming a major therapy for the treatment and control of diabetes. Some spices, like cinnamon (*Cinnamon zeylanicum),* are reported to have the medicinal value against diabetes (Khan *et al.* 1990). Botanical products can improve glucose metabolism and overall condition of persons with diabetes not only through their hypoglycaemic effect but also by improving lipid metabolism, antioxidant status, and capillary function (Broadhurst, 1997). Moreover, medicinal plants have been used for diabetes safely and with reasonable success (Duke *et al.* 1998). The present study was under-taken to evaluate the effects of alcoholic and aqueous extracts of *Cinnamon zeylanicum* in diabetic subjects on concentrations of plasma glucose, serum total cholesterol (TCh) and serum triacylglycerol (TAG), activities of serum glutamate pyruvate transaminase (SGPT) and serum glutamate oxaloacetate transaminase (SGOT).

MATERIALS AND METHODS

A Complete Randomized Design (CRD) laboratory trial with five replications on diabetic rats [Streptozotocin®(STZ) –induced] was performed between January and May 2005. Seven treatments groups were used: control (T_0), diabetic control (T_1), aqueous extract 50 mg /kg body weight (T_2), aqueous extract 100 mg /kg body weight (T_3), alcoholic extract 50 mg /kg body weight (T_4) and alcoholic extract 100 mg /kg body weight (T_5) and recommended drug Dibenol (5 mg glibenclamide) was used @ 600 μg /kg body weight (T_6).

The mixed adult albino (long Evan's male Strains) rats were collected from the International Centre for Diarrhoeal Diseases Research (ICDDR), Bangladesh and maintained at the departmental animal house. A diabetic inducing agent, 2-deoxy-2-(3-methyl-3-nitrosoureido)-D-glucopyranose (Streptozotocin®, Yako company, Japan) was dissolved in 0.1M citrate buffer solution having pH 4.5 (Bhuiya *et al.*, 2004). Freshly prepared solution of STZ was injected subcutaneously @ 50 mg /kg body to all rats, except control (T_0), and fasted for 18 hours to induce diabetes.

The pulverised cinnamon was soaked in distilled water and re-distilled ethanol with occasional shaking for 48 hours to prepare aqueous and alcoholic extracts respectively. The filtrates were evaporated in a rotary vacuum evaporator at 45°C and the concentrated slurry was freeze-dried and stored in desiccators until used. Suspension of alcoholic aqueous extract of cinnamon barks or solution of aqueous extract, or glibenclamide were administered orally to the rats through forced- feeding for 21 days by micro tubes with the help of micropipette.
Up to 21 days, at every 7th day one rat from each group was anesthetized and sacrificed and blood was collected with a syringe directly from the heart. Each blood sample was immediately divided into two test tubes. One was used for the preparation of blood plasma by adding EDTA at the rate of 10 mg /mL of blood and subsequently centrifuged at 2500 rpm for 10 minutes. Another test tube was kept at normal temperature for 20 minutes and then centrifuged for 7 minutes at 2500 rpm. The blood plasma and serum was carefully collected by microtubes and preserved in an eppendorf vials and refrigerated. The plasma and serum were used for the determination of glucose, total cholesterol (Tch), triacylglycerol (TAG), Serum Glutamate Pyruvate Transaminase (SGPT) Serum Glutamate Oxaloacetate Transaminase (SGOT) activity as described by Trinder (1969), Allian *et al.*, (1974), Fossati and Prencipe (1982), and Gella *et al.*, (1985), respectively.

RESULTS AND DISCUSSION

After STZ- injection, rats on all groups exhibited weight loss. But, at the 21st day of treatment in groups T_2, T_3, T_4, T_5 and T_6, a significant ($p < 0.01$) increase of body weight was observed. However, there was no significant variation between groups (Table 1). This indicates that the rats became obese, a characteristic of type-II diabetes caused by the STZ. It appears from Fig 1 that after 7 and 14 days of treatment, groups T_3 and T_6 showed the biggest decrease in blood glucose concentrations. At the end of the experiment (21 days), all the treated rats showed a decreasing trends of their plasma glucose, but T_3 (122.2 mg/dl) and T_6 (120.8 mg/dl) were superior to other treatments and not significantly different.

There was a significant ($p < 0.01$) decrease in concentrations of total cholesterol in treatment groups (Table 2). After 7 days of treatment, all the treated groups showed a gradual decrease in the total cholesterol concentration, but after 14 days T_3, T_4, T_5 and T_6 showed the best performance. On the 21st day of the experiment, treatments T_2, T_4, T_5, and T_6 showed a significant ($p < 0.01$) reduction of blood cholesterol concentrations (60.6, 54.0, 61.0, 56.0 and 55.4 mg/dL, respectively), but T_3, T_5 and T_6 had better responses and they were not significantly different from each other.

Table 1: Body weight (g) of normal, diabetic control (STZ) and experimental rats treated with cinnamon extracts.

Treat-ment	Body weight (g)				
	Initial (Before STZ injection (mean)	15th day of STZ injection (mean)	7th day of treatment (mean)	14th day of treatment (mean)	21st day of treatment (mean)
T_0	226.5	228.7 a	231.2 a	232.1 a	233.3 a
T_1	221.8	194.4 b	190.1 c	188.3 d	184.2 b
T_2	230.2	203.4 b	208.8 b	218.5 c	229.6 a
T_3	230.1	204.9 b	215.4 b	227.7 ab	241.6 a
T_4	222.7	200.3 b	207.0 b	220.7 bc	232.5 a
T_5	220.9	196.8 b	210.4 b	226.1 abc	231.7 a
T_6	227.9	202.3 b	213.8 b	224.4 abc	238.0 a
LS	NS	**	**	**	**
CV %	-	3.33	3.83	2.35	3.91

Note
LS = Level of significance ** = Significant at 1% level of probability
NS = Not significa

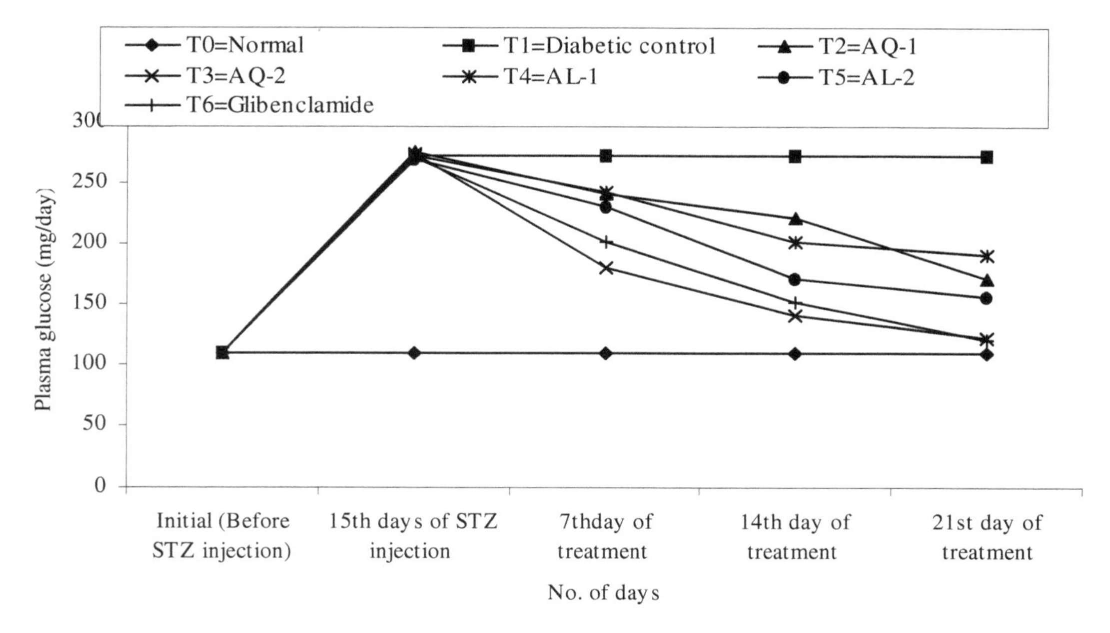

Fig. 1 Effect of *C. zeylanicum* bark extracts on plasma glucose level in STZ induced diabetic rats

Table2: Serum cholesterol (mg/dL) in normal, diabetic control (STZ) and experimental rats treated with cinnamons.

Treatment	Serum cholesterol (mg/dL)				
	Initial (Before STZ injection (mean)	15th day of STZ injection (mean)	7th day of treatment (mean)	14th day of treatment (mean)	21st day of treatment (mean)
T_0	70.8	72.5 b	71.4 d	69.3 c	74.4 b
T_1	71.3	92.5 a	92.1 a	91.1 a	90.3 a
T_2	69.5	93.0 a	80.5 b	75.1 b	60.6 c
T_3	70.3	95.1 a	78.2 b	64.6 de	53.9 d
T_4	67.8	93.4 a	78.0 b	64.1 de	61.0 c
T_5	73.2	90.5 a	76.8 bc	62.3 e	56.0 d
T_6	74.1	94.6 a	73.0 cd	66.0 d	55.5 d
LS	NS	**	**	**	**
CV %	-	3.68	3.89	3.26	3.34

Note

LS = Level of significance, ** = Significant at 1% level of probability, NS = Not significant

Fig 2 shows that in the beginning, all the groups maintained initial TAG concentrations at 71.3 mg /dL and on the 15th day of STZ injection a significant ($p< 0.01$) increase was observed.

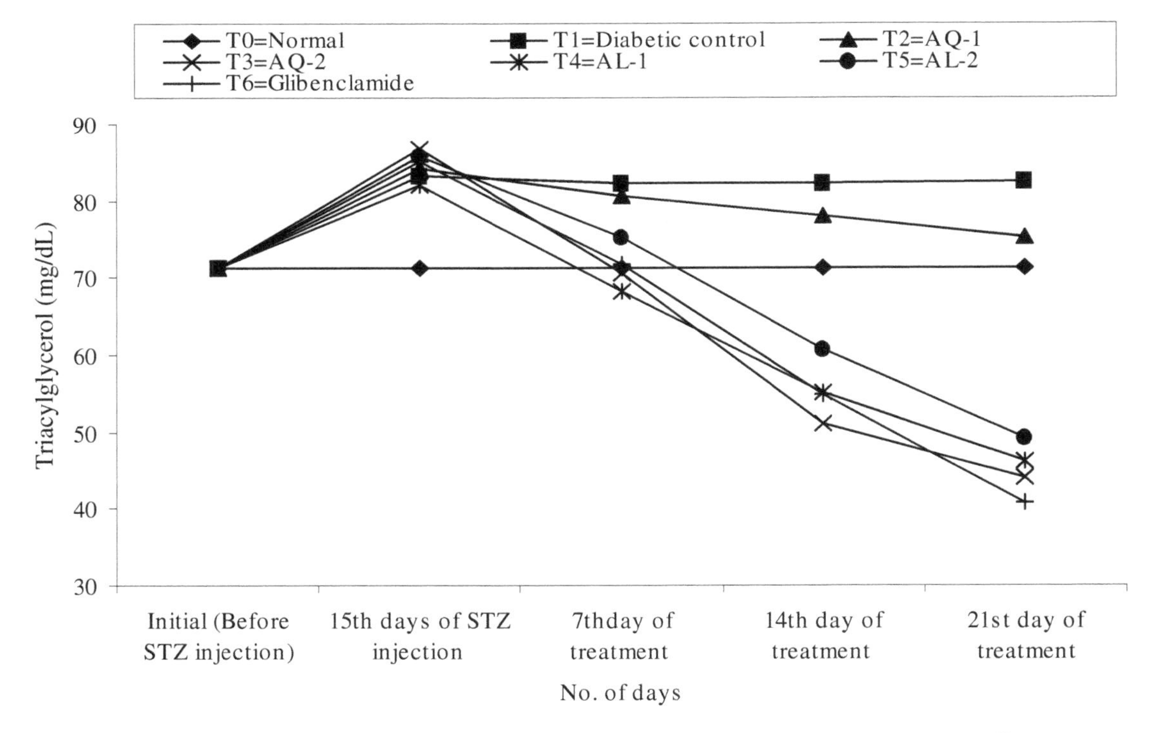

Fig. 2 Effect of C. zeylanicum bark extracts on serum triacylglycerol (TAG) level in STZ induced diabetic rats

On the 7th day of the experiment, group T_2 had decreased the TAG concentration, which fell further till the day 21. But at 14th day, treatments T_3, T_4, T_5 and T_6 showed the significant ($p< 0.01$) decrease in serum TAG concentration, which was below normal. At the end of the experiment, and treatments T_3, T_4, T_5 and T_6 exhibited the remarkable decrease in TAG concentrations. The biochemical reason for TAG increase in diabetes may be due to more circulating free fatty acids, which are the main source of TAG; moreover, lipogenesis from acetyl-CoA is depressed under this condition (Martin *et al.*, 1985).

Table 3: Serum SGPT (U/L) activity in normal, diabetic control (STZ) and experimental rats treated with cinnamon extracts.

Treatment	Serum SGPT (U/L)				
	Initial (Before STZ injection (mean)	15th day of STZ injection (mean)	7th day of treatment (mean)	14th day of treatment (mean)	21st day of treatment (mean)
T_0	10.2	11.3 b	10.9 e	12.0 e	11.5 e
T_1	10.9	55.4 a	55.4 a	54.8 a	55.1 a
T_2	9.9	61.1 a	48.0 bc	38.3 b	24.3 b
T_3	11.1	62.4 a	46.2 c	30.4 c	20.4 cd
T_4	10.9	61.5 a	50.7 b	32.2 c	25.0 b
T_5	12.0	62.3 a	49.0 b	35.2 b	21.2 c
T_6	11.9	58.4 a	24.3 d	23.0 d	18.1 d
LS	NS	**	**	**	**
CV %	-	9.92	1.91	3.75	3.72

LS = Level of significance , ** = Significant at 1% level, NS = Not significant

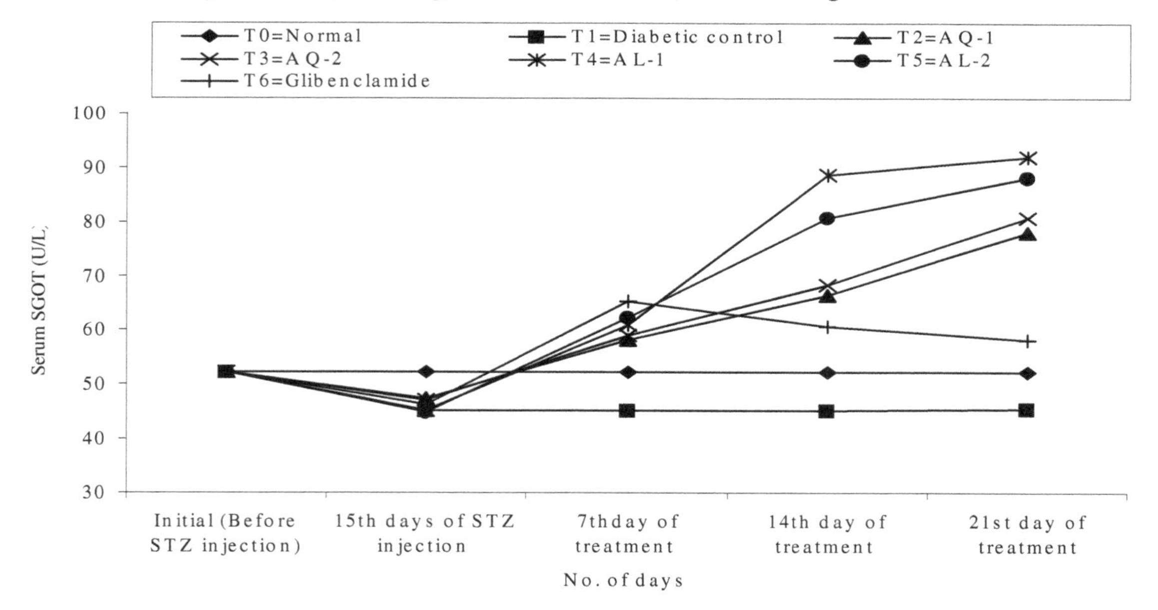

Fig. 3 Effect of C. zeylanicum bark extracts on serum SGOT (U/L) activity in STZ induced diabetic rats

Table 3 showed that all the treatments decreased the SGPT activity at the 7th day of treatment. Again, at 14th day of treatment, the best performance was found in T_3 and T_6 groups compared to control group. After 21st day of treatment, all the treated rats showed a decreasing trend, but T_3 and T_6 (20.4 U/L and 18.1 U/L) had alleviated SGPT compared to other treatments and they were statistically alike.

Figure 3 shows that on the 15th day of STZ injection, a significant ($p< 0.01$) reduction of SGOT activity in all treated groups. At 7th, 14th and 21st days, all the treatments increased SGOT activity except $T_{6,}$ which showed highest SGOT activity (65.3 U/L) at Day 7 but at Day 14 and 21 concentrations were decreased. It is noted that the aqueous and alcohol extracts of *Cinnamon zeylanicum* did not show beneficial effect on SGOT in diabetic condition of the experiment.

CONCLUSION

It may be concluded that glibenclamide, a hypoglycaemic drug, performed the best, but aqueous extract of cinnamon 100 mg /kg body weight produced similar results. It is suggested that the diabetic subjects may take cinnamon extracts in lieu of glibenclamide. However, the cinnamon extracts might be tested in human volunteers with type II diabetes before general use.

REFERENCES

1. Allian, C. C., Poon, L. S., Chan, C. S. G., Richmond, W. and Fu, P. C; Clinical Chemistry, 1974, 20, 470-475.
2. Broadhurst, C. L.; Alternative Medicine Review, 1997, 2, 378-399.
3. Burke, J. P., Williams, K., Narayan, K. M. V., Leibson, C., Haffner, S. M., and Stern, M. P.; Diabetes Care, 2003, 26,1999-2004.
4. Bhuiya, J. H., Chowdhury, B. L. D. and Hossain, M. A.; Bangladesh Veterinary Journal, 2004, 38, 35-42
5. Duke, J. A., Beckstrom-Sternberg, S. and Broadhurst, C. L.; U.S. Department of Agriculture, Phytochemical and Ethnobotanical. 1998
6. Fossati, P. and Prencipe, L.; Clinical Chemistry, 1982, 28, 2077-2080.
7. Gella, F. J., Olivella, T., Cruz Pastor, M., Arenas, J., Moreno, R., Durban, R., et al.; Clinica Chemica Acta, 1985, 153, 241-247.
8. Hegested, D.M.; Nutritition Review, 1974, 32, 33-38
9. Khan, A., Bryden, N. A., Polansky, M. M. and Anderson, R. A. Biological Trace Element Research, 1990, 24,183-188.
10. Martin, D.W. Jr., Mayes, P.A., Rodwell, V. W. and Garnner, D.K. (). Harper's review of Biochemistry. Lange medical publications. Maruzen Co. Ltd. 20th edn., 1985, pp. 243.
11. Sayeed, M. A., Ali, L., Hossain, M. Z., Rumi, A and Khan, A. K. M.; Diabetes Care, 1997, 20, 551-555.
12. Trinder, P.; Annals Clinical Biochemistry, 1969, 6, 24-27.

BRIEF DESCRIPTION OF CHEMOMUTANT WINTER WHEAT VARIETIES AND PERSPECTIVES OF USING THEM AS FOODSTUFFS FOR CHRONIC PATIENTS

Natalya S. Eiges, Larissa I. Weisfeld, and Georgij A. Volchenko

Emanuel Institute of Biochemical Physics, Russian Academy of Sciences, 119991 Russia

Corresponding author: Larissa I. Weisfeld, E-mail: liv11@yandex.ru

Keywords: Winter wheat, chronic diseases and functional foods

ABSTRACT

Previously, we studied mutant varieties and samples of common winter wheat created by the method of chemical mutagenesis at the Laboratory of Mutational Selection and Protection of Environment of the Emanuel Institute of Biochemical Physics. These varieties and samples possess of characters, which are of interest as foodstuffs for people with chronic diseases such as diabetes, obesity, rectum cancer, and cardiovascular diseases since they possess properties that facilitate removing harmful toxins from organism. The aim of this work is to study the associated valuable characters and their combinations in mutants and mutant varieties and the possibility of cultivating them in various ecological niches.

INTRODUCTION

Chronic diseases such as diabetes, obesity, and cardiovascular and cancer affect an increasing number of people. Higher intake of vegetables vitamins, antioxidants, and flavonoids could prevent some of these diseases. Hence, cultivation of new valuable plants that are rich in antioxidants, vitamins and minerals is important [1]. One such valuable plant is winter wheat. By using the chemical mutagenesis technique, we developed hereditarily changed samples – mutants with new valuable characters that are not inherent in winter wheat.

MATERIALS AND METHODS

In this work, with the method of chemical mutagenesis [2] using optimum doses of a supermutagen - ethylene imine, whose mutagenic properties had been discovered by Iosif Abramovich Rapoport we obtained a wide spectrum of mutations that contain a large diversity of mutant characters, which determines various uses of these utilizations [3], e.g., in medicine. The main section of this work concerning this aspect is devoted to investigating the properties of starch and the content of amylose polysaccharide in it.

RESULTS AND DISCUSSION

The first screening of a small number (13) of varieties and samples showed a differentiation for the content of amylose in starch [4]. By this character, the mutants are divided into several groups.

Group I includes five varieties: Sibirskaya niva, Imeni Rapoporta, Opytnyi (Belaya), Volzhanka Elena, and Ritza. These varieties are characterized by the normal (22-26%) amylose content. This amylose content is inherent in most varieties and samples of winter wheat. Among the wheat varieties with the normal amylose content, two varieties – Sibirskaya niva and Imeni Rapoporta – are region-adapted (or included in the State Register of Selection Achievements and permitted for using). Other three varieties passed the State Grade-Testing but were not included in the State Register.

Five varieties and samples, i.e. more than one third of those studied have high or elevated amylose content. Group II includes varieties and samples with high amylose content in starch. The Bulava variety has the amylose content 39.5%; the promising samples №639N and №5B have the amylose contents 37.0 and 42.1%, respectively. The Bulava variety has been adapted for the East Kazakhstan region, №639N is under competitive grade-testing in the Moscow area, and №5B is under competitive grade- and production-testings in the Moscow area.

Group III includes varieties and samples with elevated amylose content in starch. These are promising samples: №615N (31.0%) and the initial variety PPG 186 (30.0%). The wheat-agropyron hybrid PPG186 had been widely used in the non-chernozem zone. Currently, we revive this variety as a very promising one because of its elevated amylose content and as the initial material for mutational selection.

Group IV includes a variety with low amylose content. The only representative of this group, the Beseda variety contains 11.0% of amylose in starch. This variety passed the State Grade-Testing by the criterion of productivity but was not included into the State Register because its amylose content was not taken into consideration.

Group V includes varieties with lowered amylose content; these are the Botovskaya 1 (21.5%) and Moskovskaya 39 (21.0%) varieties. The latter variety was selected at the Research Institute of Agriculture of the Central Region of Non-Chernozem Zone; we have taken this variety as a reference standard.

The varieties and samples with both high and low amylose contents in starch are of great value.

High amylose content determines enzyme-resistant properties of starch. Due to these properties, starch is not subjected to the action of enzymes in the larger part of the gastro-intestinal tract and does not give off glucose. Therefore, bio-degraded foodstuffs, which may be produced from the varieties and valuable mutant samples with high amylose content, will be useful both for diabetes patients and as a prophylaxis of this disease. These foodstuffs may inhibit complications in diabetes patients such as diabetes-provoked cardiovascular diseases.
Low amylose content in foodstuffs is associated with low fat content in these foodstuffs. The latter factor prevents from obesity and development of cardiovascular and other obesity-induced chronic diseases, e.g., diseases of the locomotor apparatus. Low amylose content in foodstuffs reduces the risk of rectum oncology. These diseases, like diabetes, are widespread in the world.
As was mentioned above, the first screening of small number of the mutants and mutant varieties as to the amylose content in starch yielded positive results.

With regard to a great diversity of characters revealed in our collection of mutants, we expect that further screenings will reveal mutants with a higher (than in Bulava and №5B) and a lower (than in Beseda) amylose content, which may be of great importance for medical purposes.

In the 1990s, the countries of West Europe used low-amylose starch of some lines of corn with amylose content 5% [5]. Then, in the USA, Canada, and Japan, using the traditional methods of selection and chemical mutagen ethylmethane sulfonate, low-amylose varieties of potato and wheat were obtained. The amylose content of these varieties is 0.2-1.3%, i.e., lower than that of low-amylose corn lines and Beseda. In this work, we obtained for the first time in Russia low-amylose starch of wheat – in the Beseda variety.

By further screenings, we expect revealing samples with the amylose content in starch higher than that in the Bulava variety and №5B sample, i.e., approximately that of some corn lines (80%) [5].

Wheat flour from the Bulava variety and other high-amylose samples of our collection may be used without adding expensive high-amylose corn flour to the dough, what cost very much.

Along with the screening, it is important to test the Beseda and Bulava varieties and №615, №639, and №5B samples for low and high amylose content with respect to different years of growing (in this work we report on the harvest of 2001) and different soil and climate conditions. The experience will show how stable are the characters of low and high amylose contents in starch of our varieties and samples.

If the given varieties and samples are processed and used in production of curative products, it is important to know how the desired amylose content associates with other properties. For example, high amylose content of the Bulava variety is associated with sufficiently high gluten content and good baking quality (on the level of the reference standard Moskovskaya 39), with regard to the standard baking quality assessed as 4.5 points (Table 1). Table 1 shows a comparison of baking qualities of our other varieties: Sibirskaya niva, Imeni Rapoporta, and Ritza with the normal amylose content. The Bulava variety is not worse than these varieties by its baking quality.

Table 1. Baking properties of flour and quality of bread from winter wheat mutant varieties grown in the Podolsk region of the Moscow area. The harvest of 2001.

Index	Bulava	Sibirskaya niva	Imeni Rapoporta	Ritza	Beseda	Botovskaya 1	Mskovskaya 39 standard
Content of raw gluten, %	32.1	33.7	31.6	27.7	30.9	32.2	27.6
Quality group	II	I	II	I	I	I	I
Dough elasticity	121	109	73	57	86	71	108
Dough elasticity/stretchability	1.6	1.2	0.7	0.4	0.9	0.5	1.5
Specific work of dough deformation *W*	346	452	292	275	342	342	345
Dough fluidity	80	0	20	20	0	0	30
Valorimetric estimation	61	98	80	65	67	78	61
Volume of bread from 100 g of flour, ml	1290	1250	1250	1190	1320	1200	1190
Total baking quality, points	4.5	5.0	5.0	4.3	4.6	4.2	4.5

Note: The samples were obtained at the Laboratory of Mutational Selection and Protection of Environment of the Emanuel Institute of Biochemical Physics with an exception of the Moskovskaya 39 variety selected at the

Research Institute of Agriculture of the Central Region of Non-Chernozem Zone. Assessment of the baking qualities was performed at the All-Russian Center for Assessment of quality of agricultural crops.

It is possible to extract simultaneously high-amylose starch and gluten from the Bulava variety grains.

The high-amylose №5B sample grown in the Central Region of Russia is of value as green fodder for cattle; thus, two objectives may be achieved.

There are varieties and samples with high content of alimentary fibers. Introduction of this material into foodstuffs is a prophylaxis for chronic constipations and rectum cancers and an aid to remove toxic substances from organism.

All the varieties and samples show high and stable annual harvests due to their high adaptive properties [6, 7]. The Bulava variety when grown in the severe East Kazakhstan area with cold winters with little snow and hot dry summers yields high stable annual harvests, which exceeds those of the standard Mironovskaya 808 and Komsomolskaya 56 varieties by the average 0.92 ton/hectare (Table 2)

Table 2. Crop harvest (ton/hectare) of the Bulava variety grown in the East Kazakhstan area.

Bulava				Mironovskaya 808 –standard I				Komsomo-lskaya 56 – standard II
1989	1990	1991	Average	1989	1990	1991	Average	1991
4.64	6.28	5.54	5.49	4.150	4.73	4.84	4.57	4.77
+ to standard I								
0.49	1.55	0.70	0.92					
HCP_{05}								
0.26	0.41	0.40	0.36					
+ to standard II								
–	–	0.77	–					

Note: Data on the harvests were obtained by G.A. Olesin and V.I. Sukhovetskii, researchers at the Ust-Kamenogorsk station of the Institute of Cytology and Genetics, Siberian Division, Russian Academy of Sciences.

The Bulava variety when grown in the Central Volga area under severe conditions in 1994-1995 (in 1994, winter was especially cold; during the whole period, summers were extremely hot and dry) yielded the crop harvest, which exceeded that of all other varieties and samples grown here and exceeded the harvest of the standard Donshchina variety by 0.41 ton/hectare (Table 3).

The Bulava variety was zoned for the East Kazakhstan region and superseded the Mironovskaya 808 and Komsomolskaya 56 varieties, which had been adapted to this region earlier. The years of testing cited in Tables 2 and 4 preceded the submission of this variety to the State Grade-Testing.

At present, the Laboratory of Mutational Selection and Protection of Environment of the Emanuel Institute of Biochemical Physics has commenced the reproduction of this variety in the Central Region of Russia so that to provide the country with this high-amylose winter wheat variety, which possesses the properties of high baking quality and high content of gluten.

Table 3. Harvest and adaptive properties of the Bulava variety as compared with various reference standards under testing in the Kamyshin region of the Volgograd area (average for 1994-1995).

Variety	Harvest, ton/hectare	Winter-resistance, points	Drought-resistan-ce, points		
	Average	± to standard Don 85	± to stan-dard Don-shchina		
Local standard Don 85	2.52	0.0	+0.13	3	3
Local standard Donshchina	2.39	–0.13	0.0	3	3
Bulava	2.80	+0.28	+0.41	4–	3+
Imeni Rapoporta - standard	2.73	+0.21	+0.34	4	4

Note: Data on the harvests were obtained by A.A. Petunya and P.A. Smutnev, researchers at the State Selection Station of the Nizhnevolzhsk Research Institute of Agriculture.

The Bulava variety when grown in the Tula area (RS Venyov Company Ltd.) in 2006 yielded the crop harvest of about 5 ton/hectare, which exceeded those of all others by 0.3-1.0 ton/hectare.

High adaptive properties of the Bulava variety and its lodging-resistance are listed in Table 4.

Table 4. Winter-, drought-, and layering-resistance of the Bulava variety

Index	Bulava				Mironovskaya 808 - standard			
	19 89	1990	1991	Average	1989	1990	1991	Ave-rage
Overwintering,%	90. 0	23.0	72.6	61.9	75.0	19.0	72.0	55.3
+ to standard	15.	4.0	0.6	6.6				
Drought-resis-tance, points	5	5	5	5	4	4	4	4
+ to standard	1	1	1	1				
Resistance to layering, points	4.5	4.0	5.0	4.5	4.0	3.0	2.5	3.2
+to standard	0.5	1.0	2.5	1.3				

Note: Data on the harvests were obtained by G.A. Olesin and V.I. Sukhovetskii, researchers at the Ust-Kamenogorsk station of the Institute of Cytology and Genetics, Siberian Division, Russian Academy of Sciences.

The Bulava variety is wheat of the intense type, low-growing, and resistant to layering (Table 4). It is resistant to brown rust, possesses highly productive bushiness, and high regenerative ability. When grown with the use of adequate agricultural technology acceptable to the intense-type varieties, the Bulava variety ensures high stem-density and suppresses weeds. Hence, volumes of agricultural toxic chemicals applied are reduced, environmentally safe grains are obtained, and protection of environment is ensured. These are favorable factors that affect human health.

At the present time, we commence also the reproduction of the low-amylose Beseda variety. There are perspectives of growing varieties with high and low amylose content in the Central and other regions of Russia.

CONCLUSION

Mutant varieties and samples of winter wheat that are of interest from the viewpoint of high- and low amylose content, feature a set of other selection-valuable characters: high and stable harvest when grown in different years and under different soil and climate conditions, high adaptive properties, and high baking quality. Therefore, these are promising varieties that can be grown in other regions.

REFERENCES

1. Kononkov P.F., Pivovarov V.F., Gins M.S., Gins V.K., Introduction and selection of non-traditional cultures for elevated content of BAS and antioxidants. Non-traditional agricultural, curative, and decorative plants, 2003; 1: P. 11-16. (in Russian)
2. Rapoport I.A. Action of oxide ethylene, glycide and glycol on gene mutations. Report. AS SU (Докл. АН СССР), 1948; 61(4): 713-715. (in Russian)
3. Eiges N.S., Volchenko G.A., Collection of mutants of winter wheat created by Rapoport's method chemical mutagenesis. Cytology (Цитология), 2004; 46(9): 891-892. (in Russian)
4. Eiges N.S., Weisfeld L.I., Volchenko G.A. // Rapoport's chemical mutagenesis in the aspect of specific valuable mutation on the winter wheat applicable in medicine. Functional foods for cardiovascular diseases. Richardson: 2005, P. 207—214.
5. Bocharnikova I.I., Eiges N.S., Weisfeld L.I., et al., Assessment of properties of wheat varieties obtained by chemical mutagenesis for the starch-molasses and baking production. Part I. Primary screening of wheat varieties. Agricultural raw material - storage and processing, 2004; 4: 45-48. (in Russian)
6. Eiges N.S., Weisfeld L.I., Volchenko G.A. // Chemical mutagenesis as an effective method of production of highly adaptive initial material and winter wheat varieties. Nontraditional and rare plants, natural compounds, and perspectives of using. Belgorod: 2006, 2, P. 7-12. (in Russian)
7. Eiges N.S., Volchenko G.A, Weisfeld L.I. // Method of chemical mutagenesis in production of mutants and mutant varieties of winter wheat with stable harvest and of high quality. Problems of enhancement of quality and stabilization of productivity in natural and anthropogenic systems. Nalchik: 2006. P. 189-197. (in Russian)

EFFECTS OF `ATAULFO` MANGO FRUIT (*MANGIFERA INDICA* L.) INTAKE ON MAMMARY CARCINOGENESIS AND ANTIOXIDANT CAPACITY IN PLASMA OF N-METHYL-N-NITROSOUREA (MNU)-TREATED RATS

Pablo Garcia-Solis, Elhadi M. Yahia, and Carmen Aceves

Facultad de Ciencias Naturales, Universidad Autónoma de Querétaro, and Instituto de Neurobiología, Universidad Nacional Autónoma de México, Campus UAQ-UNAM, Juriquilla, Querétaro, 76230

Corresponding author: Elhadi M. Yahia, E-mail: yahia@uaq.mx

Keywords: mango fruit, breast cancer, mammary carcinogenesis, plasma antioxidant capacity

INTRODUCTION:

Breast cancer is the most common worldwide neoplasia in women[1]. The totality of etiology factors of breast cancer is unknown and thus an effective preventive strategy has not been developed[2]. Risk factors associated to breast cancer can be grouped into three broad categories: a) family history factors, b) endocrine and reproductive factors and c) environmental and life-style factors including diet[2]. Several authors have postulated that high intake of fruits and vegetables could participate in the prevention of cancer because their phytochemical antioxidants prevent the oxidative damage produced by free radicals[3]. Free radicals oxidize lipids, proteins and DNA and are involved in the initiation and promotion/progression of carcinogenesis[4]. Epidemiological research has not clarified if the intake of fruit and vegetables are associated with a reduction in the risk of breast cancer. In 1997, World Cancer Research Fund and American Institute for Cancer Research concluded that there was convincing evidence that high intake of vegetables decreases the risk of cancers of mouth and pharynx, esophagus, lung, stomach, colon, and rectum, and probably decrease the risk of cancer of larynx, pancreas, bladder and breast[5]. However, several recent cohort-studies have indicated no associations between fruit and vegetable consumption and risk of breast cancer[6] and total cancer[7]. However, studies with animals have shown that several phytochemical compounds prevent chemical-induced mammary carcinogenesis. The chronic intake of soy flavonoid, genistein, prevents mammary carcinogenesis-induced by 7,12-dimethylbenzo[a]anthracene (DMBA)[8]. Isothyocyanates and glucosinolates extracted from broccoli sprouts also prevented DMBA-induced mammary cancer[9]. However, phytochemicals such as the carotenoid, lycopene failed to prevent mammary cancer induced by N-methyl-N-nitrosourea (MNU) [10]. On the other hand, it has been proposed that the health benefits of vegetable and fruits may result from multiple combined effects of their phytochemicals rather than from the effects of a single active ingredient[11]. Recent reports using more complex formulations of phytochemicals such as of whole extracts of apple (Red Delicious) or purple grape juice have reported the prevention of mammary carcinogenesis in DMBA treated rats[12, 13]. Moreover using *in vitro* approaches it has been shown that several whole extracts of fruits and vegetables have antioxidant and antiproliferative activities on cancer cell lines[11, 14]. On the other hand, Mango, one of the most consumed fruits, is rich in phytochemical antioxidants such as vitamin C, carotenoids and phenolic compounds[15]. Botting et al. [16] showed that mango have antimutagens using the salmonella typhimurium mutagenicty assay against heterocyclic amine 2-amino-3-methylimidazo [4,5-f]quinoline. Recently Percival et al.[11] showed that whole mango juice was able to inhibit cell proliferation in leukemic cell line HL-60 and inhibit neoplastic transformation of BALB/3T3 cells. The presence of antioxidant and antimutagenic activities in mango as well as its antineoplastic effects using mammalian *in vitro* systems suggested important anticancer activity *in vivo*. In this study we investigated the effect of the intake of mango (*Mangifera indica* L. Cv. Ataulfo) juice in preventing mammary carcinogenesis in rats treated with the carcinogen MNU and on the antioxidant capacity in plasma.

METHODS:

Female Sprague-Dawley rats, 3 weeks of age, four per cage, were housed in a temperature-controlled room (21±1°C) with a 12-h light/dark schedule. They were provided with food (Laboratory Rodent Diet 5001; LabDiet Purina,St. Louis, MO) and water ad libitum, aand all of the animal procedures followed de Official Mexican Norm NOM-062-ZOO-1999. Seven weeks old rats were anesthetized with a ketamine and xylazine (Cheminova, Mexico) mixture (30 mg and 6 mg, respectively, per Kg of body wt), and treated with a single intraperitoneal injection of 50 mg of MNU (Sigma, St. Louis, MO) per Kg body weight (bw). MNU was dissolved in 0.9% saline, pH 5.0, and heated to 50-60°C[17]. Negative control groups received only a saline injection.

Ripe mango (*Mangifera indica* L., cv. Ataulfo) fruit were purchased in the local market of Queretaro, Mexico, were of uniform size (223.99±12.69 g), and without physical and pathological defects, and were processed immediately after their arrived to the laboratory. They were cleaned and dried, and analyzed objectively for external (peel) and internal (pulp) color on basis of the CIELAB color system (L*, a*, b*) with a Minolta CM-2002 spectrophotometer (Minolta Co., Japan) and the program Spectra Match 3.3.7, where coordinate a* indicates positive values for reddish colors and negative values for greenish ones, whereas b* indicates positive values for yellowish colors and negative values for bluish ones, and L* is an approximate measurement of luminosity, which is the property according to which each color can be considered as equivalent to a member of the gray scale, between blank and white within the range 0-100. After which seed and peels were removed from the edible portion of mangoes, and the pulp was cut, pureed and stored in aliquots at -70°C. The color of fresh aliquots of puree pulp was measured as describe above. Fresh aliquots of puree pulp had values of total soluble solids of 17.5% (°Brix), indicating that the mango used is ripe.

Five weeks old rats were randomly sorted into 4 experimental groups using a randomization process, and mango treatment was started. Mango was provided in the drinking water by diluting the mango pulp with purified water in three different concentrations. The experimental groups were: a) Control (water with 1.2% sucrose), b) Mango-1 (0.02 g of mango/mL of water); c) Mango-2 (0.04 g of mango/mL of water) and d) Mango-3 (0.06 g of mango/mL of water). All control and mango-containing fluids were adjusted with sucrose to have the same °Brix (sugar content) as well as control solution. Mango solutions were prepared every day and both food and liquid intake were quantified. At the age of 8 weeks mango treatment was stopped for half of the rats which were injected with MNU and was substituted with control treatment (short-term intake of mango). Also at this time half of the rats of the negative control were sacrificed. Total phenolic and caroteniod content were monitored in Mango-3 solution every 15 days during 4 months. In addition, total phenolic and caroteniod contents were analyzed in Mango-3 juice after 24 and 48 h incubation in the vivarium conditions descried previously.

Rats were weighed and palpated for tumors every week beginning 4 weeks after carcinogen exposure and until 22 weeks after MNU administration. A tumor was defined as a discrete palpable mass recorded for at least 2 consecutive weeks. Tumor incidence was calculated as the percentage of animals with one or more palpable tumors per treatment. Tumor multiplicity was calculated as the average number of tumors per animal in each treatment group. The mean latency of tumor onset for each treatment group was calculated as the mean time interval (in weeks) from MNU injection to the appearance of the first palpable tumor. When the tumors had grown to $\geq$ 2 cm in diameter, rats were anesthetized with the ketamine and xylazine mixture and the tumors were surgically removed and processed for the different analyses. Tumor sizes were weighed and measured using a caliper, and the volumes were calculated by the ellipsoid formula[18]. At the end of the experiment the rats from each experimental group were sacrificed by decapitation, and mammary tumors were collected and processed for different analyses. After tumor surgery or animal sacrifice the mammary tumors were fixed in 10% neutral buffered formalin, and liver was frozen in dry ice.

Fixed MNU-induced mammary tumors were embedded in paraffin blocks. Sections of 5-um thickness were cut from each block and placed on slides. Sections were deparaffinized in xylene, rehydrated in descending grades of ethanol, and stained with hematoxylin and eosin. Tumors were classified according to criteria[19].

The extraction and analysis of total carotenoids was done according to Soto-Zamora et al. [20]. Quantification of carotenoid content in mango juice were made using a B-carotene dissolved in hexane as standard, and results were expressed as ug of B-carotene equivalents/mL. The contents of total free phenolic compounds in mango juice samples were analyzed by the method Folin-Ciocalteu colorimetric method[21], with some modification. Mango-3 juice was centrifuged at 10, 000 g for 5 min and then 30 uL of centrifuged mango-3 juice were oxidized with 150 uL of Folin-Ciocalteu reagent and after 5 min the reaction was neutralized with 120 uL sodium carbonate. The absorbance of the resulting blue color was measured at 620 nm after 2 h using a MRX Dynex plate reader (Dynex technologies, Inc., Chantilly, VA). Gallic acid was used as standard, and results were expressed as ug of gallic acid equivalents/mL. Total phenolics compounds content in plasma were measured by the Folin-Ciocalteau method modified to remove protein interferences[22]. Total phenolic compounds were determined by extraction/hydrolysis, and precipitation of protein with 0.75 M metaphosforic acid. For hydrolyzing the conjugated forms of polyphenols, 500 uL of 1 M HCl was added to 250 uL of plasma, vigorously vortexed for 1 min and incubated at 37°C for 30 min. Later, 500 uL of 2 M NaOH in 75% methanol was added, and the resulting mixture vortexed for 3 min and incubated at 37°C for 30 min. This step breaks the links of polyphenols with lipids and provides a first extraction of polyphenols. Then, 500 uL of 0.75 M of methaphosforic acid was added after mixing for 3 min to remove plasma proteins, and the sample was centrifuged at 1500 X g for 10 min. The supernatant was removed and kept on ice in the dark, while polyphenols were extracted again by adding 500 uL of a 1:1 (v/v) solution of acetone: water and centrifuged for 10 min at 2700 X g. The two supernatants were combined and filtered through 0.45 mm filter paper. Samples of 30 mL were assayed for total polyphenols with the Folin-Ciocalteu reagent as explained earlier.

FRAP was measured using the method reported by Benzie and Strain[23] with some modifications. Sodium acetate buffer (300 mM, pH 3.6) was prepared by dissolving 3.1 g of sodium acetate trihydrate in 950 mL H_2O, adding 16 mL of glacial acetic acid, and brought to a total volume of 1 L. This solution was stored at room temperature until use. A 2, 4, 6-tripyridyl-s-triazine (TPTZ) solution (10 mM) was prepared by dissolving 0.312 g of TPTZ in 100 mL of HCl (40 mM) and stored in the dark at room temperature. A 20 mM ferric chloride solution was prepared by dissolving 0.270 g of ferric chloride hexahydrate in 50 mL of H_2O. A FRAP reagent was prepared fresh prior to each analysis by combining 300 mM sodium acetate buffer, 10 mM TPTZ and 20 mM solution at the proportion of 10:1:1. Ascorbic acid standard series were prepared fresh prior to analysis. FRAP assay was carried out using MRX Dynex plate reader (Dynex technologies, Inc., Chantilly, VA) and absorbance readings were taken at 620 nm. The plates were manually loaded with sodium acetate buffer (10 uL), ascorbic acid standards (10 uL) or samples (10 uL) into the respective wells, 290 uL of freshly prepared FRAP reagent were added into each well, and readings were taken 1 hr thereafter.

A modified rapid-screening of total oxyradical scavenging capacity (TOSC) assay was used to determine total antioxidant activity in plasma. TOSC assay is based on the generation of peroxyl radicals generated by of thermal homolysis of 2-2`-azo-bis-(2 methyl ropioamidine)-dihydrocholoride (ABAP) to oxidized alpha-keto-y-methiolbutyric acid (KMBA) to ethylene, which was monitored by headspace gas chromatographic analysis[24]. The final assay conditions were 0.2 mM KMBA, 20 mM ABAP in 100 mM potassium phosphate buffer, pH 7.4. The reactions were carried out in 10 mL rubber septa-sealed vials in a final volume of 1 mL. The reactions were initiated by injection of 100 mL of 200 mM ABAP in water directly through the rubber septum. The vials were incubated for 90 min at 39ºC. Ethylene production was measured by gas-chromatographic analysis in 100 uL aliquots taken directly from the headspace of the reactions vials. Analysis were performed in a HP 5890 Serie II gas chromatograph (Agilent Technologies, CA) equipped with HP-PLOT Q capillary column (30 m x 0.53 mm x 40 um) and flame ionization detector (FID). The oven, injection and FID temperatures were 130, 180, and 220ºC, respectively. Nitrogen was used as the carrier gas at a flow rate of 24 mL/min. The plasma was diluted in the potassium phosphate buffer (1:20) and 100 uL were used for the assay. Ethylene production was measured at a single time point as proposed by MacLean et al. [25].
Retinol in liver was analyzed as described by Hosotani and Kitawaga with some modifications. 500 mg of liver were homogenized with 5 mL of water, and 300 uL of 10% homogenized liver was mixed with 500 uL of 25% sodium ascorbate solution and 2 mL ethanol. After incubation at 70ºC for 5 min, samples were saponificated with 10 M KOH (1 mL) and heated at 70ºC for 30 min. After cooling, n-hexane (4 mL) was added, and samples were then vigorously shaken for a 2 min and centrifuged at 10,000 g for 10 min. This extraction process was repeated four times. Then n-hexane extracts were evaporated to dryness and the residues were dissolved in methanol. For the determination of retinol a Hewlett Packard 1100 series HPLC system with DAD (detection at 325 nm) was used. Retinol was separated on a Symetry Â® C-18 (3.5 um, 150 X 4.6 mm) column at 30ºC with a mobile phase of acetonitrile-dichloromethane-methanol-1-octanol (90:15:10:0.1) at a flow rate of 1.0 mL/min. In all cases 25 uL was injected with an isocratic elution to the HPLC system.

The effects of dietary treatments on mammary cancer incidence were analyzed using 2 X 2 contingency tables and a one-tail Fisher test. The effects of treatments on tumor multiplicity, tumor latency, and tumor size counts, plasma antioxidant capacity, total phenolic compounds in plasma and retinol liver levels were analyzed using ANOVA.

The consumption of fruits and vegetables could reduce the risk of chronic disease such as cardiovascular diseases and cancer. Mango fruit is rich in several phytochemicals such as carotenoids and phenolic compounds, and has been reported to have *in vitro* antioxidant and antineoplastic effects suggesting important anticancer activity *in vivo*. We studied the effect of mango (*Mangifera indica* L., cv. Ataulfo) consumption on chemically-induced mammary carcinogenesis and plasma antioxidant ability in rats treated with the carcinogen N-methyl-N-nitrosourea (MNU). Mango was administered in the drinking water (0.02-0.06 g/mL) during both short-term and long-term periods to rats treated or not with MNU. Rats treated with MNU did not show differences in mammary carcinogenesis (incidence, latency and number of tumors), nor differences in plasma antioxidant capacity measured by both ferric reducing capacity in plasma (FRAP) and total oxyradical scavenging capacity assays. However, we have noticed a dose dependent increase in plasma antioxidant capacity with FRAP assay in animals which did not receive MNU and had a LT intake of mango. This suggests that mango consumption by normal subjects may increase antioxidants due to either a high intake or through the induction of metabolic changes in endogenous antioxidants.

RESULTS AND DISCUSSION

We have used the same mango puree pulp to make the mango dilutions, and therefore we only analyzed total carotenoids and phenolic compounds content in mango-3 juice. Mango-3 juice contained 2.27 + 0.31 ug/mL of total carotenoids whereas the phenolic compounds were 50.99 ± 3.24 ug/mL, indicating that mango-1 and mango-2 juices contained a mean of 0.75 and 1.50 ug/mL of total carotenoids, and 17 and 34 ug/mL of total phenolic compounds, respectively. The analysis every 15 days during 4 months did not show significant changes in the content of total carotenoids and phenolic compounds of mango-3 juice prepared from mango puree pulp stored at -70°C (data not shown). When we analyzed Mango-3 juice after 24 and 48 h of incubation in the vivarium conditions (23± 1°C with a 12-h light/dark schedule) we noted a slight but significant ($p<0.05$) reduction with respect to fresh pulp in both total carotenoids and phenolic compounds. After 24 h of incubation the reduction in total carotenoids was 6% whereas after 48 h was 13%. In the case of phenolic compounds the reduction after 24 and 48 h were 11 % and 14%, respectively. Animal weight gains and both liquid and food intake were similar in all groups studied. Food intake was about 16g per day per rat with a minimum intake of 12g at the first weeks and a maximum of 19g at the last weeks. As regard to mango intake in MNU-treated rats, at the beginning the intake of mango per Kg of body weight was high but decreased later. Mango-3 juice treated rats consumed a mean of 4.5±0.9 g of mango/Kg body (minimum: 3.6; maximum 7.2), whereas the mean of mango intake in mango-2 and 3 treated rats were 8.5±1.4 g/Kg body (minimum: 7.2; maximum 12.2) and 12.8±2.7 g/Kg body (minimum: 10.6; maximum 21.8), respectively. The mean consumption of mango by rats in the 3 different concentrations was comparable to human consumption of two (about 293 g of mango pulp /70 Kg of body weight), four and six mangoes per day, respectively. The consumption of total carotenoids and phenolic compounds when mango-3 juice was used were higher in the first 4 weeks and then decreased and stabilized.

The present study is, to our knowledge, the first to test in vivo antineoplastic effect of mango. Previous studies showed that mango have in vitro antioxidant, antimutagenic and antineoplastic activities[11,16]. However, we have shown here that short and long-term consumption of mango does not prevent mammary carcinogenesis in rats treated with the carcinogen MNU. This null effect of short-term and long-term intake of mango indicated that the kind and quantity of `Ataulfo` mango phytochemicals do not seem to be able to inhibit the MNU-induced mammary cancer neither at the initiation nor at the promotion steps of carcinogenesis. Some fruits have been shown to have an effective protective effect on chemically-induced mammary carcinogenesis[12, 13]. The fact that mango intake of MNU-treated rats did not modify antioxidant capacity of plasma suggests that MNU impairs uric acid metabolism.

CONCLUSION

Ataulfo mango intake does not prevent mammary carcinogenesis or increase plasma antioxidant capacity in MNU-treated rats. These findings of the present study should not be interpreted as a lack of the health benefits from regular consumption of mango or other fruits and vegetables.

REFERENCES

1. Parkin DM, Bray F, Ferlay J and Pisani P: Global cancer statistics, 2002. *CA Cancer J. Clin* 55, 74-108, 2005.
2. Bray F, McCarron P and Parkin DM: The changing global patterns of female breast cancer incidence and mortality. *Breast Cancer Res* 6, 229-239, 2004.
3. Syngletary KW, Jackson SJ and Milner JA. Non-nutritive components in foods as modifiers of the cancer process. In *Preventive Nutrition: the comprehensive guide for health professionals*, Bendich A and Deckelbaum RJ (eds). Totowa, NJ: Humana Press, 2005, pp 55-88.
4. Ray G, Batra S, Shuka NK, Deo S, Raina V, et al.: Lipid peroxidation, free radical production and antioxidant status in breast cancer. *Breast Cancer Res Treat* 59, 163-170, 2000.
5. World Cancer Research Fund, American Institute for Cancer Research expert panel. Food, nutrition and the prevention of cancer: a global perspective. Washington, DC. WCRF/AICR, 1997.
6. van Gils CH, Peeters PH, Bueno- de-Mezquita HB, Boshuizen HC, Lahmann PH, et al.: Consumption of vegetables and fruits and risk of breast cancer. *JAMA* 293, 183-193, 2005.
7. Hung H-C, Joshipura KJ, Jiang R, Hu FB, Hunter D, et al.: Fruit and vegetables intake and risk of major chronic disease. *J Natl Cancer Inst* 96, 1577-1584, 2004.
8. Lamartiniere CA, Cotroneo MS, Fritz WA, Wang J, Mentor-Marcel R and Elgavish A: Genistein chemoprevention: timing and mechanism of action in murine mammary and prostate. *J Nutr* 132 (Suppl), 552S-558S, 2002.
9. Fahey JW, Zhang Y and Talalay P: Broccoli sprouts: An exceptionally rich source of inducers of enzymes that protect against chemical carcinogens. *Proc Natl Acad Sci USA* 94, 10367-10372, 1997.
10. Cohen LA, Zhao Z, Pittman Ba and Khachik F: Effect of dietary lycopene on *N*-methylnitrosourea-induced mammary tumorigenesis. *Nutr Cancer* 34, 153-159, 1999.

11. Percival SS, Talcott ST, Chin ST, Mallak AC, Lound-Singleton A and Pettit-Moore: Neoplastic transformation of BALB/3T3 cells and cell cycle of HL-60 cells are inhibited by mango (*Mangifera indica L.*) juice and mango juice extract. *J Nutr* 136, 1300-1304, 2006.
12. Liu RH, Liu J and Chen B: Apples prevent mammary tumors in rats. *J Agric Food Chem* 53, 2341-2343. 2005.
13. Jung KJ, Wallig MA and Singletary KW: Purple grape juice inhibits 7,12-dimethylbenz[a]anthracene (DMBA)-induced rat mammary tumorigenesis and in vivo DMBA-DNA adduct formation. *Cancer Lett* 233, 279-88, 2006.
14. Sun J, Chu YF, Wu X and Liu RH: Antioxidant and antiproliferative activities of common fruits. J Agric Food Chem 50, 7449-7454, 2002.
15. Rocha-Ribeiro SM, Queiroz J, Lopes M, Campos F, Pinheiro H, et al.: Antioxidant in mango (*Mangifera indica L.*) pulp. *Plant Foods Hum Nutr* 62, 13-17, 2007.
16. Botting KJ, Yong MM, Pearson AE, Harris PJ and Ferfuson LR: Antimutagens in food plants eaten by Polynesians: micronutrients, phytochemicals and protection against bacterial mutagenicity of heterocyclicic amine 2-amino-3-methylimidazo [4,5-*f*]quinoline. *Food Chem Toxicol* 37, 95-103, 1999.
17. Thompson HJ. Methods for the induction of mammary carcinogenesis in the rat using either 7,12-dimethylbenz[a]antracene or 1-methyl-1-nitrosourea. In *Methods in mammary gland biology and breast cancer research, 8th*, Ip M and Asch BB (eds). New York: Kluwer Academic/Plenum Publishers, 2000, pp 19-29.
18. Teelmann K, Tsukaguchi T, Klaus M and Eliason JF: Comparison of the therapeutic effects of a new arotinoid, Ro 40-8757, and all-*trans* and 13-*cis* retinoic acids on rat breast cancer. *Cancer Res* 53, 2319-2325, 1993.
19. Russo J and Russo IH: Atlas and histological classification of tumors of the rat mammary gland. *J Mammary Gland Biol Neoplasia* 5, 187-200, 2000.
20. Soto-Zamora G, Yahia EM, Brecht JK and Gardea A: Effects of postharvest hot air on the quality and antioxidant levels in tomato fruit. *LWT* 38, 657-663, 2005.
21. Singleton VL, Orthofer R and Lamuela-Raventós RM: Analysis of total phenols and other oxidation substrates and antioxidants by means of Folin-Ciocalteu reagent. *Methods Enzymol* 299, 152-178, 1999.
22. Serafini M, Maiani G and Ferro-Luzzi A: Alcohol free red wine enhances plasma antioxidant capacity in humans. *J Nutr* 128, 1003-1007, 1998.
23. Benzie IFF and Strain JJ: The ferric reducing ability of plasma (FRAP) as a measured of "antioxidant power": The FRAP assay. *Anal Biochem* 239, 70-76, 1996.
24. Winston GW, Regoli F, Dugas AJ Jr, Fong JH and Blanchard K. A rapid gas chromatographic assay for determining oxyradical scavenging capacity of antioxidants and biological fluids. *Free Radical Biol Med* 24, 480-493, 1998.
25. MacLean DD, Murr DP and DeEll JR. A modified total oxyradical scavenging capacity assay for antioxidants in plant tissues. *Postharvest Biol Technol* 29, 183-194, 2003.

SENSORY EVALUATION OF VEGETABLE AMARANTH POWDER SUPPLEMENTED LOCAL PREPARATIONS OF HIMACHAL PRADESH (INDIA)

Sonika, S.R. Malhotra[1] and Renu Bala[2]

[1]Department of Food Science and Nutrition, [2]Department of Chemistry and Biochemistry
CSK Himachal Pradesh Krishi Vishvavidyalaya, Palampur-176062, India.

Corresponding author: Sonika Banyal, E-mail: sonika_banyal@rediffmail.com

Keywords: vegetable amaranth, essential nutrients, amaranth leaf powder

ABSTRACT

Amaranth is one of the green leafy vegetables grown successfully under mid hill conditions of H.P. The present study was undertaken to see the potential of vegetable amaranth powder in locally consumed preparations of the state. Four locally consumed preparations were selected, standardized and supplemented with vegetable amaranth powder. Thereafter rated on 9 Point Hedonic Scale for sensory qualities viz. colour, chewiness/crispness, taste, flavour, texture and overall acceptability. In all the preparations the levels of supplementation was found to be in the range of 'liked slightly' to 'liked very much'. The colour was the most affected character and other characters were less affected with the supplementation.

INTRODUCTION

Green leafy vegetables occupy an important place among the crops as these provide adequate amounts of protective nutrients such as vitamins and minerals for human nutrition. Green leafy vegetables are highly seasonal and in abundant supply during the season resulting in spoilage of large quantities if not utilized. Therefore, green leafy vegetables should be utilized during the season, however, there is great potential for preservation of green leafy vegetables to reduce wastage and make them available in the lean season.

All the preparations wherein fresh green leafy vegetables are normally used can be prepared with their powders too. These powders can also be used to enrich other food preparations provided the sensory qualities of the products are not affected (Lakshmi and Vimala, 2000). The potential of vegetable amaranth in various preparations has been reviewed by a number of researchers (Devadas et al.1982; Dahiya and Kapoor, 1994) and reported the products acceptable.

Amaranthus a green leafy vegetable successfully grown under mid hill conditions of H.P. is a rare example of a vegetable where all the essential nutrients are combined in one. This paper reports the sensory acceptability of amaranth leaf powder supplemented to some selected commonly consumed preparations of H.P.

MATERIAL AND METHODS

Edible leaf samples of the vegetable amaranth cultivar AV-76 were grown and procured from the Department of Plant Physiology, College of Basic Sciences, CSK HPKV, Palampur. The leaves were washed and dried to a constant weight in a tray drier (60± 5°C), ground to powder form and preserved in polythene bags until required. All the raw materials for products development viz. flours (maize flour, rice flour, and refined wheat flour), potato, seasonings and fat/oil were procured from the local market.

PRODUCT DEVELOPMENT

Four commonly consumed preparations of Himachal Pradesh i.e. maize flour-*roti*, rice flour-*babroo*; refined wheat flour-*mathri* and potato-*tikki* were selected. Different preparations were standardized with different proportions of amaranth leaf powder (ranging from 1.5 to 7.5 percent levels of supplementation) and prepared in the department laboratory. Finally, four proportions of supplementation for each recipe as evaluated by a panel of judges were optimized for supplementation purpose.

SENSORY EVALUATION

Cooked samples of the preparations were subjected to sensory evaluation by a taste panel of 10 trained judges. The panel was presented with the samples for rating colour, chewiness/crispness, taste, flavour, texture and overall acceptability on a Hedonic Rating Scale using values ranging from 1 to 9, where 1 represented 'Disliked Extremely' and 9 represented 'Liked Extremely'. The Hedonic Ratings were treated statistically by analysis of variance at 5 percent level of significance.

RESULTS AND DISCUSSION

The results in Table 1 shows that scores of the control *roti* for colour, chewiness, taste, flavour and overall acceptability were higher than the supplemented *roti,* however, the scores for the texture were found same for control and supplemented *roti*.

Table 1. Effect of supplementation of Amaranth powder on the organoleptic properties of *maize roti*

Proportions	Colour	Chewiness/ Crispness	Taste	Flavour	Texture	Overall acceptability
100:0	8.5	8.3	8.4	8.4	8.3	8.3
95:5	8	7.9	7.9	7.8	8.3	8
90:10	7.2	7.4	7.2	7.2	8	7.5
85:15	7	7.2	6.8	7	7.7	7.2
CD (0.05)	0.94	0.66	0.81	0.93	NS	0.64

Supplementation affected all the organoleptic characters of the *roti* significantly at 5 percent level of significance except for texture, which was non-significantly affected with the supplementation. Colour was affected the most with the supplementation followed by flavour, taste, chewiness and overall acceptability. Lakshmi and Vimala (2000) reported the acceptability of *misi roti* to vary from average to good supplemented with a blend of leafy vegetable powders containing 75 percent amaranth powder. Bhavani and Kamini (1998) reported the acceptability of maize-leaf powder blends by preschool children.

Table 2. Effect of supplementation of Amaranth powder on the organoleptic properties of *babroo*

Proportions	Colour	Chewiness/ Crispness	Taste	Flavour	Texture	Overall Acceptability
100:0	8.4	8	8	8	8	8.1
98.5:1.5	8.1	8.4	8.4	8.2	8.3	8.3
97.5:2.5	7.5	7.4	7.6	7.8	7.3	7.7
95.5:5	6.8	7.5	7.7	7.6	7.4	7.5
CD (0.05)	0.83	0.59	NS	NS	0.82	0.54

The results in Table 2 shows that rice flour *babroo* (control) rated superior only in colour, however, *babroo* supplemented with 1.5 percent level rated superior for all other characters viz. crispness, taste, flavour, texture and overall acceptability. Substantial reduction only in the colour and texture of *babroo* was observed with the increase in the level of supplementation; however, supplementation affected little with regard to crispness and overall acceptability of *babroo*. The supplementation affected non-significantly to the taste and flavour of the babroo. Lakshmi and Vimala (2000) reported the acceptability of *biriyani* supplemented with dried vegetable powder blends in terms of colour, texture, taste and overall acceptability.

Table 3. Effect of supplementation of Amaranth powder on the organoleptic properties of *mathri*

Proportions	Colour	Chewiness/ Crispness	Taste	Flavour	Texture	Overall Acceptability
100:0	8.2	8.1	8.1	8.1	7.9	8.1
97.5:2.5	8	7.9	8.3	8.2	8.2	8.2
95.5:5	7.4	7.6	7.9	7.9	8	7.8
92.5:7.5	7	7.8	7.5	7.6	7.7	7.5
CD (0.05)	0.71	NS	NS	NS	NS	NS

The results in Table 3 shows that the effect of supplementation of amaranth leaf powder to the *mathri* was found significant only in the colour at 5 percent level of significance, however it was found non significant for all other characteristics of the *mathri.* The superiority of *mathri* (control) was observed for colour, crispness; however, supplemented *mathri* was superior in taste, flavour, texture and above all in overall acceptability. Lakshmi and Vimala (2000) reported the acceptability of leafy vegetable powders (blends) supplemented deep fried products as *vada, pakori* and bread rolls to vary from good to excellent for colour, taste, texture and overall acceptability.

Table 4. Effect of supplementation of Amaranth powder on the organoleptic properties of *Tikki*

Proportions	Colour	Chewiness/ Crispness	Taste	Flavour	Texture	Overall Acceptability
100:0	8.4	8.1	8.2	8.1	7.7	8.1
98.5:1.5	8	7.2	8.3	8.3	8.3	8.2
97.5:2.5	7.3	7.6	7	7	7.3	7.2
95.5:5	6.5	6.8	6.4	6.5	7.1	6.7
CD (0.05)	0.75	0.83	0.88	0.79	0.90	0.65

Table 4 shows that all the organoleptic characters were affected significantly with the supplementation (0-5%) of amaranth leaf powder to the potato *tikki*. The control *tikki* was rated superior only in colour; however, supplemented tikki (1.5%) was rated superior for all other characters. Considering overall acceptability, *tikki* supplemented at 1.5 percent level ranked first followed by control, 2.5 percent level of supplementation. Lakshmi and Vimala (2000) reported the acceptability of amaranth powder supplemented potato curry.

CONCLUSIONS

The sensory qualities of all the recipes were affected with the increase in the level of supplementation of amaranth leaf powder to these. In all the recipes the least levels of supplementation were most acceptable in terms of overall acceptability. Colour was the most affected character; others were less affected with the supplementation. Considering overall acceptability, all the supplementation levels were found in the range of 'Liked Slightly' to 'Liked Very Much'.

GENERAL DESCRIPTION OF THE PREPARATIONS

Name of Recipe	**Description of Recipe**
Roti	Maize flour dough made, rounded into circular *roti* and baked on hot griddle.
Babroo	Seasoned rice flour batter, flattened on greased hot griddle.
Mathri	Stiff, seasoned refined wheat flour dough was made with added ghee. Rolled, cut the pieces and deep fried to light brown.
Tikki	Boiled and mashed potatoes. Seasoned, made small ball. Flattened and shallow fried on hot griddle to golden brown.

REFERENCES

1. Bhavani, K.N. and Kamini, D.(1998). Development and acceptability of a ready to eat ß-Carotene rich maize based supplementary products. Plant Foods Human Nutrition 52(3): 271-278.
2. Dahiya, S. and Kapoor, A.C.(1994). Development, nutritive content and shelf life of home processed supplementary foods. Plant Foods Human Nutrition 45(4): 331-342.
3. Devadas, R.P., Chandrasekher, V. and Bhooma, N. (1982). Acceptability of diets based on low cost locally available foods for various target groups. Ind. J. Nutr. Dietet. 19:1-7.
4. Lakshmi, B. and Vimala, V.(2000). Nutritive value of dehydrated green leafy vegetable powders. J. Food Sci. Technol. 37(5): 465-471.

PART THREE
REVIEWS

PREVENTION AND TREATMENT OF CARDIOVASCULAR DISEASES BY NUTRITIONAL FACTORS

Undurti N. Das

UND Life Sciences, Shaker Heights, OH 44120, USA

Corresponding author: Undurti N. Das, E-mail: undurti@hotmail.com

Keywords : Nitric oxide, hypertension, endothelium, free radicals, folic acid, ω-3 fatty acids, endothelial dysfunction, lipoxins, resolvins, prostaglandins, prostacyclin.

ABSTRACT

Decreased endothelial nitric oxide (eNO) activity, a marker of endothelial dysfunction, is seen in cardiovascular diseases. Folic acid reduces plasma homocysteine levels, enhances eNO synthesis, stimulates endogenous tetrahydrobiopterin (H4B) regeneration, a co-factor needed for eNO synthesis, inhibits intracellular superoxide generation, increases tissue concentration of ω-3 polyunsaturated fatty acids (PUFAs), and augments prostaglandin I3 synthesis. H4B enhances NO generation and L-arginine transport into the cells. Vitamin C augments eNO synthesis by increasing intracellular H4B and stabilization of H4B. ω-3 fatty acids enhance the action of insulin, and insulin in turn stimulates both H4B synthesis and metabolism of ω-3 and ω-6 fatty acids. ω-3 fatty acids and insulin suppress tumor necrosis factor-α and superoxide anion generation and enhance eNO formation. ω-3 fatty acids inhibit the activities of HMG-CoA reductase and angiotensin-converting enzymes (ACE), enhance parasympathetic tone and form precursors to anti-inflammatory molecules: lipoxins (LXs) and resolvins; and have anti-arrhythmic action. Thus, a rational combination of folic acid, H4B, vitamin C, ω-3 fatty acids, and L-arginine is of benefit in the prevention and treatment of cardiovascular diseases.

INTRODUCTION

Adequate intake of folic acid, vitamin B6 and vitamin B12 decreases homocysteine levels, which is an independent risk factor for atherosclerosis, coronary heart disease, and venous thromboembolism1-3. Plasma homocysteine concentrations are dependent on the dietary intake of folate (the term folate refers to the natural form in the human body and folic acid refers to the synthetic compound which is used in pills) and vitamin B12. Folic acid is not only involved in homocysteine metabolism but also enhances nitric oxide (NO) metabolism, increases the availability of tetrahydrobiopterin (BH4), reduces superoxide anion generation, and thus, improves endothelial dysfunction and possess anti-inflammatory actions.

HOMOCYSTEINE INCREASES OXIDANT STRESS

Homocysteine enhances oxidative stress4, which is due to its auto-oxidation leading to the formation of homocystine, homocysteine-mixed disulfides, and homocysteine thiolactone.

During this process, superoxide anion (O2-.) and hydrogen peroxide (H2O2) are generated, which cause endothelial cytotoxicity and dysfunction. Homocysteine converts normal antithrombotic endothelium to a more prothrombotic phenotype by increasing Factor V and Factor XII activity, decreasing protein C activation, inhibition of thrombomodulin expression, induction of tissue factor expression, suppression of heparan sulfate expression, reducing the binding of tissue-type plasminogen activator to its endothelial cell receptor: annexin II, reducing the production of NO and prostacyclin (PGI2), events that facilitate thrombotic tendency4-6. Normal endothelial cells detoxify homocysteine by releasing NO or a related S-nitrosothiol, which causes the formation of S-nitroso-homocysteine7, a potent vasodilator and platelet anti-aggregator. This S-nitrosation of homocysteine (thus forming S-nitroso-homocysteine) attenuates sulfhydryl-dependent generation of H2O2. However, continued exposure of endothelium to homocysteine compromises the production of adequate amounts of NO that ultimately leads to unopposed homocysteine-mediated injury to the endothelium and initiation of atherosclerosis and/or thrombus formation or acceleration of existing atherosclerosis. Homocysteine enhances the production of O2-. that, in turn, inactivates NO. Normally, a balance is maintained between NO and O2-.. Hence, enhancement of NO production prevents the thrombotic action of homocysteine and restores the antithrombotic properties of endothelium.

Homocysteine inhibits glutathione peroxidase (GP) activity *in vitro*. Inhibition of GP activity by homocysteine is responsible for its (homocysteine) vascular toxicity8. Homocysteine induces smooth muscle cell migration and proliferation by increasing cyclin D1 and cyclin A mRNA expression9, and inhibits NO production, which contribute to its pro-atherosclerotic and pro-thrombotic actions. Homocysteine up regulated vascular cell adhesion molecule-1 expression in human aortic endothelial cells and enhanced monocyte adhesion. Cyclo-oxygenase (COX) inhibitors completely prevented homocysteine-induced monocyte adhesion, whereas scavenging reactive oxygen species and elevation of NO caused only partial inhibition10. These results suggest that pro-inflammatory actions of homocysteine are mediated by prostaglandins.

Homocysteine enhanced the activity of HMG-CoA (3-hydroxy-3-methylglutaryl coenzyme A) reductase enzyme in human umbilical vein endothelial cells (HUVECs), and enhanced cellular cholesterol content. Cell-permeable superoxide dismutase (SOD) mimetic, Mn-TBAP, reversed homocysteine-induced expression of HMG-CoA reductase activity. Simvastatin, an HMG-CoA reductase inhibitor, reduced HUVECs cholesterol content and prevented homocysteine-induced suppression of NO production in a dose dependent manner11. Thus, homocysteine participates in the pathogenesis of atherosclerosis by enhancing cholesterol synthesis.

Both folic acid and tetrahydrobiopterin (H4B) abolished homocysteine-induced intracellular superoxide production in culture12. Oral folic acid supplementation (10 mg/once a day) to healthy volunteers not only restored NO synthesis to normalcy but prevented nitrate tolerance by restoring/stimulating endogenous regeneration of H4B13.

FOLATE/FOLIC ACID, H4B, INSULIN, VITAMIN C, AND L-ARGININE IN CARDIOVASCULAR DISEASES

Folic acid, H4B, and insulin suppress O2-. production and thus, prolong the half-life of NO4, 14-16 that, in turn, preserves endothelium-dependent vasodilation. Folate/folic acid, H4B and insulin possess anti-inflammatory actions both by suppressing O2-. generation and enhancing

NO production17. Folic acid and H4B restore and attenuate cholesterol-induced endothelial dysfunction and coronary hyperreactivity to endothelin18 respectively. Since folic acid restores the tissue stores of H4B, it is important that both folic acid and H4B should be given together to obtain their maximum benefit. H4B stimulated endothelial cell proliferation, whereas NG-mono-methyl-L-arginine, a NO synthase inhibitor, attenuated H4B-induced endothelial cell proliferation19. This suggests that H4B levels regulate proliferation of normal endothelial cells and that its deficiency impairs NO-dependent proliferation of endothelial cells. H4B augments NO generation by enhancing arginine transport in cardiac myocytes20. This suggests that it is necessary to provide L-arginine, folic acid or 5-methyltetrahydrofolic acid, the active form of folic acid, H4B, and vitamin C together in adequate amounts to stimulate eNO synthesis. Vitamin C enhances eNOS activity by increasing intracellular levels21, 22, and via chemical stabilization of H4B and it also enhances the release of NO23 and thus, suppresses the formation of total S-nitrosothiols and S-nitrosoalbumin.

ASYMMETRICAL DIMETHYLARGININE AND HYPERTENSION

One endogenous factor that inhibits NO synthesis is asymmetrical dimethylarginine (ADMA). Increased plasma ADMA concentrations have been shown to prospectively determine cardiovascular and overall mortality in patients with end-stage renal disease. A strong association between ADMA concentrations in serum with risk of acute coronary events has been described. Endothelial dysfunction in hypercholesterolemic individuals can be related to plasma concentrations of ADMA24. Plasma concentrations of ADMA were found to be high in hypertension and pre-eclampsia25, 26. Increased serum concentrations of ADMA were associated with an increased risk of acute coronary events indicating that endothelial dysfunction occurs in coronary heart disease (CHD)27. Endothelial dysfunction in offspring of patients with essential hypertension was found to be due to impaired basal production of NO, which could be restored to normalcy by intra-brachial L-arginine28, 29. This suggests that impairment in NO production precedes the onset of hypertension.

LONG-CHAIN POLYUNSATURATED FATTY ACIDS AND HYPERTENSION

Dietary linoleic acid (LA, 18:2 ω-6) is converted to gamma-linolenic acid (GLA, 18:3 ω-6), dihomo-GLA (DGLA, 20:3 ω-6), and arachidonic acid (AA, 20:4 ω-6), and α-linolenic acid (ALA, 18:3 ω-3) to eicosapentaenoic acid (EPA, 20:5 ω-3), and docosahexaenoic acid (DHA, 22:6 ω-3) by specific enzymes, which are controlled by genetic, hormonal, and nutritional factors30-34. GLA, DGLA, AA, EPA, and DHA are called as long-chain polyunsaturated fatty acids (LCPUFAs). Saturated and trans-fatty acids interfere with the formation of vasodilator prostaglandins (PGs), PGE1 and PGI2, elevate blood pressure, and exacerbate spontaneous hypertension. LA and DGLA enhance the formation of PGE1 and PGI2 vasodilators and prevent increase in blood pressure induced by feeding of saturated fats. Fish oil, a rich source of ω-3 eicosapentaenoic acid (EPA, 20:5) and docosahexaenoic acid (DHA, 22:6), reduces blood viscosity and lower blood pressure30, 35. EPA and DHA inhibit the formation of thromboxane A2 (TXA2), a potent vasoconstrictor and platelet aggregator, and enhance that of PGI3, a vasodilator and platelet anti-aggregator. In addition, EPA lowers the tissue levels of AA and enhances those of DGLA, the precursor of PGE135, 36. Hence, when adequate amounts of ω-3 and ω-6 fatty acids are provided blood pressure remains normal. Furthermore, AA, EPA, and DHA also give rise to anti-inflammatory compounds such as lipoxins (LXs), resolvins, and neuroprotectin D1 that are essential for would healing and protect neurons and other cells from the toxic action of various endogenous and exogenous molecules.

LCPUFAs inhibit ACE activity30, 37 and augment the synthesis of eNO38. L-arginine-NO system up regulates the metabolism of LCPUFAs in experimental animals. Hence, whenever the tissue concentrations of LCPUFAs are low, the synthesis and release of NO will also be low, and *vice versa*. Since endothelial cells are the major source of NO, LCPUFA content of endothelial cells could influence the synthesis and release of NO, explaining the ability of AA, EPA, and DHA to enhance the synthesis of NO30. In patients with hypertension, type 1 and type 2 diabetes mellitus, and CHD the plasma concentrations of various LCPUFAs and those of LA and ALA were found to be low30, 39. This suggests that abnormalities in NO, O2-. and LCPUFAs are closely associated with each other in various cardiovascular diseases.

LCPUFAs suppress the synthesis of tumor necrosis factor-α (TNF-α) and other pro-inflammatory cytokines40, 41. Healthy Indian Asians had higher C-reactive protein (CRP), a marker of systemic inflammation, than did Europeans and have higher levels of TNF-α, which explains the high incidence of insulin resistance and other features of metabolic syndrome X in them41, 42. Weisinger et al43 reported that DHA deficiency in the perinatal period could raise blood pressure later in life, even when animals were subsequently replete with this fatty acid. Animals raised on an LCPUFA-deficient diet under drank water and over ingested sodium, features that are somewhat similar to high salt intake-induced hypertension.

EPA and DHA inhibited the development of proteinuria and suppressed hypertension in stroke-prone spontaneously hypertensive rats, and prevented the exaggerated growth of vascular smooth muscle cells from these animals by suppressing the synthesis of TGF-β44, 45. LCPUFAs also interact with other nutrients and L-arginine-NO-O2- system, suppresses ACE enzyme, and lower blood pressure30, 46, 47, 48, 49.

SUMMARY AND CLINICAL IMPLICATIONS

It is evident from the preceding discussion that folic acid, H4B, vitamin C, L-arginine, and LCPUFAs augment NO synthesis and inhibit/prevent platelet aggregation, atherosclerosis, thrombosis, and thus, CHD. It is interesting to note that folic acid increases concentration of ω-3 polyunsaturated fatty acids, and decreases vitamin K-dependent coagulation factors, which could reduce risk of thrombosis50-52. Evidence suggests that H4B controls NOS activity in the human placenta and a defect in this interaction between H4B and NOS activity may play a role in the pathogenesis of pre-eclampsia53. Interaction between folic acid and PUFAs could be one potential mechanism by which folic acid brings about some of its clinical benefits.

Based on these evidences, it is reasonable to suggest that a rational combination of folic acid, vitamin B12, B6, vitamin C, L-arginine, H4B, PUFAs (especially EPA and DHA) could be useful in the prevention and treatment of cardiovascular diseases.

REFERENCES:

1. Welch GN, Loscalzo J. Homocysteine and atherothrombosis. N Engl J Med 1998; 338: 1042-1050.

2. Quere I, Perneger TV, Zittoun J, Bellet H, Gris J-C, Daures J-P, Schved J-F, Mercier E, Laroche J-P, Dauzat M, Bounameaux H, Janbon C, de Moerloose P. Red blood cell methylfolic acid and plasma homocysteine as risk factors for venous thromboembolsim: a matched case-control study. Lancet 2002; 359: 747-752.
3. Boushey CJ, Beresford SA, Omenn GS, Motulsky AG. A quantitative assessment of plasma homocysteine as a risk factor for vascular disease: probable benefits of increasing folic acid intakes. JAMA 1995; 274: 1049-1057.
4. Loscalzo J. The oxidant stress of hyperhomocyst (e) inemia. J Clin Invest 1996; 98: 5-7.
5. Heinecke JW, Rosen H, Suzuki LA, Chait A. The role of sulfur-containing amino acids in superoxide production and modification of low-density lipoprotein by arterial smooth muscle cells. J Biol Chem 1987; 262: 10098-10103.
6. Wall RT, Harlan JM, Harker LA, Striker GE. Homocys(e)ine-induced endothelial cell injury in vitro: a model for the study of vascular injury. Thromb Res 1986; 18: 113-121.
7. Stamler JS, Osborne JA, Jaraki O, Rabbani LE, Mullins M, Single D, Loscalzo J. Adverse vascular effects of homocysteine are modulated by endothelium-derived relaxing factor and related oxides of nitrogen. J Clin Invest 1993; 91: 308-318.
8. Upchurch GR, Welch GN, Freedman JE, Loscalzo J. Homocys(e)ine attenuates endothelial glutathione peroxidase and thereby potentiates peroxide-mediated cell injury. Circulation 1995; 92: I-228.
9. Tsai J.-C, Perrella MA, Yoshizumi M, Hseih CM, Haber E, Schlegel R, Lee ME. Promotion of vascular smooth muscle cell growth by homocyst(e)ine: a link to atherosclerosis. Proc Natl Acad Sci USA 1994; 91: 6369-6373.
10. Silverman MD, Tumuluri RJ, Davis M, Lopez G, Rosenbaum JT, Lelkes PI. Homocysteine upregulates vascular cell adhesion molecule-1 expression in cultured human aortic endothelial cells and enhances monocyte adhesion. Arterioscler Thromb Vasc Biol 2002; 22: 587-592.
11. Li H, Lewis A, Brodsky S, Rieger R, Iden C, Goligorsky MS. Homocysteine induces 3-hydroxy-3-methylglutaryl coenzyme A reductase in vascular endothelial cells. Circulation 2002; 105: 1037-1043.
12. Doshi SN, McDowell IFW, Moat SJ, Lang D, Newcombe RG, Kredan MB, Lewis MJ, Goodfellow J. Folic acid improves endothelial function in coronary artery disease: an effect mediated by reduction of intracellular superoxide? Arterioscler Thromb Vasc Biol 2001; 21: 1196-1202.
13. Gori T, Burstein JM, Ahmed S, Miner SE, Al-Hesayen A, Kelly S, Parker JD. Folic acid prevents nitroglycerin-induced nitric oxide synthase dysfunction and nitrate tolerance: a human in vivo study. Circulation 2001; 104: 1119-1123.
14. Das UN. Is insulin an anti-inflammatory molecule? Nutrition 2001; 17: 409-413.
15. Das UN. Possible beneficial action(s) of glucose-insulin-potassium regimen in acute myocardial infarction and inflammatory conditions: a hypothesis. Diabetologia 2000; 43: 1081-1082.
16. Das UN. Hypothesis: can glucose-insulin-potassium regimen in combination with polyunsaturated fatty acids suppress lupus and other inflammatory conditions? Prostaglandins Leukot Essen Fatty Acids 2001; 65: 109-113.
17. Guidot DM, Hybertson BM, Kitlowski RP, Repine JE. Inhaled nitric oxide prevents IL-1 induced neutrophil accumulation and associated acute edema in isolated rat lungs. Am J Physiol 1996; 271: 1225-1229.

18. Verma S, Dumont AS, Maitland A. Tetrahydrobiopterin attenuates cholesterol induced coronary hyperreactivity to endothelin. Heart 2001; 86: 706-708.
19. Marinos RS, Zhang W, Wu G, Kelly KA, Meininger CJ. Tetrahydrobiopterin levels regulate endothelial cell proliferation. Am J Physiol Heart Circ Physiol 2001; 281: H482-H489.
20. Schwartz IF, Schwartz D, Wollman Y, Chernichowski T, Blum M, Levo Y, Iaina A. Tetrahydrobiopterin augments arginine transport in rat cardiac myocytes through modulation of CAT-2 mRNA. J Lab Clin Med 2001; 137: 356-362.
21. Huang A, Vita JA, Venema RC, Keaney JF Jr. Ascorbic acid enhances endothelial nitric oxide synthase activity by increasing intracellular tetrahydrobiopterin. J Biol Chem 2000; 275: 17399-17406.
22. Baker TA, Milstien S, Katusic ZS. Effect of vitamin C on the availability of tetrahydrobiopterin in human endothelial cells. J Cardiovasc Pharmacol 2001; 37: 333-338.
23. Tyurin VA, Liu S-X, Tyurina YY, Sussman NB, Hubel CA, Roberts JM, Taylor RN, Kagan VE. Elevated levels of S-nitrosoalbumin in preeclampsia plasma. Circulation Res 2001; 88: 1210-1215.
24. Boger RH, Bode-Boger SM, Szuba A, Tsao PS, Chan JR, Tangphao O, Blaschke TF, Cooke JP. Asymmetric dimethylarginine : a novel risk factor for endothelial dysfunction. Its role in hypercholesterolemia. Circulation 1998; 98: 1842-1847.
25. Vallance P. Importance of asymmetrical dimethylarginine in cardiovascular risk. Lancet 2001; 358: 2096-2097.
26. Surdacki A, Nowicki M, Sandmann J, Tsikas D, Boeger RH, Bode-Boeger SM, Kruszelnicka-Kwiatkowska O, Kokot F, Dubiel JS, Froelich JC. Reduced urinary excretion of nitric oxide metabolites and increased plasma levels of asymmetric dimethylarginine in men with essential hypertension. J cardiovasc Pharmacol 1999; 33: 652-658.
27. Valkonen V-P, Paiva H, Salonen JT, Lakka TA, Lehtimaki T, Laakso J, Laaksonen R. Risk of acute coronary events and serum concentrations of asymmetrical dimethylarginine. Lancet 2001; 358: 2127-2128.
28. Taddei S, Virdis A, Mattei P, Ghiadoni L, Sudano I, Salvetti A. Defective L-arginine-nitric oxide pathway in offspring of essential hypertensive patients. Circulation 1996; 94: 1298-1303.
29. McAllister AS, Atkinson AB, Johnston GD, Hadden DR, Bell PM, McCance DR. Basal nitric oxide production is impaired in offspring of patients with essential hypertension. Clin Sci (Lond) 1999; 97: 141-147.
30. Das UN. Nutritional factors in the pathobiology of human essential hypertension. Nutrition 2001; 17: 337-346.
31. Das UN. Minerals, trace elements, and vitamins interact with essential fatty acids and prostaglandins to prevent hypertension, thrombosis, hypercholesterolemia, and atherosclerosis and their attendant complications. IRCS Med Sci 1985; 13: 684-688.
32. Das UN. Essential fatty acids: biology and their clinical implications. Asia Pacific J Pharmacol 1991; 6: 317-330.
33. Das UN, Horrobin DF, Begin ME, Huang YS, Cunnane SC, Manku MS. Clinical significance of essential fatty acids. Nutrition 1988; 4: 337-342.
34. Das UN. Essential fatty acids in health and disease. J Assoc Physicians India 1999; 47: 906-911.

35. Das UN. Beneficial effect(s) of n-3 fatty acids in cardiovascular diseases: but, why and how? Prostaglandins Leukot Essen Fatty Acids 2000; 63: 351-362.
36. Das UN. Biological significance of arachidonic acid. Med Sci Res 1987; 24: 1485-1489.
37. Kumar KV and Das UN. Effect of cis-unsaturated fatty acids, prostaglandins, and free radicals on angiotensin converting enzyme activity in vitro. Proc Exp Biol Med 1997; 214: 374-379.
38. McVeigh GE, Brennan GM, Johnson GD, McDermott BJ, McGrath LT, Henry WR, Andrews JW, Hayes JR. Dietary fish oil augments nitric oxide production or release in patients with type 2 (non-insulin dependent) diabetes mellitus. Diabetologia 1993; 36: 33-38.
39. Das UN. Essential fatty acid metabolism in patients with essential hypertension, diabetes mellitus, and coronary heart disease. Prostaglandins Leukot Essen Fatty Acids 1995; 52: 387-391.
40. Kumar GS, Das UN. Effect of prostaglandins and their precursors on the proliferation of human lymphocytes and their secretion of tumor necrosis factor and various interleukins. Prostaglandins Leukot Essen Fatty Acids 1994; 50: 331-336.
41. Das UN. Obesity, metabolic syndrome X, and inflammation. Nutrition 2002; 18: 430-432.
42. Das UN. Metabolic syndrome X is common in South Asians, but why and how? Nutrition 2002; 18: 774-776.
43. Weisinger HS, Armitage JA, Sinclair AJ, Vingrys AJ, Burns PL, Weisinger RS. Perinatal omega-3 deficiency affects blood pressure in later in life. Nature Med 2001; 7: 258-259.
44. Miyazaki M, Takemura N, Watanabe S, Hata N, Misawa Y, Okuyama H. Dietary docosahexaenoic acid ameliorates, but rapeseed oil and safflower oil accelerate renal injury in stroke-prone spontaneously hypertensive rats as compared with soybean oil, which is associated with expression for renal transforming growth factor-beta, fibronectin and rennin. Biochim Biophys Acta 2000; 1483: 101-110.
45. Nakayama M, Fukuda N, Watanabe Y, Soma M, Hu WY, Kishipka H, Satoh C, Kubo A, Kannatsuse K. Low dose of eicosapentaenoic acid inhibits the exaggerated growth of vascular smooth muscle cells from spontaneously hypertensive rats through suppression of transforming growth factor-beta. J Hypertens 1999; 17: 1421-1430.
46. Mohan IK, Das UN. Prevention of chemically induced diabetes mellitus in experimental animals by polyunsaturated fatty acids. Nutrition 2001; 17: 126-151.
47. Mohan IK, Das UN. Effect of L-arginine-nitric oxide system on the metabolism of essential fatty acids in chemical-induced diabetes mellitus. Prostaglandins Leukot Essen Fatty Acids 2000; 62: 35-46.
48. Das UN. Can perinatal supplementation of long-chain polyunsaturated fatty acids prevent hypertension in adult life? Hypertension 2001; 38: e6-e8.
49. Baur LA, O'Connor J, Pan DA, Kriketos AD, Storlien LH. The fatty acid composition of skeletal muscle membrane phospholipids: its relationship with the type of feeding and plasma glucose levels in young children. Metabolism 1998; 47: 106-112.
50. Pita ML, Delgado MJ. Folic acid administration increases N-3 polyunsaturated fatty acids in rat plasma and tissue lipids. Thromb Haemost 2000; 84: 420-423.
51. Durand P, Prost M, Blache D. Pro-thrombotic effects of a folic acid deficient diet in rat platelets and macrophages related to elevated homocysteine and decreased n-3 polyunsaturated fatty acids. Atherosclerosis 1996; 121: 231-243.

52. Andriamampandry M, Freund M, Wiesel ML, et al. Diets enriched in (n-3) fatty acids affect rat coagulation factors dependent on vitamin K. C R Acad Sci III 1998; 321: 415-421.
53. Kukor Z, Valent S, Toth M. Regulation of nitric oxide synthase activity by tetrahydrobiopterin in human placentae from normal and pre-eclamptic pregnancies. Placenta 2000; 21: 763-772.

ECDYSTEROIDS: USAGE IN MEDICINE, SOURCES, AND BIOLOGICAL ACTIVITY (REVIEW)

N.P. Timofeev

Collective Farm BIO; Koryazhma, 165650, Russia,

Corresponding author: N.P. Timofeev, E-mail: timfbio@atnet.ru

Keywords: ecdysteroids, 20-hydroxyecdysone, ponasterone, muristerone, Rhaponticum carthamoides, Bioinfusin

ABSTRACT

In this paper, the actual level of scientific investigations on ecdysteroids is reviewed: fields of application, medical importance, major representatives, sources, and biological activity. Historical retrospection views and the achieved research levels are shown while the world flora screening, identification of most active formulas, and study of practical application possibilities. It is emphasized that chemically isolated ecdysteroids are extremely expensive and asked for mainly in science intensive investigations. To satisfy the mass demand for ecdysteroids in the pharmaceutical industry, non-purified or weakly purified plant formulations from super producer species with null toxicity rate that do not require very expensive processing technologies have good prospects.

In the case of Russia, cultivating Rhaponticum carthamoides (Willd.) Iljin and Serratula coronata L. plants. Regarding the former species, there has appeared an industrial growing technology and a new class of pharmaceutical preparations from its aboveground shoots is being developed. The effective biological activity rate of extracts from Rhaponticum carthamoides grown with a special technology in agropopulations accounts for 10-11...10-13 M, that is 3-4 orders of magnitude higher than the activity rate of highly purified individual ecdysteroids (0.5-10 microgram/kg against 5-50 milligram/kg).

CONTENTS

MEDICAL SIGNIFICANCE

One of the most important achievements of science in the past few years is providing technologies on the plant-synthesized ecdysteroids' usage to control the growth and development parameters of different life organisms. In addition to well-known adaptogenic and immuno-modulating properties of ecdysteroid-containing preparations used in medicine (http://insectsciense.org/3.7), this last discovery becomes even more significant and urgent for people's health. Presenting ligands for intracell and membrane receptors, their regulating elements, ecdysteroids have an ability to change the homeostasis processes in an organism, influencing cell growth, differentiation, and programmed death, production of specific products of metabolism.

In practical medicine, ecdysteroid-containing formulations are applied to prevent illness and preserve the immune status of healthy people [1-3]; they take an important part in sport, cosmic, and military medicine as adapting and work efficiency increasing drugs under limiting factors, e.g. when overcoming extreme physical and psychical efforts [4-6]. They are usable for the acceleration of regeneration of tissues human organs and skin, to increase the hair growth, cure wounds, ulcers, and burns; they improve sexual function, stimulate libido and obviate difficulties in sexual life [7-12].

The ecdysteroid molecules (Fig. 1), presenting a group of lipophilic poly-hydroxylated steroids, participate in the life activity sustainability of practically all classes of organisms, while fulfilling numerous functions. The question on their role in living nature is still open. Certain is only the fact that one of the major ecdysteroid representatives, *20-hydroxyecdysone, makisterone C* and *25-deoxyecdysone*, is a true molting hormone of arthropoda (insects and crustaceans) and initiates changes that occur at different developing stages from larva to chrysalis and then to a grown-up insect [13, 14].

Since appeared several thousand million years ago, ecdysteroids have participated in a complex co-evolution pathway of ecosystems development and adaptation to the environment. The presence of ecdysteroids is characteristic, together with flowering plants, of such ancient organisms as ferns, mushrooms, mosses, algae, gymnosperms [15-16].

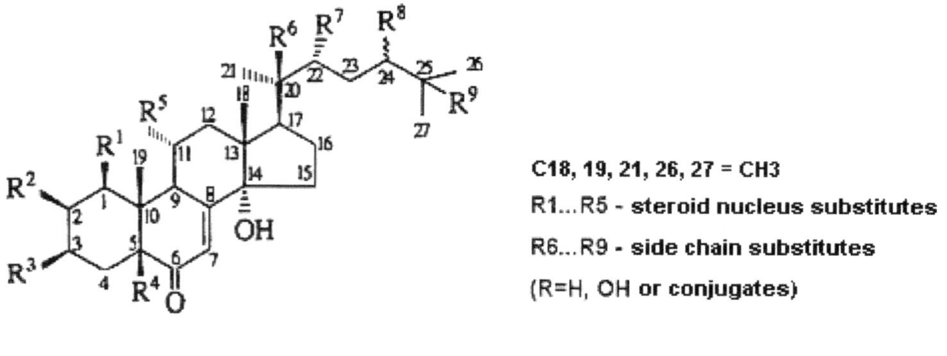

C18, 19, 21, 26, 27 = CH3

R1...R5 - steroid nucleus substitutes

R6...R9 - side chain substitutes

(R=H, OH or conjugates)

Figure 1. The structural formula of ecdysteroids

In the 60-ies years of XX century, the discovery of colossal amounts of molting hormones in plants (million times higher than in insects) caused a great scientific sensation. This discovery was proposed to contribute to an ecologically safe and quite effective method for the insect population control. Nevertheless, as the more detailed research has shown, most of insects are resistant to phytoecdysteroids (http://www.sciteclibrary.ru/rus/catalog/pages/4723.html) or learnt how to detoxify the hormones [17-19] penetrating into their organisms through the mouth and started synthesis of their own zooecdysteroids (ecdysones) using other metabolic pathways differing from those of plants.

However, a 20-year-old investigation work in a field of cell and molecular biology, ecological genetics and physiological sciences has led to even more significant discoveries: ecdysteroids represent natural and absolutely safe ligands in molecular systems of gene switching [20-22]; mechanisms on ecdysone- (ecdysteroid-) induced gene expression systems like those existing in insect cells are applicable for mammals, including humans [23-26].

The importance of the last discoveries becomes greater in the post-genome medical era. Upon completion of the human genome library sequencing, it is proposed that switch genes will allow switching off cells which produce organism-destroying structures (e.g. cancerous growth) and bringing to a stop diseases, incurable by conventional treatment modes (many inherited diseases) [27-28]. Analogically, it will be possible to implant and point wise switch on genes not present in the host cells but responsible for producing target therapeutic agents, as well as to set off the regeneration factors of damaged tissues [29].

Though the new directions on ecdysteroid application seem to be quite unusual, ecdysone-induced systems are not only created and patented but also realized for commercial purposes (http://www.invitrogen.com). Moreover, important aspects on clinical application of ecdysteroids are their participation in numerous non-genome effects. Despite the mechanisms on interconnection of ecdysteroids with membrane receptors as signal molecules, which activate secondary messengers, have been just recently put under investigation [28, 30], ecdysteroid-containing preparations are broadly used in practical medicine when curing cardio-vascular, nervous, and reproduction system diseases, whatever disorders of the whole homeostasis processes [1-2, 31-32].

That is why, for today we need those sources of ecdysteroid molecules or ecdysteroid-related compounds that would act in small quantities, be highly active, non-toxic, resistant to decomposing influence, quickly removable from the organism, cost effective, and could be produced in large dimensions [3, 25, 33].

ALL-IMPORTANT ECDYSTEROIDS

The first investigations on ecdysones aimed at the isolation of insect hormones, which began in the early 30's, were conducted by German scientists. In 1954, they managed to isolate 25 mg of weakly purified substance from 500 kg of silkworm chrysalides (*Bombix mori*) and to crystallize it [34]. In 1963, as its general structure was discovered, *α-ecdysone* was related to steroids (with molecular weight M=464). In 1965, the molecular structure of *α-ecdysone* was determined by the X-ray structure analysis [35-36]. These studies themselves were familiar only to a limited number of specialists and, possibly, the things would not have changed for a long time, if it were not for the concurrence of circumstances elicited a great interest and big investments in the investigations dealing with the world flora screening and research of new molecules' properties.

The discovery of ecdysones in plants happened by a lucky chance when the scientist Karel Slama (Czecho-Slovakia) went to the USA to cultivate a soil insect *Pyrrhocoris apterus* L. on filter paper. Here he was surprised – the insect metamorphosis was disturbed, and he could not obtain puration in the last larval stage. The key to explaining this phenomenon lay in filter paper:

in that case it was produced from balsam fir (*Abies balsamea*). Other papers did not influence the metamorphism. The extraction procedure allowed for isolation of juvabione, structurally related to a juvenile hormone, which has selective effects on this single insect. Testing any other plants revealed that they contain numerous compounds with insect hormone activity. The biotest method was later modified in B_{II}-biotest and to date is broadly used for the primary screening of vegetal ecdysteroids, together with the radio-immune analysis (RIA).

Taking into account the economical and biological importance of phytoecdysones, the last three decades were characterized by making significant efforts by screening the world flora to discover species producing most ecdysones, identify most active formulations, and to study possibilities on practical usage of ecdysones in different biology and medicine fields.

Ponasterone was a first phytoecdysterone (fig. 2) isolated and described in 1966 by the Japanese scientists from the conifer *Podocarpus nakaii* [37]. Afterwards, it was identified in the affined species *Podocarpus macrophyllus* and *Podocarpus reichei* with concentration 100 g/kg dry wt, and in the yew-trees *Taxus canadasis, T. chinensis, T. cuspidata* (50-80 mg/kg). In the late 70s and early 80s, *ponasterone* was detected in cancroids, while in 1995 in flat-cap mushrooms (*Paxillaceae*) with concentration up to 50 mg/kg.

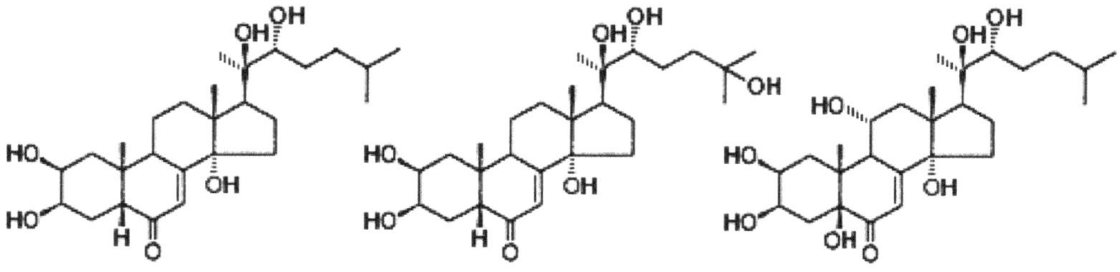

Ponasterone A

Ecdysterone

Muristerone A

Figure 2. Ecdysteroids, found wide practical application

Ecdysterone (*β-ecdysterone, 20-hydroxyecdysone, 20E*) was first isolated in 1966 from the crustacean *Jasus calandei* with a quantity of 2 mg/t and, therefore, named *crustedysone* [38]. Then, it was found in insects, silkworms *Bombix mori* and *Antharea pernyi*, and extracted with 200 mg 31 kg chrysalides [39]. In the same year the structure was assigned, and *ecdysterone* was isolated from conifers and ferns: first 50 mg/kg of *Podocarpus elatus,* further 10 g/kg *Polipodium vulgare* rhizomes.

Lately, *ecdysterone* was found in most plants [40], including cereals (maize *Zea mays*) and crucifers (*Arabidopsis thaliana).* The concentrations differ by 1 million to 1 milliard times (20-300 ng to 20-30 g/kg). One of the main sources of reception on ecdysterone industrial production is the perennial plant *Rhaponticum carthamoides* (Willd.) Iljin (*Leuzea carthamoides* DC.) and *Serratula coronata* L., induced in different regions of Russia.

Muristerone A, the most active, rare and extremely expensive ecdysteroid for today, was found in 1972 by the German scientists in endemic plants' seeds, genus *Ipomoea* [41], which occupy southern slopes of the Himalayas. The *Ipomoea* presents the most mysterious ecdysteroid source. Since the discovery of *muristerone A*, more than 1.5 thousand articles have been published on various aspects of its research but only some very first papers mention the source. It appeared because the *Ipomoea* nomenclature is highly looped: it can mean completely diverse plants, often endemic [42]. Only 30 years later, there occurred new messages on *muristerone* isolation from sequoia-trees, not scientifically proven yet (http://www.sequoiasciences.com).

ECDYSTEROID SOURCES

3.1. *Plants.* Actually, ecdysteroids were identified to compose not only higher flowering plants but also gymnosperms, ferns, mushrooms, algae, mosses, as well as insects, crustaceans, and nematodes. According to the last investigations, almost all terrestrial and water higher plants have ecdysteroid-synthesizing genes [15-16, 40-43].

Today we know the structure of more than 310 ecdysteroid molecules (www.ecdybase.org). Besides, angiosperms exceed in the diversity of kinds of ecdysteroids they contain. Among all various ecdysteroid molecules the mammals contain, the following three: *ponasterone A*, *muristerone A*, and *ecdysterone* are most active. The first two ecdysteroids do not occur in the higher flowering plants. *Ponasterone A* is present in single ferns (including eagle fern – *Pteridium aquilinum*), mushrooms of the *Paxillaceae* family (*Paxillus atrotomentosus*), and relict plants of the *Podocarpaceae* and *Taxaceae* families. *Muristerone A* is typical of the *Ipomoea* genus (morning-glory) from the *Conovolvulaceae* family. Less active *ecdysterone* is wide-spread among flowering plants.

Plants possess ecdysteroids in the form of water-soluble conjugates: conjugates with inorganic acids – sulphates, phosphates; conjugates with organic (carbon, fatty, phenolic) acids – acetates, benzoates, cumarates; conjugates with sugars – glucosides, galactosides, xylosides etc.

Apart from the above-mentioned main ecdysteroids, all the studied objects contain in trace amounts other structure analogues and their derivates (the so called minor ecdysteroids) numbering to 30-40 and more units. Some endemic and rare species, as well as those growing in specific ecological and geographical conditions, include ecdysteroids of unusual or abnormal structure, which do not occur in the most studied objects. The 90s are notable for isolation of ecdysteroids with new structures (*polyporusterone A...G*) from the Chinese bracket fungus *(Polyporus umbellatus)* in a quantity 0.1-3.0 mg/kg [44-45]. It was also in this period when a new type of ergostane ecdysteroids (*paxillosterone, atrotosterone, malakosterone)* and their derivates from the mushrooms *Tapinella panuoides* and *Paxillus atrotomentosus* were obtained [46-47].

None of the mammal species contains ecdysteroids. The artificial chemical synthesis is only possible in the case of the secondary, biologically inactive or weakly active products *via* chemical transformation of major ecdysteroids. For this purpose, *ecdysterone* is most often used [1]. Just recently the artificial photochemical transformation method was discovered allowing for structures being uncharacteristic of chemical transformation, e.g. dimers [48].

By the origin, it is accepted to divide ecdysteroid sources into phyto-, zoo-, and myco-ecdysteroids (i.e. plants; insects, cancroids, nematodes; mushrooms). Zooecdysteroids cannot be used for industrial production because of their utterly low concentrations in anthropoda. The value of this or that plant or mushroom species as a raw material source depends on its uniqueness grade formed by such indices as the biological activity, end use, concentration in biomass, availability, cost expediency [49].

The most important ecdysteroid sources for industrial isolation are plants, which, by their ability to biosynthesize ecdysteroids, can be classified as follows (on a dry weight basis):

I. 1-30 g/kg (0.1-3.0 %) – superconcentrator species;
II. 0.1-1 g/kg (0.01-0.1 %) – species with high ecdysteroid content;
III. 10-100 mg/kg (0.001-0.01 %) – species with mean ecdysteroid content;

IV. 0.5-10 mg/kg (0.00005-0.001 %) – species with low ecdysteroid content;
V. 0.1-0.5 mg/kg and less – species with trace ecdysteroid content.

In general, the differences among ecdysteroid concentration levels in plants are huge values – 8-9 orders (from 20-300 ng/kg to 20-30 g/kg). Normally, it is an extremely small value (thousandths and hundredths of a per cent on a dry weight basis). But there are plants which single organs in a restricted age and vegetation diapason can concentrate essential amounts of ecdysteroids. On average, among several thousands of other species there is one concentrator species. The species Rhaponticum carthamoides (Willd.) Iljin and Serratula coronata L. belong to the most important ecdysteroid-containing plants serving as industrial ecdysteroid sources. These species are considered to be highly promising in developing new classes of pharmaceutical preparations and biologically active food additives, and also ecologically safe products against pests [2].

Among angiosperms there is an insignificant number of other species with high content of ecdysteroids (mainly ecdysterone) in single organs, which are of interest for scientific purposes. Detailed investigations of the European northeast flora [43] showed the distribution of ecdysteroid-containing plants by taxons corresponding with the analogous distribution in other regions. The presence of these plants was identified in most species, whereas only 4% of them, represented by species with mean and high ecdysteroid content, developed a positive response in the radio-immune activity biotest. These data agree with the works done by other scientists where essential ecdysteroid concentrations belonged to 5-6% of plants.

The species of secondary importance in the Russian flora are: some Silene and Lychnis varieties; Coronata flos-cuculi L.; Helleborus purpurascens and Helleborus caucasicus; Paris guadrifolia L.; Ajuga reptans; Sagina procumbens L.; Potamogeton natans and Potamogeton perfoliatus; Pulmonaria officinalis; Butomus umbellatus; Androsace filiforms etc.

Unfortunately, all these plants have some negative features, which do not make their industrial use possible. The main limiting factor is that they are difficult of access, grow in scattered groups or lonely, only wild and cannot be cultivated. Often they are low-height, creeping, rosulate, forest, meadow or water plants, and poisonous or weakly toxic. They are met with in flood land thickets of meadow shrubs, forest edges and felled areas, peat-bogs, waste plots of land, along road and trench sides, at riverbanks and lake coasts or hill foots at high places. The life strategy of these plants is combined growing together with other species under the forest canopy, within meadows or as ruderal plants at cultivated fields. In most cases, introduction was not tried at all or represents serious difficulties.

3.2. ***Methods of biotechnology.*** Since isolation and purification of ecdysteroids from plant biomass is a complicated procedure heightening the prime cost of the end products, there have developed ecdysteroid production technologies using biotechnological methods (cultures of cells, tissues, and transformed roots). Since ecdysteroid-biosynthesizing genes are present in all plant organs, callose cultures (cultures multiplying in the artificial nutritional medium) can be obtained from every quick-growing tissue: seed-lobs, hypocotyl, leaves, shoots, buds, and roots. Ecdysterone and some other secondary important components of genera Ajuga, Serratula, Rhaponticum, Pteridium, Polypodium can be synthesized in cell culture. However, most active ecdysteroids, muristerone and ponasterone, cannot be synthesized in artificial conditions.

On the whole, ecdysteroid content in cell culture is significantly lower than that in nature. A long-term cultivation lowers total content and changes the proportion ratio between individual compounds. Moreover, not identified inactive ecdysteroids are synthesized. Ecdysteroid synthesis in suspension cultures has somewhat higher results than in cell culture, but these results are instable and ecdysterone concentration increases very slowly [15].

The most promising biotechnology method is a transformed roots' culture. Inoculating sterile seedlings by the Agrobacterium rhizogenes strains causes an infection and agrobacterial transformation of roots to a hairy form. In case with Rhaponticum carthamoides the characteristic roots and swellings appear a month ago, they start spontaneous regeneration of modified plants. Within 4 weeks, hairy roots increase their mass by 4-6 times. However, ecdysteroid fraction in culture, 0.02-0.03 % in total, differs from that of natural plant roots [50].

The method of ecdysteroid production via hairy roots culture has its positive and negative features. Its advantage, compared to field culture, consists in a continuous reproduction source, high growth and regeneration rates of specimens; there is no need in external growth hormones as in the case with cell and tissue cultures. The nutrition source is saccharose. The method is applicable in respect to many cultures, modified roots are characterized by a high growth rate and genetic stability, large concentrations of secondary metabolites, comparable with that of natural plants. In addition to phytoecdysteroids, the secondary metabolites of hairy roots also involve alkaloids, polyacetilene compounds, glycosides, polyphenols, tannins, flavonoids, saponins etc.

However, the commercial use of such systems is restricted due to significant defects, which find a detailed treatment in the assay of Giri and Narasu [51]. The main restrictions arise from need for specially designed patterns of bioreactors with automatic control models to provide a free vertical culture development; these cultures need careful selection of optimal nutritional medium, temperature, and illumination conditions. One of the most important parameters, optimal roots morphology, influencing the density and aeration degrees of specimen is difficult to meet because many of various morphologies existing are in connection with different plasmid strains.

Uniform aeration and intermixing procedures also provide certain difficulties, which lead to stagnant zones and fermentation, tissues necrosis and vital capacity loss. The end product accumulation is inhibited by its concentration saturation. It is necessary to filtrate and renew the liquid medium constantly. Also there are some difficulties appearing by crop collection and treatment, i.e. partially root removing from the reservoir etc.

Generally, ecdysteroid production by the biotechnology methods did not find large acceptance. Modified secondary ecdysteroids obtained by methods of biotechnology comprise less ecdysteroid activity indices, compared to nature. Therefore, such systems are used only to get chemically pure ecdysteroids. Besides, as the world experience readily shows, it is not enough just to have an ecdysteroid, it is to be highly active already in minimal concentrations, as muristerone A and ponasterone A are. Otherwise, chemically pure substances do not meet a ready sale.

ACTIVITY

4.1. *Chemically pure ecdysteroids.* The activity rate of isolated ecdysteroids is evaluated by biotesting with insect cells containing natural ecdysteroid receptors (EcR). Ponasterone A, muristerone A, and ecdysterone are regarded to be most active ecdysteroids with large practical use possibilities. Each of them can show different results with different receptors, but their initial activity rates are generally the same comprising 10-9 (10-8…10-10) M [52]. There are other ecdysteroids with 5, 6, 7 or even 8 OH-groups but relatively less active in isolated form. The following decreasing activity series was obtained: muristerone A, ponasterone A, polypodine B, 20E, 22-acetate 20E, 2-deohy-20E [20]. Concerning minor components some derivatives from muristerone (kaladasterone) and ajugasterone (dachryhainansterone) appeared active in biotests [53].

The activity rate of ecdysteroids in a real organism essentially differs from that in cell cultures, the doses are tissue-specific. Ecdysone-induced systems have effective doses of muristerone A and ponasterone A as single ligands in transgenic mice being equal to 10-5…10-7 M. Scientists give more preference to muristerone despite it is a rare and expensive substance ($ 120-135 for 1 mg). The ponasterone use is complicated because of its instability: after 3 hours the receptor complex decays by 50 % in buffer solution, whereas for muristerone the figures lie by 5 % [54].

Concerning ecdysterone, though the biotests on insect cells revealed a significantly high bioactivity rate of 10-8 M [20], the ecdysone-induced systems are 2-3 orders of magnitude less in results. The activity rate of other ecdysteroids, polypodine B, ecdysterone, inokosterone, makisterone, is even more less, while α-ecdysone, 2-deoxyecdysone, 20-deoxyecdysterone, 22-acetate-ecdysterone do not have it at all [25].

Apart from EcR, none of the steroid receptors can interact with isolated ecdysteroids as ligands in mammal cells [26]. This situation seems to be a satisfactory one because it allows for avoiding negative unforeseen side effects when using ecdysteroids as switch genes in the ecdysone-induced systems.

The anabolic activity of ecdysteroids, including the ability for protein synthesis inhibition or stimulation, was experimentally proven: a) in substances with insufficient purification degree (95 % and less); b) substances isolated from the producer plant Rhaponticum carthamoides; c) via unclear activation mechanisms with secondary agents [55-58]. Numerous experiments in the fields of cell and molecular biology with individual highly purified compounds (99 %) and other ecdysteroid sources, e.g. Serratula coronata [59-60], did not detected any signs of ecdysterone, muristerone, and ponasterone anabolic activity without secondary agents.

To activate gene transcription, hybrid ecdysteroid/retinoid receptors (EcR/RxR) and their modifications with other nuclear receptors are applied, where the RxR-partner is necessary to stabilize the heterodimer complex and to provide fixation of response elements before activation of gene expression mechanisms. In a living organism both ecdysteroid agonists and ligands of the heterodimer complex second partner (retinoid receptor) can act together, which enlarge their biological activity diapason significantly [74]. Moreover, the effective doses decrease to 10-9…10-10 M under condition of a local treating a target organ [24].

The aforesaid is true only for highly purified (more than 98-99 %) formulations isolated from the primary sources. The artificial systems on their basis are extremely expensive and used mainly for scientific purposes. Mass ecdysteroid production in pharmaceutical industry can use non-purified or weakly purified plant formulations from non-toxic superproducer species, which do not require high-expensive ecdysteroid-possessing technologies.

Note: Chemically pure ecdysteroids are obtained according to the following sequential procedures: raw material fining, extraction and extract's concentration by evaporation, watering, filtration, accompanying hydrophobic substances' re-extraction by hydrocarbon solvent, ballast compounds' precipitation, phytoecdysteroids extraction from water portion with evapoconcentration, chromatographic purification in column with aluminum oxide, eluate evapoconcentration, selective elution of ecdysteroids from sorbent material with solvent and resulted fractions' concentration, re-crystallization and vacuum drying after dissolution in metanole and evapoconcentration, or filtration after freezing at -40 to -70°C.

4.2. ***Non-purified Rhaponticum carthamoides formulations.*** Chemically isolated ecdysteroid fraction (91 %, including 75 % 20-hydroxyecdysone) extracted from the overground Serratula coronata shoots demonstrated a complex and ambiguous "dose-effect" modulation activity (also called two-phase operation) in the spontaneous E-rosette-forming biotest within the concentration limits 10-4...10-12 M [61]. The effective immuno-modulating activity rate of CD2+-rosette forming with human T-lymphocytes was achieved by a concentration value of 1 μM (10-6 M) and a stimulation index of 1.132 [62].

Natural ecdysteroid-containing substances can be significantly more active than chemically isolated ecdysteroids. Cultivating lymphocyte populations in vitro in presence of the Rhaponticum carthamoides extract [63] can cause proliferation of spleen cells in concentration 10-13...10-14 M, if 20-hydroxyecdysone. The nonspecifically activating ConA (T-mytogene) and LPS (B-mytogene) agents stimulate proliferation up to 10-15 M (Fig. 3).

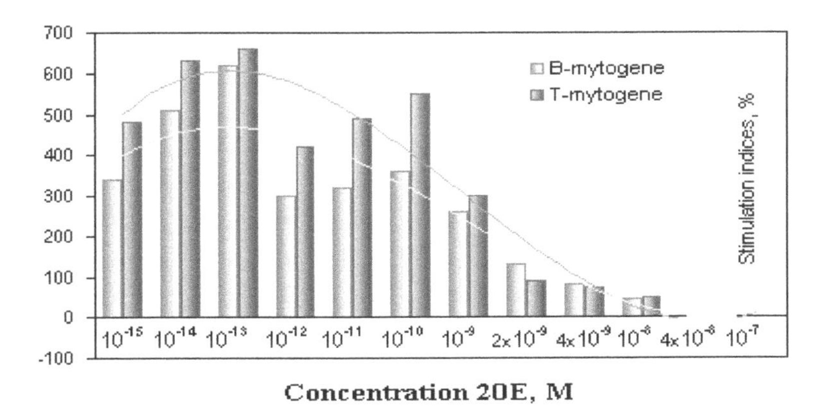

Figure 3. Cell proliferation stimulation by *Rhaponticum carthamoides* extract (according to [63] changed)

At present, a new class of ecdysteroid-containing pharmpreparations with super-low doses of active substances is being worked out. The new preparations are produced from the overground *Rhaponticum carthamoides* shoots grown according to a special technology in agropopulations [64-66]. The crude drug used in production allows decreasing today doses – by 3-4 orders of magnitude, if *20-hydroxyecdysone* [67]. For example, effective doses of the *Bioinfusin, BCL-PHYTO (BCL – BactoCelloLactin), Lipolite* and *Rapontik* pharmpreparations account for 0.5-10.0 microgram/kg biomass ($10^{-12}...2\times10^{-13}$ M), for *ecdysterone* [2, 68, 74]. This is not a mistake or misprint because an average daily dose of chemically pure *20-hydroxyecdysone* and preparations on its basis is 5-50 milligram/kg body weight [13, 55, 58-60, 69-71].

The action mechanism specificity of new preparations lies in a stimulating activity of small doses and inhibiting effects on proliferative organism processes in large doses. Even one-time introduction of such drugs can cause essential immuno-stimulating effects on cell and humoral levels [68-72]. The seven-day-long course of treatment allows for considerable immuno-stimulating aftereffects, which last for 30 days (Fig. 4). Besides, non-purified ecdysteroid formulations possess a stable industrial anabolic effect when used for mass industrial production [73] (Fig. 5).

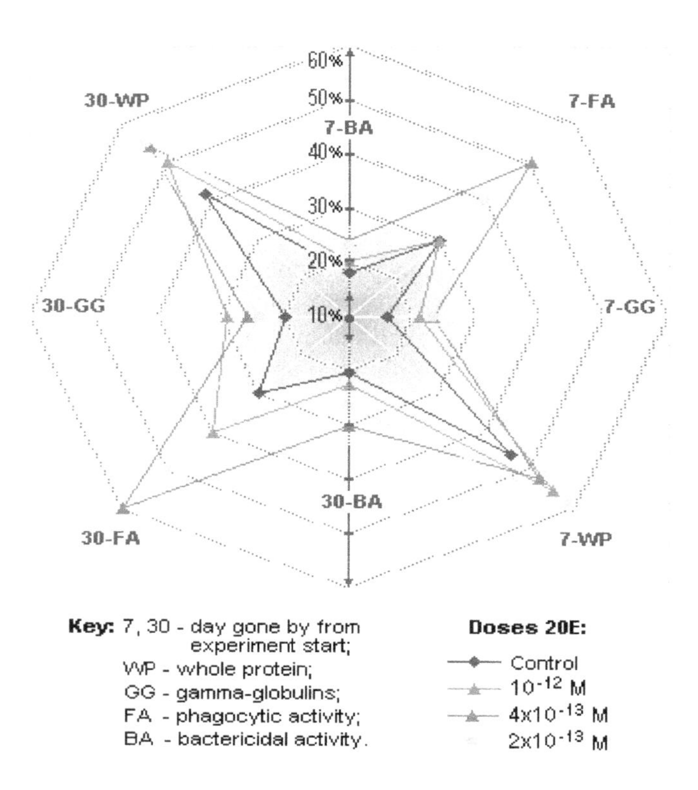

Figure 4. Immuno-modulating effect of the "Bioinfusin" preparation (course – sevenfold introduction; according to [68])

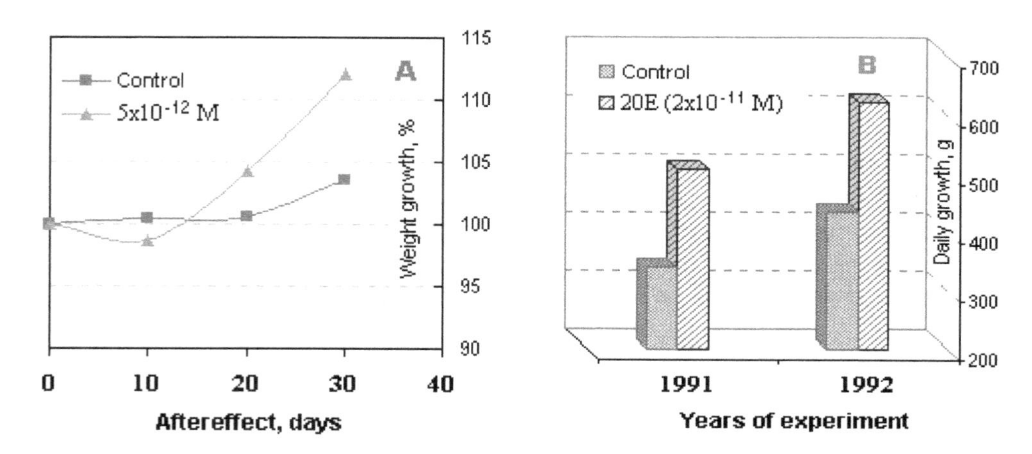

Figure 5. Anabolic effect of small ecdysteroid doses:
A –single intramuscular injection, according to [72];
B – production testing for 3 months, according to [73]

CONCLUSION

Research on ecdysteroids is a direction in biology offering rich possibilities for fundamental and practical scientific studies. Exploring the role and mechanisms of ecdysteroid biological activity allows for a real possibility to realize the boldest human projects, i.e. learn how to get control over the vital activity of different organisms manipulating the activity rate of the particular genes according to the "switch on-switch off" principle. Practically, it would help to get rid of whole number incurable diseases and refuse the chemical synthesis for the ecologically safe biological synthesis of many important substances.

Research on ecdysteroids including data on the genetics; cell and molecular biology; human, animal, and plant physiology, as well as commercial offers intended for solving real questions in chemistry, biotechnology, pharmacology, medicine, entology, and agriculture. Different states offer as their sources native plant species: ferns, convolvuluses, conifers, yew-trees, and amaranths.

In the case of Russia, cultivating Rhaponticum and Serratula plants representing super concentrator plant species is economically sound. The basic ecdysterone concentrations in plants of Rhaponticum carthamoides (Wlld.) Iljin (Leuzea carthamoides DC.) and Serratula coronata L. comprise 0.12-0.57 % and 0.31-1.15 % dry weight, respectively. Regarding the former species, there appeared an industrial growing technology and a new class of pharmaceutical preparations from its aboveground shoots is being developed. The latter is under introduction study in different Russian regions.

To satisfy the mass demand for ecdysteroids in the pharmaceutical industry, non-purified or weakly purified plant formulations from super producer species with a null toxicity rate that do not require high-expensive processing technologies have good prospects. The effective biological activity rate of extracts from Rhaponticum carthamoides grown with a special technology in agropopulations accounts for 10-11...10-13 M. It is about 3-4 orders of magnitude higher than the activity rate of highly purified individual ecdysteroids. Stable results of comparable doses were experimentally obtained in biotests, tests on laboratory animals, and in conditions of a large-scale production. What is particularly responsible for the unusually high activity rate of Rhaponticum carthamoides has to be looked for in its complex chemical composition causing the complex biological activity of ecdysteroids with other metabolites.

REFERENCES

1. Slama K., Lafont R. Insect hormones – ecdysteroids: their presence and actions in vertebrates. Eur J Entomol., 1995; 92: 355-377.
2. Timofeev N.P. Leuzea carthamoides: Introduction questions and application prospects as biologically active components // Netraditsionnye prirodnye resursy, innovatsionnye tekhnologii i produkty. Collected proceedings. Is. 5. Moscow, RANS, 2001. P. 108-134.
3. Lafont R., Dinan L. Practical uses for ecdysteroids in mammals including humans: An update. Journal of Insect Science, 2003; 3(7): 30 pp.
4. Novikov V.S., Shamarin I.A., Bortnovskiy V.N. Experience on pharmacological correction of seamen sleep disturbances during navigation.Voenno-Meditsinskiy Zhurnal,1992;8:47-49.
5. Seyfulla R.D. Use of drugs by healthy people. Eksperimental'naya i Klinicheskaya Farmokologiya, 1994; 4: 3-6.

6. Seyfulla R.D. Sport pharmacology. Moscow: Sport-Pharma Press, 1999. 120 pp.
7. Mirzaev Yu.R, Syrov V.N, Krushev S.A, Iskanderova S.D. Study of the effects of ecdysten on the sexual function under experimental and clinical conditions. Eksperimental'naya i Klinicheskaya Farmakologiya, 2000; 63: 35-37.
8. Kibrik N.D., Reshetnyak J.A. Therapeutical approaches to sexual disadaption. European Neuropsychopharmacology, 1996; 6(4): 167.
9. Meybeck A., Bonte F., Redziniak G. Use of an ecdysteroid for the preparation of cosmetic or dermatological compositions intended, in particular, for strengthening the water barrier function of the skin or for the preparation of a skin cell culture medium, as well as to the compositions. US Patent 5,609,873. March 11, 1997.
10. Kovler L.A., Volodin V.V., Pshunetleva E.A. Ecdysteroid-containing liposomes and their description // Scientific reports of the Komi SC UrD RAS. Is. 407. Syktyvkar, 1998. 18 pp.
11. Tsuji K., Shibata J., Okada M., Inaoka Y. Blood flow amount-improving agent comprising steroid derivative and cosmetic using same. US Patent 5,976,515. November 2, 1999.
12. Darmograi V.N., Petrov V.K.,Yukhov Yu.I. Theoretical and clinical bases of a phytoecdysteroid action mechanism model // Biochemistry at the end of XX century. Interregional proceedings. Ryazan, 2000. P. 489-492.
13. Akhrem A.A., Kovganko V.V. Ecdysteroids: chemistry and biological activity. Minsk: Nauka i tekhnika, 1989. 327 pp.
14. Rees H.H. Ecdysteroid biosynthesis and inactivation in relation to function. Europ J Entomol., 1995; 92(1): 9-39.
15. Lafont R. Phytoecdysteroids and world flora: Diversity, distribution, biosynthesis, and evolution. Physiologiya Rasteniy, 1998; 3: 326-346.
16. Baltaev U.A. Phytoecdysteroids – structure, sources, and biosynthesis pathways in plants. Bioorganicheskaya Khimiya, 2000; 26(12): 892-925.
17. Zeleny J., Havelka J., Slama K. Hormonally mediated insect-plant relationships: Arthropod populations associated with ecdysteroid-containing plant, Leuzea carthamoides (Asteraceae). Eur J Entomol., 1997; 94(2): 183-198.
18. Dinan L. Strategy of valuating the role of ecdysteroids as deterrents in respect to invertebrates phytophagans. Phiziologiya Rasteniy, 1998; 3: 347-359.
19. Timofeev N.P. Whether protect phytoecdysones populations Rhaponticum and Serratula from damage by insects? // Chemistry and technology of plants substances. Saratov, Institute of Biochemistry and Physiology of Plants and Microorganisms RAS, 2004. P. 189-191.
20. Wang S.F, Ayer S, Segraves W.A, Williams D.R, Raikhel A.S. Molecular determinants of differential ligand sensitivities of insect ecdysteroid receptors. Mol Cell Biol., 2000; 20: 3870-3879.
21. Carlson G.R., Cress D.E., Dhadialla T.S., Hormann R.E., Le D.P. Ligands for modulating the expression of exogenous genes via an ecdysone receptor complex. US Patent 6,258,603. July 10, 2001.
22. Jepson I., Martinez A., Greenland A. J. Gene switch. US Patent 6,379,945. April 30, 2002.
23. Suhr S.T., Gil E.B., Senut M-C., Gage F.H. High level transactivation by a modified Bombyx ecdysone receptor in mammalian cells without exogenous retinoid X receptor. PNAS, 1998; 95: 7999-8004.
24. Albanese C., Reutens A.T., Bouzahzah B., Fu M., D'Amico M., Link T., Nicholson R., Depinho R.A., Pestel R.G. Sustained mammary gland-directed, ponasterone A-inducible expression in transgenic mice. FASEB J., 2000; 14: 877-884.

25. Saez E., Nelson M.C., Eshelman B., Banayo E., Koder A., Cho G.J., Evans R.M. Identification of ligands and coligands for the ecdysone-regulated gene switch. Proc Natl Acad Sci USA., 2000; 97: 14512-14517.
26. Evans R.M., Saez E. Formulations useful for modulating expression of exogenous genes in mammalian systems, and products related thereto. US Patent 6,333,318. December 25, 2001.
27. Vegeto E., McDonnell D.P., O'Malley B.W., Schrader W.T., Tsai M.J. Mutated steroid hormone receptors, methods for their use and molecular switch for gene therapy. US Patent 5,935,934. August 10, 1999.
28. Wolter S., Mushinski J.F., Saboori A.M., Resch K., Kracht M. Inducible expression of a constitutively active mutant of mitogen-activated protein kinase kinase 7 specifically activates c-JUN NH2-terminal protein kinase, alters expression of at least nine genes, and inhibits cell proliferation. J Biol Chem., 2002; 277(5): 3576-3584.
29. Patrick C.W., Zheng B., Wu X., Gurtner G., Barlow M., Koutz C., Chang D., Schmidt M., Evans G.R. Muristerone A induced nerve growth factor release from genetically engineered human dermal fibroblasts for peripheral nerve tissue engineering. J Tissue Eng., B, 2001; 7(3): 303-311.
30. Constantino S., Santos R., Gisselbrecht S., Gouilleux F. The ecdysone inducible gene expression system: unexpected effects of muristerone A and ponasterone A on cytokine signaling in mammalian cells. European Cytokine Network, 2001; 12(2): 365-367.
31. Syrov V.N. Phytoecdysteroids: biological effects in organism of higher animals and prospects of use in medicine. Experimentalnaya i Klinicheskaya Farmakologia, 1994; 5: 61-66.
32. Kholodova Yu.D. Phytoecdysteroids: biological effects, application in agriculture and complementary medicine. Ukrainskii Biokhimicheskii Zhurnal, 2001; 73: 21-29.
33. Rossant J., McMahon A. Meeting review "Cre"-ating mouse mutants a meeting review on conditional mouse genetics. Genes & development, 1999; 13(2): 142-145.
34. Butenandt A., Karlson P. Über die isolierung eines metamorphose-hormones der insekten in kristallisierten form. Z Naturforsch, 1954; 9b: 389-391.
35. Huber R., Hoppe W. Zur chemie des ecdysons. VII. Die kristall- und molekül-strukturanalyse des insektenverpuppung hormons ecdyson mit der automatisierten falt-molekülmethode. Chem Ber., 1965; 98: 2403-2424.
36. Karlson P., Hoffmeister H., Hummel H., Hocks P., Spiteller G. Zur chemie des ecdysons. VI. Reaktionen des ecdysonmoleküls. Chem Ber., 1965; 98: 2394-2402.
37. Nakanishi K., Koreeda M., Sasaki S., Chang M.L., Hsu H.Y. Insect hormones. The structure of ponasterone A, an insect-moulting hormone from the leaves of Podocarpus nakaii Hay. J Chem Soc Chem Commun., 1966; 24: 915-917.
38. Hampshire F., Horn D.H.S. Structure of crustecdysone, a crustacean moulting hormone. J Chem Soc Chem Commun., 1966; 37-38.
39. Hocks P., Wiechert R. 20-Hydroxyecdyson, isoliert aus insekten. Tetrahedron Lett., 1966; 26: 2989-2993.
40. Dinan L., Savchenko T., Whiting P. On the distribution of phytoecdysteroids in plants. Celluar and Molecular Life Sci., 2001; 58(8): 1121-1132.
41. Canonica L., Danieli B., Weisz-Vincze G., Ferrari G. Structure of muristerone A, a new phytoecdysone. Journal of the Chemical Society Chemical Communications, 1972; 1060-1061.
42. Austin D.F., Huaman Z. A synopsis of Ipomoea (Convolvulaceae) in the Americas. Taxon, 1996; 45: P. 3-38.
43. Volodin V., Chadin I., Whiting P., Dinan L. Screening plants of European North-East Russia for ecdysteroids. Biochemical Systematics and Ecology, 2002; 30(6): 525-578.

44. Ohsawa T., Yukawa M., Takao C., Muruyama M., Bando H. Studies on constituents of fruit body of Polyporus umbellatus and their cytotoxic activity. Chemical and Pharmaceutical Bulletin, 1992; 40(1): 143-147.
45. Ishida H., Inaoka Y., Shibatani J., Fukushima M., Tsuji K. Studies of the active substances in herbs used for hair treatment. II. Isolation of hair regrowth substances, acetosyringone and polyporusterone A and B, from Polyporus umbellatus Fries. Biol Pharm Bull., 1999; 22(11): 1189-1192.
46. Vokac K., Budesinsky M., Harmatha J., Pis J. New ergostane type ecdysteroids from fungi. Ecdysteroid constituents of mushroom Paxillus atrotomentosus. Tetrahedron, 1998; 54(8): 1657-1666.
47. Vokac K., Budesnsky M., Harmatha J., Kohoutova J. Ecdysteroid constituents of the mushroom Tapinella panuoides. Phytochemistry, 1998; 49(7): 2109-2114.
48. Harmatha J., Dinan L., Lafont R. Biological activities of a specific ecdysteroid dimer and of selected monomeric structural analogues in the B (II) bioassay. Insect Biochem Mol Biol., 2002; 32(2): 181-185.
49. Timofeev N.P. Industrial ecdysteroid sources: Part I. Ponasterone and muristerone. Netraditsionnye prirodnye resursy, innovatsionnye tekhnologii i produkty. Collected proceedings. Is. 9. Moscow, RANS, 2003. P. 64-86.
50. Orlova I.V., Zakharchenko N.S., Semenyuk E.G., Nosov A.M., Volodin V.V., Buryanov Ya.I. Development the Rhaponticum carthamoides transformed roots culture. Physiologiya Rasteniy, 1998; 45(3): 397-400.
51. Giri A., Narasu L.M. Transgenic hairy roots: recent trends and applications. Biotechnology Advances, 2000; 18: 1-22.
52. Harmata J., Dinan L. Biologocal activity of natural and synthetic ecdysteroids in the BII bioassy. Archives of Insect Biochemistry and Physiology, 1997; 35: 219-225.
53. Bourne C., Whiting P., Dhadialla T.S., Hormann R.E., Girault J-P., Harmatha J, Lafont R., Dinan L. Ecdysteroid 7,9(11)-dien-6-ones as potential photoaffinity labels for ecdysteroid binding protein. Journal of Insect Science, 2002; 2(11): 11 pp.
54. Landon T.M., Sage B.A., Seeler B.J., O'Connor J.D. Characterization and partial purification of the Drosophila Kc cell ecdysteroid receptor. J Biol Chem., 1988; 263(10): 4693-4697.
55. Syrov V.N., Ayzikov M.I., Kurmukov A.G. The influence of ecdysterone on protein, glycogene, and fat concentrations in the liver, heart, and muscle of white rats. Reports of the Uzbek SSR Academy of Sciences, 1975; 8: 37-38.
56. Slama K., Koudela K., Tenora J., Mathova A. Insect hormones in vertebrates: anabolic effects of 20-hydroxyecdysone in Japanese quail. Experientia, 1996; 52(7): 702-706.
57. Syrov V.N., Kurmukov A.G. To the anabolic activity rate of phytoecdysone-ecdysterone isolated from Rhaponticum carthamoides (Willd.) Iljin. Pharmokologiya i Toksikologiya, 1976; 6: 690-693.
58. Todorov I.N., Mitrokhin Yu. I., Efremova O.I., Sidorenko L.I. The impact degree of ecdysterone on protein and nucleic acid biosynthesis in mice organs. Khimiko-Pharmatsevticheskiy Zhurnal, 2000; 34(9): 3-5.
59. Pchelenko L.D., Metelkina L.V., Volodina S.O. The adaptogenic effect of Serratula coronata L. ecdysteroid fraction. Khimiya Rastitel'nogo Syrya, 2002; 1: 69-80.
60. Zaynullin V.V., Mishurov V.P. Punegov V.V., Starobor N.A., Bashlykova L.A., Babkina N. Yu. Biological efficiency rate of two folder components containing Serratula coronata L. ecdysteroids. Rastitel'nye Resursy, 2003; 39(2): 95-103.

61. Trenin D.S., Volodin V.V., Beykin Ya.B., Shlykova A.B. Ecdysteroid fraction of overground Serratula coronata L. in the spontaneous E-rosette-forming reaction and agaromigration test in vitro. Eksperimental'naya i Klinicheskaya Farmokologiya, 1996; 1: 55-57.
62. Trenin S., Volodin V.V. 20-hydroxyecdysone as a human lymphocyte and neutrophil modulator: in vitro evaluation. Archives of Insect Biochemistry and Physiology, 1999; 41: 156-161.
63. Zelenkov V.N., Timofeev N.P., Kolesnikova O.P., Kudaeva O.T. Revealing the biological activity rate of water extracts isolated from Rhaponticum carthamoides (Willd.) Iljin leaves in vitro // Topical questions on innovations concerning production of phytoproducts from untraditional plant resources and their use in phytotherapy. Moscow, RANS, 2001. P. 59-62.
64. Timofeev N.P., Volodin V.V., Frolov Yu.M. Some aspects of the production of ecdysteroid containing raw material from aerial part of Rhaponticum carthamoides (Willd.) Iljin // International Workshop on Phytoecdysteroids. Syktyvkar, Komi SC UrD RAS, 1996. P. 89.
65. Timofeev N.P., Volodin V.V., Frolov Yu.M. Distribution of 20-hydroxyecdysone in the structures of the above ground biomass of Rhaponticum carthamoides (Willd.) Iljin, under condition of agrocenosis in Komi Republic. Rastitel'nye Resursy, 1998; 34(3): 63-69.
66. Timofeev N.P. Growth and accumulation of ecdysteroids in the Rhaponticum carthamoides Iljin overground organs depending on the age and environmental conditions, Innovatsionnye tekhnologii i produkty. Collected proceedings. Is. 6. Moscow, RANS, 2002. P. 94-107.
67. Timofeev N.P. New preparations development from Rhaponticum carthamoides (Willd.) Iljin ("Bioinfusin" and "BCL-PHYTO") // Innovatsionnye tekhnologii i produkty. Collected proceedings. Is. 4. Novosibirsk, ARIS, 2000. P. 26-36.
68. Ivanovskiy A.A. Bioinfusin impact degree on several immunity indices. Veterinariya, 2000; 9: 43-46.
69. Kurakina I.O., Bulaev V.M. Ecdystene as a tonic in tablets 0.005 g each. Novye lekarstvennye preparaty, 1990; 6: 16-18.
70. Gadzhieva R.M., Portugalov S.N., Panyushkin V.V., Kondratyeva I.I. Anabolic effect comparative study on the plant preparations ecdystene, levetone, and "Prime-Plus". Eksperimental'naya i Klinicheskaya Farmokologiya, 1995; 5: 46-48.
71. Koudela K., Tenora J., Bajer J., Mathova A., Slama K. Simulation of growth and development in Japanese guails after oral administration of ecdysteroid-containing diet. Eur J Entomol., 1995; 92: 349-354.
72. Timofeev N.P., Ivanovskiy A.A. Anabolic effect of small doses Rhaponticum carthamoides preparations // International Workshop on Phytoecdysteroids. Syktyvkar, Komi SC UrD RAS, 1996. P. 132.
73. Timofeev N.P. Results on practical application Rhaponticum carthamoides to pig-breeding as ecdysteroid source // Untraditional plants selection, cultivation and processing technologies. Simferopol, 1994. P. 166-167.
74. Timofeev N.P. Phytoecdysteroids: Pharmacological use and activity. Medical sciences, 2005; 4

DIABETES MELLITUS: A METABOLIC SYNDROME

Sharon Rabb

Alternative Healing, Dallas, TX, USA

Corresponding author: Sharon Rabb, E-mail: drsrabb@yahoo.com

Keywords: diabetes mellitus, nutrient starvation, whole food vitamins and minerals, herbal therapy

ABSTRACT

Diabetes mellitus, the most common non-communicable disease in the U.S., is a chronic digestive disorder as well as a metabolic disorder. This disease is a syndrome involving almost every system in the body from the endocrine to the cardiovascular. Insulin replacement and limited dietary counseling alone are insufficient to meet the complicated needs of the diabetic. Because of the significant lag time between probable onset and diagnosis, prevention and intervention in at risk individuals are key factors in a successful program. The medical establishment has been totally focused on insulin replacement and has missed other important variables. Fat and protein digestion and other endocrine dysfunctions are also important factors. It has been known since the early twentieth century that diabetes is a disease of nutrient starvation and needs to be addressed as such. An effective program includes:

A nutrient dense diet
Whole food vitamins and minerals
Glandular products
Herbal therapy
Exercise
Psychological counseling

Because each person is unique there is no single program that works for everyone. Individual differences play a vital role in designing an effective diabetic protocol.

INTRODUCTION

Diabetes mellitus is a chronic metabolic and digestive disorder involving the assimilation not only of carbohydrates but also fats and proteins. The result is a defective or deficient production of insulin by the beta cells in the islets of Langerhans, specialized cells in the pancreas, and an imbalance of other enzymes and hormones. This leads to impaired glucose use or hyperglycemia as well as impaired fat and protein metabolism due in large part to impaired production of enzymes, their cofactors, and hormones.

Diabetes is now the most common non-communicable disease and the 6th leading cause of death in the U.S. More people die of diabetes than car fatalities. The prevalence of diabetes is reaching epidemic proportions, and individuals with diabetes are at high risk of developing severe complications leading to other chronic conditions. Obviously, insulin therapy and limited dietary counseling are not adequate. Then, what is the problem, and what can be done? Because the medical establishment has been totally focused on a drug that will supply insulin, it has failed to grasp the underlying causes of diabetes and the adequate support of diabetic needs.

According to Drs. Gyr, Beglenger and Stalder, the pancreas has both exocrine and endocrine components. The exocrine component manufactures, stores, and packages digestive enzymes for digestion of food. The endocrine secretes hormones that regulate the metabolism and utilization of the absorbed nutrient components. Both functions are closely related both anatomically and functionally. It has been shown that the endocrine part exerts a profound effect upon the exocrine functions of the pancreas and that diabetes severely affects both components of the gland.

Dr. Harry Harrower said as much as early as 1932 in his work Practical Endocrinology. He stated that no gland works in isolation and that until we understand the interactions of the endocrine system, many diseases would remain a mystery.

Diabetes is a syndrome of a multitude of factors and needs to be approached as such. Diet and nutrition are, of course, major players, but just what constitutes adequate diet? What should be avoided and what added and why? This discussion will attempt to answer these and other pertinent questions.

Diabetes can be classified into either Type 1 or Type 2:

Type 1: Is the autoimmune destruction of the pancreatic islet beta cells with total loss of insulin secretion. This type accounts for about 8% of the population with diabetes. Type 1 is also known as IDDM (insulin dependent diabetes mellitus).

Type 2: Is a progressive chronic illness that usually is present for 4 to 7 years before diagnosis. The symptoms are less acute than Type 1. This disease is known as NIDDM (non-insulin dependent diabetes mellitus). Most of this article will discuss Type 2, but much information can be applied to Type 1.

Contributing Factors to the Syndrome Diabetes Mellitus

Lack of functional insulin (see note).
Lack of digestive enzymes for fats, carbohydrates and proteins.
Lack of whole food vitamins and other cofactors.
Lack of healthy bile salts and insufficient lecithin and other factors to metabolize fats.
Imbalances in endocrine function not limited to the pancreas.
Chronic disorder of carbohydrate, fat and protein metabolism and assimilation.
Hyperglycemia - Hyperlipidemia.
Glycosylation of proteins leading to a number of complications.
A build up of sorbitol.
Metabolic or cardiac Syndrome X.
Toxicity because of poor metabolism and digestion resulting in acidosis among other problems.
Sympathetic nervous system imbalances contributing to nerve imbalances and emotional issues.
Chronic infective disorders resulting from immune deficiencies.
Vascular changes and cardiovascular disturbances
Liver dysfunction and toxicity.

Note

Diabetes can result from either the lack of sufficient insulin or the overproduction of insulin due to the fact that the cells in the body have become insulin resistant. The function of insulin is to aid in the transport of glucose and other nutrients into the cell. When the body becomes resistant to insulin, it requires much more insulin to transport glucose, and it is not done effectively. Insulin resistance, coupled with obesity, elevated cholesterol, low HDL cholesterol, high triglycerides and high blood pressure have become known as Syndrome X. In this syndrome it can require 300 to 400% more insulin to maintain normal blood sugar. There is some speculation that heavy metal toxicity and environmental chemicals might play a role in diabetes.

Hyperglycemia results in both micro-vascular and macrovascular damage known as the complications of diabetes.

Micro-Vascular Damage

retinopathy
neuropathy
nephropathy

Macro-Vascular Damage

cardiovascular disease
cerebral vascular disease

Other Complications

Fatigue
brain fatigue
irritability
depression
reduced immunity to infections
ketoacidosis
gastric dysmobility
probably predisposes to cancer, heart disease, arthritis and other chronic diseases.
gangrene
obesity
abnormal lipid metabolism
fatty liver

Processed Foods and Diabetes

Diabetes is, for the most part, a disease of consuming copious amounts of synthetic toxic chemicals going under the alias as "food" and not eating adequate fresh whole foods.

Fraudulent Foods

Artificial chemical sweeteners
High fructose corn syrup (HFCS) or dextrose - synthetic sugar
Synthetic hydrogenated fats and old rancid oils
Bleached enriched white flour, processed white sugar
Synthetic and/or isolate vitamin fragments sold as vitamins
Chemical preservatives, flavors and dyes
Pasteurized milk
The products of one or all of these
Chlorinated foods and water and fluoridated water
Genetically hybridized foods
Too much animal based protein and other meat toxins (not necessarily a fraudulent food but one that contributes to diabetes)
Caffeine taken with meals increases blood sugar

It is the intake of these counterfeit foods as well as the lack of nutritious whole foods that are causing in large part the rapid increase in the incidence rates of diabetes mellitus (primarily Type 2).

Don Harkins in his article in the Idaho Observer, 2000 Nov 26, considers that most individuals are unaware that the artificial sweetener aspartame becomes formaldehyde in the body. Formaldehyde is so toxic that neither the FDA nor the EPA has identified a safe level of ingestion. Dr. James Bowen considers it a neurotoxin and a catalyst for polychemical hypersensitivity syndrome (PCS). According to Dr. Bowen, "The Persian Gulf Syndrome is largely PCS from massive NutriSweet (aspartame) exposure experienced by our men in combat units in the Persian Gulf". The saddest joke on the American consumer is that aspartame – sweetened sodas are marketed as "diet" drinks. The truth is that aspartame suppresses the production of serotonim, which makes a person crave carboyhydrates which in turn causes them to gain weight". There is a now a class action lawsuit against several individuals and companies over the many diseases believed connected to aspartame.

Diabetes is caused in part by what we eat and what we fail to eat. Most people don't consider fat metabolism in relation to diabetes, but some of the major complications are directly related to fat metabolism (more later).

Statistically, there is a lag time of between 4 to 8 years from the date or probable onset of diabetes to the date of diagnosis. So, it behooves us to correct our dietary intake as soon as we can and as soon as we know how. There is much more to it than what you normally hear from a dietician or read in most books. We once had a certified dietician at the Master's level tell us that a pepperoni pizza was a completely balanced meal because it contained all the four food groups. As you will read, this is not accurate information.

The real problem of diabetes is a lack of whole food vitamins and other whole food nutrients found in fresh whole foods and a surfeit of junk masquerading as food.

High Fructose Corn Syrup (HFCS)

Numerous studies abound in the medical literature from early 1900s to recently linking HFCS to diabetes (and other diseases as well). HFCS is used as a food filler not just a sweetener because it is not only cheap but subsidized by the U.S. government. It is ubiquitous in everything from bread to baby food. Today Americans consume more HFCS than sugar, according to a study by J.E. Swanson (Metabolic effects of dietary fructose in healthy subjects; *American Journal of Clinical Nutrition*, 55(4), 1992: 851-56).

Abstracted from L. S. Gross, and L. Li, May 26, 2004, "Increased consumption of refined carbohydrates and the epidemic of Type 2 diabetes in the United States: An ecologic assessment" in May 2004 issue of the *American Journal of Clinical Nutrition:*

HFCS is produced by processing corn starch to yield glucose, and then processing the glucose to produce a high percentage of fructose. Two enzymes used to make HFCS, alpha-amylase and glucose-isomerase, are genetically modified to make them more stable. The ubiquitous nature of HFCS (used in almost everything, from jams to condiments to soft drinks to so-called "health foods" also makes those trying to avoid genetically engineered foods even more difficult. Today Americans consume more HFCS than sugar.

According to the study published in the *American Journal of Clinical Nutrition*, corn syrup's ubiquity in our food has now been linked to Type 2 Diabetes.

Seeking to examine the correlation between consumption of refined carbohydrates and the prevalence of type 2 diabetes in the United States, researchers conducted an ecologic correlation study. They examined the per capita nutrient consumption in the United States between 1909 and 1997 obtained from the US Department of Agriculture and compared that with the prevalence of type 2 diabetes obtained from the Centers for Disease Control and Prevention. After conducting a multivariate nutrient-density analysis, in which total energy intake was accounted for, corn syrup was positively associated with the prevalence of type 2 diabetes. Fiber was negatively associated with the prevalence of type 2 diabetes.

These results led the researchers to conclude, "intakes of refined carbohydrate (corn syrup) concomitant with decreasing intakes of fiber paralleled the upward trend in the prevalence of type 2 diabetes observed in the United States during the 20th century."

The question begs to be answered--why is this information being kept secret and why are food manufacturers allowed to continue--and--why is corn subsidized? The answer, of course, is money. HFCS is cheap and corn is highly subsidized.

Since the early 1900's HFCS has been suspected to cause diabetes in humans as well as test animals. In 1907, Dr. Harvey Wiley, then the head of the FDA, tried unsuccessfully to ban it from commercial use and failing that to at least acquire honest labeling. HFCS is processed so extensively that it becomes an exogenous or synthetic glucose which over-stimulates the pancreas among other problems. It was the only sugar to cause diabetes in test animals. The following is a quote from Dr. Royal Lee (Lectures of Dr. Royal Lee, I, pp. 198-99, 1958).

We will find that there are available vitamin concentrates that will often in minutes erase the heart reactions of various kinds that follow the use of the refined sugar, and bleached flour, the foods that kill seven hundred thousand people every year.

You may ask, "Why are these unfit foods permitted on the market if they are so dangerous?" May I refer to Dr. Harvey W. Wiley's attempt to get synthetic sugar--glucose--properly labeled as a synthetic sugar substitute instead of being permitted to masquerade under the phony cognomen of "corn syrup." He predicted that we would become unduly afflicted with diabetes if we consumed much of this synthetic, counterfeit sugar, and tried his best to get it at least properly described and labeled instead of being palmed off on us as a natural food. Forty years after his unsuccessful attempts to enforce honest labeling of the product, Drs. Lukens and Dohan, at the University of Pennsylvania, confirmed their fears by showing that corn syrup--dextrose--was the only sugar known to science that was capable of causing diabetes in test animals when fed in substantial amounts.

The average American consumes as much as 170 lbs. of sugar a year! More than 1/4 the total caloric intake! HFCS combined with processed white sugar is one of the primary reasons for all chronic disease in this country. Bleached white flour along with its isolate vitamins and hydrogenated synthetic oils is another.

Dr. Wiley was not the only doctor to warn against HFCS. Dr. David Quigley, an oncologist, was another. Dr. Quigley in his book, *The National Malnutrition*, states that 75% of foods eaten in large cities have their vitamin and mineral content removed. This book was banned from public libraries in Rochester, New York around 1940. Dr. Quigley also wrote a book, *Notes on Vitamins and Diets* (1933) where he comments on cancer and HFCS stating that he could not help people clear cancer without abstaining from HFCS.

More recently, Dr. Bray in his paper on the epidemic of obesity and changes in food intake, believes that the consumption of HFCS is the largest factor in the epidemic of diabetes. Peter Jennings of ABC News went on Prime Time Live on December 8, 2003 condemning HFCS and linking it to obesity. He claimed that two thirds of the population is overweight due in large part to HFCS. Since the 1970's the use of HFCS has increased by 4,000 percent. The role of corn sweeteners is to lower the cost of processed foods because corn is subsidized by the US government. The program states: "You have to bike for one hour just to burn the calories in the average soda".

Dextrose or HFCS is rapidly absorbed and overloads the pancreas, causing diabetes in test animals. This overloading also imbalances other pancreatic functions and other endocrine glands leading to imbalances in the whole body.

Bleached White "Enriched" Flour

HFCS is not the only problem for the diabetic. Bleached white "enriched" flour is another. White flour has been robbed of its vital nutrients because of the short shelf life of fresh ground wheat flour. Processing it and bleaching it renders it bug free and stable for long periods of time (bugs won't touch it.), thus robbing the consumer of vital nutrients and minerals. This is bad enough, but when the isolate and/or synthetic vitamins are added, you have a very toxic substance. In the 1940's, Dr. Agnes Fay Morgan showed that "fortified white flour was much more toxic to animals than just plain bleached flour." This is because the manufacturer is not adding real vitamins.

The bleach in the flour acts on Xanthine, a component of some bodily tissues, to produce alloxan. It also acts on the germ in wheat as well. Alloxan is used in research to cause diabetes in test animals by killing the beta cells in the pancreas.

The following is from Lectures of Dr. Royal Lee, VI, pp. 177, 194 (1955):
The next great advance in flour technology was the invention of the bleaching system (about 1895). It is somewhat coincidental (maybe) that the first case of coronary thrombosis was seen in an autopsy in 1898. But it is no coincidence that all flour bleaches so far used convert the

xanthine of flour (xanthine is one of the germ components that have vitamin value--it cooperates with vitamin E) into a highly potent poison, alloxan. The special characteristic of alloxan is that it destroys the islets of Langerhans of the pancreas, and specifically causes diabetes. But now we know that one of the vitamin co-factors known as XANTHINE is capable of being oxidized by any oxidizing bleach chemical (and that is why they all act as bleaches) which converts it into ALLOXAN.

ALLOXAN is a very potent DIABETOGENIC POISON. This ALLOXAN is another dangerous synthetic substance created in the flour by CHEMICAL MEDDLING with a natural food product.

Flour has been bleached now for over fifty years, but this fact of the possible conversion of xanthine to alloxan is very new to most of us. It was reported, however, in Thorpe's Dictionary of Chemistry as far back as 1918. The diabetogenic nature of alloxan was not discovered until 1943.

A more recent article by Bosakowski and Leven (1987) states:

Both Halocitrates produced a similar diabetes-like syndrome (hyperglycemia, glycosuria) mediated by a significant hyperglucagonemia and slight hypoinsulinema. Chlorocitrate was more potent in its effect with a much greater buildup of plasma lactate. In contrast, fluorocitrate produced a severe life-threatening hypocalcemia. Both halocitrates had a similar depressive effect on circulation as evidenced by hypothermia, bradycardia and elongalion of the QT-interval. These changes were considered to be the result of lactic acidosis and the ongoing ion imbalance since heart levels were not depleted.

This study indicates a causative relationship between oxidizening halocitrates (chlorine and floride) and diabetes.

Even though HFCS, bleached white enriched flour, white sugar, and other synthetic foods and chemicals contribute to diabetes and obesity, the main problem is one of whole food vitamin and mineral deficiencies.

The problem of diabetes (at the core of the onion) is one of vitamin starvation. (Remember real vitamin complexes cannot be bought over the counter.) Particularly important are the whole vitamin B complexes and whole vitamin C complex. However, all whole vitamin complexes play a role. Deficiencies in these vitamins result in starvation of the organs and glands, particularly the endocrine glands. The pancreas is an endocrine gland, but contrary to established belief, it is not the only one responsible for diabetes. Through proper nutrition and whole food vitamin concentrates, it is possible to rebalance these glands if they are not already dead. The endocrine glands are a living dynamic system that support and are being supported by each other. Imbalances affect all endocrine glands and the organs they support.

We have found that most, if not all, chronic illness is nutritionally based and the same is true for diabetes. Remember, that not only does diet profoundly affect diabetes after diagnosis, but that diet also plays a causative role in bringing about diabetes.

However, you can start now in cleaning up your diet (and not with bleach!). If you have diabetes or suspect that you are at risk, there are definite steps you can take to live a healthier lifestyle. Keys to recovery and prevention include the use of concentrates of whole food vitamins and herbs as well as dietary modifications.

Fat Metabolism

One of the most serious consequences of diabetes is high levels of blood and liver fat. This results from both impaired digestion and metabolism. In the process of digestion, fat is reduced to fatty acids and glycerin before it is converted to glycogen in the liver. The rate of conversion from fat to glycogen is dependent upon the concentration of blood glucose. If blood glucose remains high, the fat is not converted to glycogen and is "backed up" in the blood and liver. Many of the complications of diabetes and indeed diabetic coma is a result of high blood and liver fat. It has long been accepted that diabetic coma is a direct result of high blood fat rather than high blood sugar. (Joslin, New Eng. Med. Jol. V. 209, p. 525). This is also the primary cause of obesity in diabetes because the excess fat is stored in the tissue.

Mahay and Adeghate (Mol Cell Biochem 2004 Jun; 261 (1-2), pp. 175-81) also reported that fat digestion is compromised in the diabetic. They also stated that diabetes is associated with numerous conditions to include hypo-secretion of digestive enzymes. The end result of this study was that the parotid glands of the diabetic person become extensively infiltrated with various sized lipid droplets. "This indicates that DM (diabetes) can elicit changes in the morphology, secretory function and acyl fatty acid quantity of isolated rat parotid glands. The levels of acyl lipids was greatly reduced in diabetic rats. This indicates that fat digestion is compromised in the diabetic."

Fats remain high because, in part, they are not converted to glycogen. The conversion of fats to glycogen is dependent upon a pancreatic factor that is a complex of whole B vitamins and betaine.

In experimental animals, removal of the pancreas is not fatal if betaine (as well as insulin) is given. Insulin is needed in the metabolism of sugar and betaine is needed in the metabolism of fat. Betaine is associated in food with the whole vitamin B complex. Whole vitamin B is needed in carbohydrate as well as fat metabolism and in the kreb cycle where glucose is converted to energy.

Fat digestion and metabolism is a very complicated process involving many steps from healthy bile, available lecithin, hormonal factors and enzymes to available whole vitamin cofactors and minerals. It is beyond the scope of this paper to delve into the exact steps involved which are varied and complicated. A major concern for the diabetic is obtaining high quality fats and abstaining from highly processed oils, rancid fats, high levels of saturated fats and trans fats. Every cell in the body depends upon high quality oils to function properly. Each cellular membrane is composed of phospholipids that give the cell its ability to transport substances in and out of its environment. Poor quality fats greatly reduce this ability and lead to disease and debilitation. High quality oils are also needed in nerve health, hormone production, and the metabolism of many vital nutrients. It is essential that a diabetic consume high quality oils and supplementation can greatly facilitate this issue.

Fats to avoid:
Trans fats
Rancid oils
Margarine
Low fat chemical substitutes
Fried foods

Highly processed foods
Cheap oils
Red Meat

Fresh ground flax seeds plus supplementation with gamma-linoleic acid (GLA) helps the diabetic obtain a good source of essential fatty acids (Omega 3 plus Omega 6). GLA is necessary because the diabetic has difficulty in forming GLA from linoleic acid. The body needs a good monounsaturated fat such as olive oil and some saturated fat preferably from butter, yogurt, or cheese. Raw nuts also provide a very good source of good fats.

Impaired fat metabolism can be the result of a variety of factors which include:

Toxicity of the liver
Lack of functional bile
Lack of whole B vitamins and the betaine factor
Lack of whole B vitamins that act as methyl donors when converting homocysteine to methionine
Lack of natural lecithin in processed fats
Lack of lipase and other enzymes
Toxic fats
Lack of other whole food vitamins and minerals
Lack of secretin--a duodenal enzyme that stimulates the pancreas to produce enzymes
Lack of a good source of essential fatty acids
Imbalances in other endocrine glands (sometimes not picked up in medical tests)
Excess of toxic foods

The Cycle of Diabetes

1. The pancreas fails for whatever reason to supply effective insulin.
2. Glucose builds up in the blood and urine (hyperglycemia and glycosuria).
3. An automatic reduction in the amount of reserve fat converted to glycogen (a compensatory reaction due to reduced glycogen metabolism).
4. Fat begins to accumulate in the liver, blood and other tissues, elevating blood lipids among other complications.
5. Diabetic coma can result due to high blood fat.
6. Increased blood sugar can affect healthy bile production, a vicious cycle.

Since the digestion and metabolism of fat is complicated and involves various metabolic steps, it behooves the diabetic to determine which metabolic processes are faulty. These processes are regulated by the endocrine system.

Endocrine Factors that Affect Blood Sugar

Adrenal Glands

The effects of the adrenal glands are probably the most overlooked factor in the total picture of diabetes. They have an antagonistic effect on the pancreas. An increase in adrenalin increases the sugar index of the diabetic by decreasing insulin. Most medical tests do not pick up sensitive imbalances in the endocrine gland function. Overall health is a direct result of balances in endocrine function. Imbalances occur as a direct result of poor nutrition and whole vitamin and mineral starvation. The adrenal is part of the sympathetic nervous system – a key factor in the disease of diabetes as well as the cardiovascular system.

Wexler and McMurtry (Life Sci, 1983 Sep 12; 33 (11), PP. 1097-103) report:

The alloxin induced diabetic rats showed a slight but statistically significant increase in blood pressure, pituitary and adrenal glandular hyperplasia hyperlipidemia, hyperglycemia and increased BUN levels. The obese and non-obese rats manifested gross and microscopic degenerative changes that suggests acceleration of the normal aging process! The genetically programmed problems of diabetes, hypertension, obesity, and cushingoid pathopysiology of obesity may be due to the hyperadrenocorticism.

This study verifies the close relationship between the adrenal and the pancreas and the complications of diabetes when imbalances occur.

Adrenaline releases glycogen from the liver. Normal glucose homeostasis is tightly regulated by three interrelated processes:

Insulin secretion
Utilization of glucose by peripheral tissues
Glycogen production in the liver

While the thyroid controls the accumulation of energy in the body, the adrenals, brain and the sympathetic nervous system control the expenditure. Diabetes is a broad endocrine dysfunction—it is not merely a pancreatic beta cell dysfunction. This was recognized at least as early as 1911 by O. Minowski of Russia.

Dr. Royal Lee (Vitamin News) in 1936 reports: "Diabetes with hypotension is a more evident case of vitamin B deficiency....Diabetes accompanied with hypertension is a different matter. The hyperglycemia here may be due to the oversecretion of adrenin, which may also be causing the increased blood pressure. Adrenin has as one of its functions, the release of glycogen from the liver. Too much adrenin may be due to avitaminosis C and B, as each of these vitamins is necessary to the proper function of the adrenals...the hypertrophy often means a hypersecretion of medulla with attendant hypertension—and diabetes if the pancreas is unable to take the load...We must accept as a basic fact that most endocrine unbalance is a result of vitamin starvation."

Dr. Lee is not speaking here of isolate or synthetic vitamins that most of us think of as vitamins. Ascorbic acid is not vitamin C. Vitamins occur in nature as complexes of many factors as yet not fully understood. This is why whole foods and whole food vitamin concentrates (rare on the market) show different test results than isolate vitamin supplements in medical research.

Parathyroid gland

Blood sugar levels increase 8-fold after parathyroidectomy. The parathyroid controls calcium metabolism. (Lack of utilizable calcium can cause acidosis, a major factor in diabetes.) Even a minor weakness in this gland affects the pancreas as well as other endocrines. Otto, Buczkowska and Kokot (Endokrynol Pol 1991, 42 (3,) pp. 447-53) report:

It is known that calcitonin participates in the homeostasis of calcium and is an important regulator of insulin secretion. The results obtained suggest that calctonin may play a role both in the pathogensis of diabetes and in developing of diabetic osteopenia. Calctonin was significantly increased in two of the three groups of diabetics studied.

This study indicates the relationship between the pancreas and the parathyroid which secretes calcitonin. It also stresses the relationship between calcium metabolism and diabetes.

Calcium metabolism is a very complicated and often overlooked aspect of diabetes. It is beyond the scope of this paper but is a very important variable of the total program for diabetes.

Secretin – duodenal hormone

Secretin, a hormone from the duodenum which activates the pancreas, has been shown to be low in diabetics. Secretin was the original hormone or the "mama" hormone first observed in 1902 by W.M. Bayless and E.H, Starling of London. They called it the "pancreatic hormone" because of the remarkable effects on the pancreas.

Secretin is needed to:

Stimulate bile flow and hormone production
Stimulate enzyme production from the pancreas
Decrease blood glucose

The secetin – pancreatic relationship has long been overlooked by established medicine. Dr. Henry Harrower, a well-known endocrinologist in the 1930's had this to say in Practical Endocrinology: "I feel that one of the shortcomings in our consideration of diabetes mellitus concerns the pancreatic acinous function and the duodena-pancreatic relationship. ... The makings of secretin are deficient, hence the difficulty in completing the cycle of secretin production and pancreatin activation. ... All of this should stress more definitely the digestive phases of diabetes." Digestion is the first process that should be evaluated and facilitated in the diabetic. Combining secretin with pancreatic enzymes greatly facilitates healing. The next phase in the healing cycle should be detoxification.

Liver Function

Liver function (including toxicity) is a key ingredient in the disease of diabetes as well as other chronic conditions. We believe that it is impossible to have diabetes and not have hypohepatica and toxic liver. Both the phase I and phase II detoxification processes in the liver can be compromised. The use of certain herbs, whole food concentrates, adequate diet and fasting have been known to aid in liver function.

The toxemia of diabetes exerts a widespread detrimental effect in the entire endocrine system as well as the liver. All endocrine glands are affected by diabetes. When endocrine glands become dysfunctional, they accumulate acid waste which further disrupts the acid-base balance and liver function.

Pituitary gland

The permeability of the intestinal wall is regulated by the posterior pituitary hormone. This hormone prevents constipation and promotes the deposition of fat in the liver to be burned. Dr. Royal Lee stated that 88% of obesity in children was attributed to pituitary weakness. Because the pancreas is intact in some cases of severe diabetes and the pituitary (both anterior and posterior) secretes hormones that cause hyperglycemia, pituitary disorder may play a role in all cases of diabetes. Of course, the pituitary regulates all of the endocrine glands.

Acid-Base Balance

Acidosis is a dangerous sequel to diabetes. Ketosis, the result of utilizing fat instead of glucose for energy, also causes acid build-up. This stresses not only the liver but also the kidneys. Acidosis is the lack of functional minerals. Alkalinization is remineralization and is important because every form of cellular inaction produces acidosis which burdens the entire endocrine system and indeed the whole body. If the liver is unable to anabolize toxic wastes, which are acidic, then further acid is dumped into the system. The pH of the blood should be approximately 7.0 and the saliva 6.8.

Correcting acid- base balance is critical to the diabetic. This can be greatly facilitated by supplements with chelated alkaline ash minerals as found in organic matter.

Alkali minerals:
sodium
magnesium
calcium
potassium

The actual mineral levels as well as major mineral ratios can provide key information about health status. Mineral hair analysis is an excellent choice in determining functional mineral levels. A low sodium to potassium ratio indicates low adrenal function. Of course, every endocrine gland needs not only a broad spectrum of minerals but also is dependent on one key mineral. For the pancreas, it is chromium. It is beneficial to provide chromium in an organic form. It works closely with insulin and is known as the “glucose tolerance factor (GTF).”

However, taking isolated minerals is not the total picture because in the human body mineral metabolism functions as a transport system. The body has to be able to absorb, transport, and deposit minerals. The body cannot utilize calcium without Vitamins A and D and essential fatty acids to "unload" the calcium in the blood to the tissues. Minerals do not work in isolation—mineral ratios are the key because minerals work together in a holistic way to produce health. The kidneys are particularly affected by the alkali minerals (alkali reserves). Dr. Royal Lee in Vitamin News, Vol. 1, 1933, pp. 3-4.

We believe that vitamins have an important bearing on the synthesis of ammonium to protect the alkali reserves, and that the cause of acidosis is always kidney failure, and that such failure is a consequence of vitamin deficiency. Our reason for such a belief is the remarkable and rapid improvement that occurs when "catalyn" is given to persons having the acidosis complex which usually includes some kidney complications. (catalyn is a whole food vitamin).

Vitamins and Diabetes

For over fifty years the international research community has been aware of the link between the vitamin B complex and diabetes. Mainstream medicine has chosen to ignore it, and many people have suffered as a consequence. Dr. Royal Lee had this to say: "The first account of the successful treatment of hyperglycemia with vitamin B concentrates was mentioned in these pages (Vol. 2, No. 5, page 29, paragraph 4, 1939). Vitamin B caused a slow but definite improvement in a majority of the cases treated. At the time attention was called to the fact that diabetes with attendant hypertension also required vitamin C concentrate to get the best response according to clinical findings" (Royal Lee, Vitamin News, p. 70). Dr. Lee is referring to whole food vitamin concentrates--not the isolate or synthetic vitamins sold in stores.

Most diabetics suffer from two major problems: carbohydrate and fat metabolism and nerve degeneration. What are the broad functions of the vitamin B complex? Carbohydrate and fat metabolism and nerve health!

Here are the partial functions of the vitamin B complex:

1. Support nerve and brain function and, therefore, muscle function.
2. Support energy or glucose metabolism and also fat metabolism.
3. Support endocrine function (pancreas is an endocrine gland).
4. Support healthy heart function.

Whole vitamin B deficiency can result in either under or over-stimulation of nerve function to the endocrine glands and to other organs and glands (specifically the heart). This is one of the major complications of diabetes.

Again, you cannot buy whole food vitamin complexes over the counter anywhere! You can only buy isolate fragments of vitamins which can be natural but they are not from food sources, and not the entire complex.

Niacinamede, also known as nicotinamide, a part of the vitamin B complex was shown in the middle 1900s to prevent the development of diabetes in experimental animals (Lazarow A, Liambies L, and Tausch A.J, Protection against diabetes with nicotinamide, *J Lab Clin Med 36*, 249-258, 1950). This is only one of the many studies showing different vitamin B fragments to be effective in diabetes and also in heart disease. These studies used only one fragment of the vitamin B complex. Think what would be possible if the entire vitamin B complex from whole food sources with emphasis on a few of the Vitamin B fragments could do for diabetes; especially when the Vitamin B Complex is combined with other whole food vitamin concentrates.

Following is a synopsis of the Vitamin B and C complex. The B complex is found primarily in nuts, whole grains or legumes or soybeans. They are also found in the cruciferous vegetables (kale, cauliflower, broccoli, brussel sprouts, and cabbage).

The B vitamins are listed below. Vitamin B1 (Thiamin) – used in:
Energy production
Carbohydrate metabolism
Nerve function (needs magnesium to function)

Deficiency results in consuming significant amounts of white rice. The B1 is in the bran left behind. It is found in brown rice.

Food Sources:
nuts
soybeans
whole grains
beans
wheat germ – fresh
brown rice

Wheat germ spoils in less than 48 hours after grinding unless refrigerated, then it only keeps about one month.

B2 - Riboflavin – Used in:

1) Energy production
2) Mucous membranes
3) Regenerates glutathoine (antioxident)

Food Sources:
almonds
wheat germ
mushrooms
green leafy veggies
soybeans

B3 – Niacin - Used in:
Energy production
Fat, cholesterol and carbohydrate metabolism
Manufacture of hormones
Used particularly in Diabetes Type 1

Food Sources:
rice bran
wheat bran
peanuts
seeds and nuts

B6 – Pyridoxine – Used in:
Formation of body proteins and structural components
Chemical transmitters in the nervous system
Proper immune function
Maintain hormonal balance

Food Sources:
wheat germ
legumes

seeds
potatoes
soybeans
kale, cauliflower, brussel sprouts
spinach
avocados
Biotin - Used in:
Manufacture and utilization of fats and amino acids
General metabolism

Food Sources:
produced to some extent in the gut by bacteria (vegetarians produce more)
soybeans
nuts
rice bran
oatmeal
legumes
cauliflower
Biotin enhances the sensitivity of the body's receptors to insulin. It also increases the activity of glucokinase, an enzyme responsible for the utilization of glucose by the liver.

B5 - Pantothenic Acid (B5) –Used in:
Energy metabolism
Manufacture of adrenal hormones
Manufacture of red blood cells
Joint function
Lowering blood lipids

Food Sources:
peanuts
soybean
mushroom
whole grains
seeds
oatmeal
lentils
broccoli
Folic Acid – Used in:
Works with B12
Nervous system in fetus
Cell division - DNA synthesis
Linked to depression and other diseases
Linked to homocysteine levels

Food Sources:

legumes
wheat germ
asparagus
nuts
spinach
peanuts
cauliflower
kale
broccoli
brussel sprouts
oatmeal
Cobalamin Vitamin (B12) – Used in:
Works with foic acid
Supports nerve tissue
Cell division (DNA synthesis)
Nerve conductivity

Food Sources:
organ meats
eggs
imported cheeses
meat

Choline (works with inisitol) – Used in:
Manufacture of neurotransmitters
Cellular membrane
Fat metabolism (freeing fat from liver)
Food Sources:
egg yolk
grains
legumes
cauliflower
lettuce
soybeans
bananas
peanuts
Inisitol (works with choline) – Used in:
Cell membranes
Supports nervous tissue
Fat metabolism
Food Sources
whole grains
citrus fruit
seeds
nuts

legumes

Vitamin C

Insulin facilitates the transport of vitamin C as well as glucose into the cells. Diabetics often lack sufficient vitamin C at the cellular level. Whole food concentrates of vitamin C are needed-- not ascorbic acid in any form. The lack of whole food vitamin C at the cellular level is one of the reasons for diabetic complications.

Whole food vitamin C:
Promotes wound healing
Assists in the metabolism of cholesterol
Decreases capillary permeability
Promotes healthy immune system, increases phagocytic activity
Acts as an antioxidant for free radical activity
Promotes blood sugar control
Reduces the accumulation of sorbitol within cells
Inhibits the glycosylation of proteins leading to many complications.

Sorbitol accumulation in combination with the glycosylation of proteins leads to many diabetic complications including nerve and eye diseases. The drugs that are designed to inhibit sorbitol accumulation are extremely toxic. Whole food vitamin C is a much better alternative (again, not ascorbic acid). Sorbitol accumulation is the result of faulty glucose metabolism. Sorbitol is normally converted to fructose, in the non-diabetic and is easily excreted from the cell. In the diabetic, sorbitol accumulates and becomes a toxin disturbing the osmotic balance in the cell.

In glycosylation, glucose binds to proteins such as albumin in blood, the eye lens, and the myelin sheath of nerves. This causes a host of abnormal structures and functions leading to the diseases of diabetic complications. Vitamin C has been shown to reduce glycosylation. Glycosylation of proteins occurs in diabetics when glucose "sticks" to proteins. It sticks to proteins in the blood, nerves and the eyes. This pathological process causes much of the damage in the complications of diabetes.

No vitamin or mineral works in isolation. the body is truly a "holistic" organism with "interdependent" functions and reactions. Vitamin C is dependent on other vitamins, minerals and phytochemicals to function which in turn are dependent on other nutrients.

Sources:

peppers
red chili
sweet peppers
guavas
green leafy veggies (collard , kale, parsley, turnip, mustard greens
brussel sprouts
cauliflower and cabbage
oranges
strawberries

papayas
lemon
grapefruit
mangos
others

Herbs and Diabetes

Research is surfacing that substantiates the belief that diabetic drugs can be very toxic.

According to A. Marcus (2000): The risks associated with drug treatment are generally class-specific. Among anti-diabetic agents, sulfonglureas and insulin are associated with risk for severe hypoglycemia, metformin with risk for lactic acidosis, and troglitagone with risk for idiosyncratic hepatocellular injury.

Herbs are a much safer alternative. They not only lower blood sugar (among other benefits), they do not have the negative side effects. Several herbs have been effective for the diabetic. One of the most promising is Gymnema sylestre. It seems to lower blood glucose, increase insulin levels, lower serum lipid levels to near normal levels, regenerate the beta cells of the pancreas and suppress sweet taste cravings! Wow—what an herb! Because of the toxicity of diabetic drugs, this is nothing short of miraculous. (Gymnema montanum also shows great benefit in diabetes; however, it is an endangered species).

The following are three references on Gymnema:

Title: Use of Gymnema sylvestre leaf extract in the control of blood glucose in insulin-dependent diabetes mellitus.

Author(s): Shanmerg E. R., et al.

Source: Journal Ethnopharmacology [J Ethnopharmacol] 1990 Oct; 30(3), pp. 281-94.

Abstract: GS4, a water-soluble extract of the leaves of Gymnema sylvestre, was administered (400 mg/day) to 27 patients with IDDM on insulin therapy. Their insulin requirements, fasting blood glucose, glycosylated haemoglobin (HbAlc) and glycosylated plasma protein levels all were lowered. Serum lipids returned to near normal levels with GS4 therapy. GS4 therapy appears to enhance endogenous insulin possibly by regeneration of the beta cells of the pancreas in insulin-dependent diabetes.

Title: An overview on the advances of Gymnema sylvestre: Chemistry, pharmacology and patents.

Author(s): Porchezhian E; Dobrigal RM

Source: Dre Pharmazie [Pharmazie[2003 Jan; 58(1), pp. 5-12.

Abstract: Extracts of Gymnema sylvestre are widely used in Australia, Japan, Vietnam, and India to treat diabetes. Preparations from this plant have been shown to suppress sweet taste sensations. In other words, Gymnema helps to diminish craving for sweets. It is used in the treatment of diabetes, obesity and dental caries. Anti-allergic, antiviral, and lipid lowering effects are also reported. However, Gymnema has a very bitter taste.

Title: Effect of Gymnema montanum on blood glucose, plasma insulin, and carbohydrate metabolic enzymes in alloxan-induced diabetic rats.

Author(s): Ananthan R; Latha M; et al.

Source: Journal of medicinal food [J Med Food [2003 Spring; 6(1)], pp. 43-9.

Abstract: The effects of Gymnema montanum, an endangered plant used in ancient India, on blood glucose, plasma insulin, and carbohydrate metabolism enzymes were studied in diabetic rats. This plant administered at 200 mg/kg body weight significantly decreased blood glucose levels and significantly increased insulin levels. The antidiabetic efficacy of his herb was better than that of glebenclamide.

Diet and Diabetes

Diet is really the key in the total program for diabetes. We advocate a nutrient dense diet that is high in fiber, both soluble and insoluble, and moderate in carbohydrates. The diabetic needs to eliminate junk carbs and "bad" fats and add fresh ground grains and nuts, brown rice, and fresh fruits and vegetables free of chemicals. The use of fresh ground nut butters, eggs, cheese and legumes are excellent protein choices. The diabetic needs to at least limit red meat if not eliminate it. Chicken and fish can be used sparingly.

Riccardi and Riveliese (Diabetic Care [Diabetes Care] 1991 Dec; 14(12), pp. 1115-25) state:

There are two camps in the dietary treatment of diabetes. One that advocates a low carbohydrate diet with high fat, the other says a high carb diet. This paper indicates that dietary fiber has been left out of the equation and when a high carb/high fiber diet significantly improves blood glucose compared with a low-carb/low fiber (high fat) diet. There are indications that only water-soluble fiber is active on plasma glucose and lipoprotein metabolism in humans. The use of legumes, vegetables and fruits should be encouraged. This study advocates increasing these fiber rich foods and decreasing saturated fats.

The glycemic index is used to measure the effect of foods on blood sugar and the glycemic load is the glycemic index of a food plus the available carbohydrate per serving. These food rating systems are not effective in selecting food choices because of the great difference among foods, the way they are prepared, other foods consumed, and the great variability in individual metabolic systems. We believe that the sugar response does not tell the whole story and quality carbohydrates are the better way to healthier choices.

The Do's and Don'ts of Diabetes

Do's:

At least 7 servings of fresh pure vegetables and fruits
Freshly ground wheat and other grain products
Fresh ground raw almonds, walnuts, peanuts and sesame seeds

Organic whole milk yogurt and butter
Fresh high quality oils
Organic range fed eggs
Raw milk cheeses
Fresh ground flax seeds
Small amounts of raw honey and molasses
Fermented soy products
Spring water with lime or lemon
Herbal teas
Fresh vegetable and fruit juices
Fresh vegetable broths
Lightly cooked fresh vegetables
Brown rice, not white
Raw vegetables and fruits

Don'ts:
No white sugar or "fortified" bleached white flour (white trash)
No high fructose corn syrup (HFCS)
No chemical preservatives, no dyes
No alcohol or caffeine
No synthetic hydrogenated fats, no low fat anything
No processed foods
No low grade or old rancid oils
No beef or pork
Limit animal flesh
No chemical sugar substitutes (researchers believe aspartame to be extremely dangerous)
No chlorinated tap water
No pasteurized milk

Diabetes - Emotional

The emotional aspect of diabetes is often overlooked by standard medical treatment. Nevertheless, it is very important to consider this modality. There seems to be a sympathetic nervous problem resulting in either in or from unresolved emotional issues. Fear, worry, anger, shame, and grief affect our physical health on all levels. There are many avenues in research to approaching these issues which we use in our clinical practice, as well as diet and supplements.

Most diabetics are sympathetic dominant because the sympathetic nervous system is imbalanced in comparison to the parasympathetic. The negative emotions of anger and fear lead to this imbalance and should be considered as part of a total health care program.

Diabetes and Exercise

It is imperative that an exercise program be started in conjunction with nutrition, whole food supplements, and other lifestyle changes. It is wise to start very slowly with exercise, if you do nothing currently. Brisk walking on a regular basis is a good start. Gradually increase the distance and try to walk 5 days a week. Exercise helps speed up the metabolism and helps detox the body as well as boost many enzyme process. it is also a good way to improve emotional outlook. Dieting without exercise is like swimming upstream in molasses. After a walking program, initiate other forms of exercise.

Possible choices:

Walking
Cycling
Yoga
Tai Chi
Medicine Ball
Callanetics
Swimming
Sports
Dancing
Weightlifting
Pilates
Others

CONCLUSION

Diabetes mellitus is a broad endocrine dysfunction and not every diabetic will be exactly the same. Therefore, individual differences need to be addressed in all aspects of the program. A generalized diet can be used as a foundation where certain foods can be added or deleted. Early as well as current research indicates that diabetes is a disease of nutrient starvation leading to broad endocrine dysfunction as well as other issues and complications. We have seen many diabetics improve significantly by using a broad systems approach.

REFERENCES

1. Anathan, R., & Latha, M., et al. (2003). Effect of gymnema montanum on blood glucose, plasma insulin, and carbohydrate metabolic enzymes in alloxan-induced diabetic rats. *Journal of Medicinal Food, 6*(1), 43-49.
2. Ashero, A., & Wilett, W. (1997). Health effects of trans fatty acids. *American Journal of Clinical Nutrition, 66*(45), 10065-10100.
3. Bosakowski, T. E., & Leven, A. A. (1987). Comparative acute toxicity of chlorocitrate and fluorcitrate in dogs. *Toxicology and Applied Pharmacology, 89*(1), 97-104.
4. Bray, G. A. (2004). The epidemic of obesity and changes in food intake: The floride hypothesis. *Psysiology Behavior, 82*(1), 115-121.
5. Elvehjim, C. A. (1940). Vitamin B complex in normal nutrition. *Journal of the American Dietetic Association, 16,* 646-654.
6. Gross, L. S., Li, L., Ford, E. S., & Liu, S. (2004). Increased consumption of refined carbohydrates and the epidemic of type 2 diabetes in the United States: An ecological assessment. *American Journal of Clinical Nutrition, 79*(5), 774-779.
7. Gyr, K., Beglenger, C., & Stalder, G. A. (1985). Interaction of the endo- and exocrine-pancreas. *Schweizeresche Medizenesche Wochenschrift, 115*(38), 1299-1306.

8. Harkins, D. (2000, November 26). Researcher believes aspartame promotes hypersensitivity syndrome. *Idaho Observer.*
9. Harrower, H. R. (1932). *Practical endocrinology* (Rev. ed.). Glendale, CA: Pioneer.
10. Jennings, P. (2003, December 8). *Primetime Live* [Television series]. ABC News.
11. Joslin, E. (1935). *Treatment of diabetes mellitus*. Philadelpia: Lea and Febiges.
12. Lazarow, A., Liambres, L., & Tausch, A. J. (1950). Protection against diabetes with nicotinamide. *Journal of Laboratory and Clinical Medicine, 36*, 249-258.
13. Lee, R. (1998). *Lectures of Dr. Royal Lee*. Fort Collins: Selene River Press.
14. Lee, R. (2006). *Vitamin news* (Rev. ed.). San Diego, CA: International Foundation for Nutrition and Health.
15. Mahay, S., & Adeghate, E., et al. (2004). Streptozotocin--induced type 1 diabetes mellitus alters the secretory function and acyl contents in the isolated rat parotid salivary gland. *Molecular and Cellular Biochemistry, 261*(1-2), 175-181.
16. Marcus, A. O. (2000). Safety of drugs commonly used to treat hypertension, dyslipidema, and type 2 diabetes (the metabolic syndrome): Part 1. *Diabetes Technology Therapeutics, 2*(1), 101-110.
17. Morgan, A. F. (1941). Effects of imbalance in filtrate fraction of vitamin B. *Science News, 93*, 261-262.
18. Otto-Buczkowska, E., & Kobot, F., et al. (1991). Level of calcitonin in blood serum of children with insulin dependent diabetes. *Edobrynologia Polska, 42*(3), 447-53.
19. Porchezhian, E., & Dobrigal, R. M. (2003). An overview on the advances of Gymnema sylvestre: Chemistry pharmacology and patients. *Dre Pharmazie, 58*(1), 5-12.
20. Quigley, D. (1943). *The national malnutrition.* Wisconsin: The Lee Foundation.
21. Quigley, D. (1933). *Notes on vitamins and diets.* Wisconsin: The Lee Foundation.
22. Riccardi, G. & Rivellese, A. A. (1991). Effects of dietary fiber and carbohydrates on glucose and lipoprotein metabolism in diabetic patients. *Diabetic Care, 14*(12), 1115-1125.
23. Shanmerg, E. R., et al. (1990). Use of gymnema sylvestre leaf in the control of blood glucose in insulin-dependent diabetes mellitus. *Journal Ethnopharmacology,* (3), 281-294.
24. Swanson, J. E. (1992). Metabolic effects of dietary fructose in healthy subjects. *American Journal of Clinical Nutrition, 55*(4), 851-856.
25. Wexler, B. C., & McMurtry, J. P. (1983). Resistance to obese and non-obese spontaneously hypertensive rats to alloxon-induced diabetes. *Life Sciences, 33*(11), 1097-1103.
26. Wiley, H. (1930). The history of a crime against the pure food law. Published by himself (first director of the FDA).

PROTECTIVE ROLE OF SILICON IN LIVING SYSTEMS

[1,2,3]Biel K.Y., [1]Matichenkov V.V., and [1,2]Fomina I.R.

[1]Institute of Basic Biological Problems, Russian Academy of Sciences, Pushchino, Moscow Region 142290, Russia , [2]Biosphere Systems International, Tucson, Arizona 85755, USA , [3]Research Center for Food and Development, Hermosillo, Sonora 83000, Mexico

Corresponding author: Biel K.Y, E-mail: karlbiel@hotmail.com

Keywords: silicon, function of silicon

ABSTRACT

A brief review about the role of silicon in the soil-plant-animal system is presented, and the participation of silicon in stress tolerance promotion of organisms is discussed. It is hypothesized that the human body, as well as animals, plants and microorganisms, have a unique protective mechanism, which partially involves the mobile silicon compounds. According to our hypothesis, the function of silicon constituent can provide the additional synthesis of stress-protective molecules under genetical control but without "physical" participation of the genetic apparatus. This assumption is based on indirect experiments, and on a bases of subordination of two constituencies: a) response of genetic apparatus on stress, insuring synthesis of stress-protective compounds, such as antioxidant enzymes, stress-proteins, glutathione, phenols, and others antioxidants, etc., and b) on additional non-enzymatic formation of the same protective compounds on the matrixes of polysilicic acids. The active forms of silicon are considered as a matrix-depot for formation of compounds, which give assistance in any organism to maintain positive homeostasis under stress conditions.

INTRODUCTION

Medicine. In the early 20th century, amorphous silicic dioxide, silica, was the most commonly used pharmaceutical remedy in the world. It is well known that a silicon imbalance in the human body is accompanied by general metabolic disease and induces anemia, osteomalacia, shedding of hair, joint illness, and other diseases (Carlisle *et al.*, 1997; Voronkov *et al.*, 1978). Silicon compounds can prevent atherosclerosis development; and problems such as headaches, cardiovascular and skin diseases, sicknesses of joints and osseous tissues have been treated successfully by silicon medications (Voronkov *et al.*, 1978; Birchall and Chapell, 1988; Kaufmann, 1995*a*; 1995*b*; Jugdaohsingh *et al.*, 2002; Perez-Grandos and Vaquero, 2002; Sripanyakorn *et al.*, 2004; McNaughton *et al.*, 2005).

It was shown that the concentration of silicon in blood increased several times around the wound (Voronkov *et al.*, 1978; Mayras *et al.*, 1980), and the bandage sodden by silicic acids hastened wound repair (Kuzin *et al.*, 2000; Brown, 2002). It is also known that silicon concentration in the blood of pregnant women and in infants is increased (Voronkov *et al.*, 1978; Kaufmann 1995*a*; 1995*b*; Van Dyck *et al.*, 2000).

As early as 1878, great Louie Pasteur proposed that "*therapeutic action of silica has a great future*" (see Voronkov *et al.*, 1978). And Pasteur was right: the traditional medicineis again returned to the application of silicon preparations, which comes in handy, and is not only absolutely safe for people, but also are very effective for treatment of a wide range of diseases.

Agriculture. Protection of agricultural plants against diseases and insect attacks by silicon compounds has been used for many centuries, or possibly, several thousands of years ago because even agriculturists of the Roman Empire and Ancient China were familiar with unique capacities of silicon remedies and successfully applied them. Today, the list of scientific publications related to the application of active forms of silicon for protection of plants against various stressors (nematodes, funguses, negative climatic conditions, etc.) amounts to over 4000 references (see ref. in Voronkov *et al.*, 1978; Savant *et al.*, 1997; Matichenkov *et al.*, 2001*a*; 2001*b*; Ma and Takahashi, 2002; Matichenkov and Bocharnikova, 2004*a*; 2004*b*; and others).

Under stress conditions, the active forms of silicon have positive influences on the health of animals, fishes and birds (Carlisle, 1997). For example, small dietary supplement by Diatomaceous Earth, zeolite, silicon water, or other forms of active silicon harshly provides disease tolerance of the domestic animals, and decreases their death-rate (Carlisle, 1972; 1984; D'yakov *et al.*, 1990; Carlisle *et al.*, 1997). At the same time, the presence of active silicon supplements in food does not increase the level of silicon concentration in the blood of robust animals (Iler, 1979; Mayras *et al.*, 1980).

Previously, it was observed that wild animals aspire not only to take, as food diet, various clays, Diatomaceous Earth, and zeolite, but also to take up their residences close to wedging out high-silica soft minerals. The animals thus used these minerals as nutrition.

In the 1970's, the scientists of the Soviet Union and America proved the importance of silicon for improving the natural developing of the domestic and agricultural animals and birds; at least 6% of their fodder should be amorphous silicon dioxide (Carlisle, 1972; Voronkov *et al.*, 1978; D'yakov *et al.*, 1990; Carlisle *et al.*, 1997). It was also demonstrated that silicon promotes disease- and stress-tolerances of the domestic animals (Carlisle, 1984; Bgatov *et al.*, 1987; Anokhin *et al.*, 1997; Vladimirov *et al.*, 1998), disease-tolerance of fishes (Exley *et al.*, 1997; Camilleri *et al.*, 2003), and increases the rate of cancroids growth (Pellenard, 1969).

Microorganisms and plants. Silicon-rich compounds have a positive effect on growth and development of microorganisms (Al Wajeeh, 1999; Wainwright, 2005). In particular, silicon is one of the basic nutrient elements for phytoplankton, diatoms and cyanobacteria (Siegel and Siegel, 1973; Cha *et al.*, 1999; Brummer, 2003; Martin-Jezequel and Lopez, 2003; Velikova *et al.*, 2005; Tuner *et al.*, 2006). At the same time, silicon is a very important nutrient for organisms, which do not have as much silicon as diatoms and cyanobacters do. For example, the cultures of microorganisms are developing in the glass Petri dishes better than in plastic ones (pers. obs. and cons. with microbiologists).

The literature data testifies that silicon has a positive effect on the growth of a number of microorganism species (Mohanty *et al.*, 1990; Yoshon, 1990; Wainwright *et al.*, 1997; Soomro, 2000). We also observed increasing populations of various soil microorganisms when active silicon was applied to the grounds or to the soils, contaminated by oil or oil derivatives (Bocharnikova *et al.*, 1999).

The positive influence of silicon on plant growth and development has been known since Justius von Leibigh published his work (Leibigh, 1840) about mineral nutrition of plants. Today, the positive effects of silicon nutrition are estimated on various soils for the following cultivated plants: rice (*Oriza sativa* L.), sugar-cane (*Saccharum officinarum* L.), barley (*Hordeum vulgare* L.), wheat (*Triticum aestivum* L.), oats (*Avena sativa* L.), rye (*Secale cereale* L.), sorghum

(*Sorghum vulgare* L.), corn (*Zea mays* L.), sunflower (*Helianthus annuus* L.), beans (*Vicia faba* L.), soy (*Glycine max* L.), trefoil (*Trifolium pretense* L.), lucerne (*Medicago sativa* L.), millet (*Panicum miliaceum* L.), tomato (*Lycopersicum esculentum* L.), cucumber (*Cucumis sativus* L.), marrow (*Cucurbito pero* L.), lettuce (*Lactuca sativa* L.), tobacco-plant (*Nicotiana tabacum* L.), sugar-beet (*Beta vulgaris* L.), lemon tree (*Citrus x limon* L., Burm. f.), tangerine-tree (*Citrus reticulate* L.), vine (*Vitis vinifera* L.), apple-tree (*Malus silvestris* L.), melon (*Cucumis melo* L.), and others (D'yakov *et al.*, 1990; Savant *et al.*, 1997; Epstein, 1999; Matichenkov *et al.*, 2001; Ma and Takahashi, 2002; Matichenkov and Bocharnikova 2004*b*).

Summarizing the data about the influences of silicon fertilizers on plant productivity and soil fertility allows us to conclude:

1. Silicon enrichments provide the protective functions of plants on mechanical, physiological, and biochemical levels (Aleshin *et al.*, 1987; Savant *et al.*, 1997; Ma and Takahashi, 2002; Belanger, 2005).
2. Agricultural activity results in deficiency of active silicon forms in plants and drives to the degradation soil processes (Matichenkov and Bocharnikova, 1994; Bocharnikova *et al.*, 1995; Matichenkov *et al.*, 1999*a*).
3. Silicon fertilizers and silicon-rich soil amendments increase and retain phosphorus in accessible forms and results in facilitatation absorption of phosphorus by plants (Matichenkov, 1990; Matichenkov and Ammosova, 1994; Matichenkov, Bocharnikova, 2001).
4. Silicon nutrition could be applied as a lime material in acid soils (D'yakov *et al.*, 1990; Savant *et al.,* 1997; Ma and Takahashi, 2002).
5. Silicon-rich soil amendments are able to optimize the physical properties of soils (Emadian and Newton, 1989; Matichenkov and Ammosova, 1994; Matichenkov and Bocharnikova, 2001).

Thus, the known facts show the positive influence of silicon on organisms throughout a certain general mechanism, which mostl likely connects with the self-protective function of living systems. Furthermore, it is logical to presume that the biochemical basis of this mechanism is uniform for all organisms.

FORMS OF SILICON

Silicon is the second most abundant element in the earth's crust, after oxygen, and its concentration is increased by a decrease of the lithosphere layer depth, and achieves the maximal value in the surface of soil (Reimers, 1990; Bashkin, 2005). In soil, the main part of silicon is presented as a silicon dioxide and various forms of aluminosilicates (Sokolova, 1985). Quartz, crystalline variety of silica, is characterized by stability to eolation (Orlov, 1985; Olier, 1990) and together with other macrocrystalline silicates [feldspar, plagioclase, pyroxenes] and secondary or clay silicon-containing minerals [kaolinite, vermiculite, smectite, and others] forms the skeleton of soil (Orlov, 1985).

Stability of silicon minerals to eolation is reflected in the classification of soil elements on the basis of their mobility, where the silicon is considered as an inert element (Polynov, 1952). However, in the same classification, silicon is also attributed to the group of mobile elements. In soil, solutions are permanently present in the mono- and polysilicic acids and organo-silicon compounds, which have high chemical and biological activity (Matichenkov, 1990; Matichenkov,

Calvert *et al.*, 2001; Matichenkov and Bocharnikova, 2004*a*; 2004*b*). Thus, the soil silicon consists of two main groups – inert elements and biogeochemically active compounds.

Figure 1 presents the classification of silicon compounds, in which special attention has been focused on biogeochemically active forms of silicon from the solid and liquid soil's phases (Matichenkov *et al.*, 1999*a*; 1999*b*). Traditionally (Sokolova, 1985), the silicon compounds in the solid phases of soil are classified on the bases of mineralogical structure, and are a priori considered as inert. But silicon compounds of the liquid phases are considered as active because they have numerous physico-chemical and biological properties. Undoubtedly, these active silicon forms are presented in the tissues of living organisms.

Monosilicic acid is a weak inorganic acid with a slight buffering capacity at pH ≈ 7.0. This acid can react with ions of aluminum, iron, calcium, manganese, magnesium and heavy metals (Bocharnikova *et al.*, 1995; 1999; Matichenkov and Bocharnikova, 2001). Monosilicic acid is also able to replace phosphate anions from the calcium, iron, and aluminum phosphates (Matichenkov, 1990; Matichenkov and Ammosova, 1994); this process results in facilitating phosphate uptake by plants and soil microorganisms (Yoshida, 1975; Trevors, 1997).

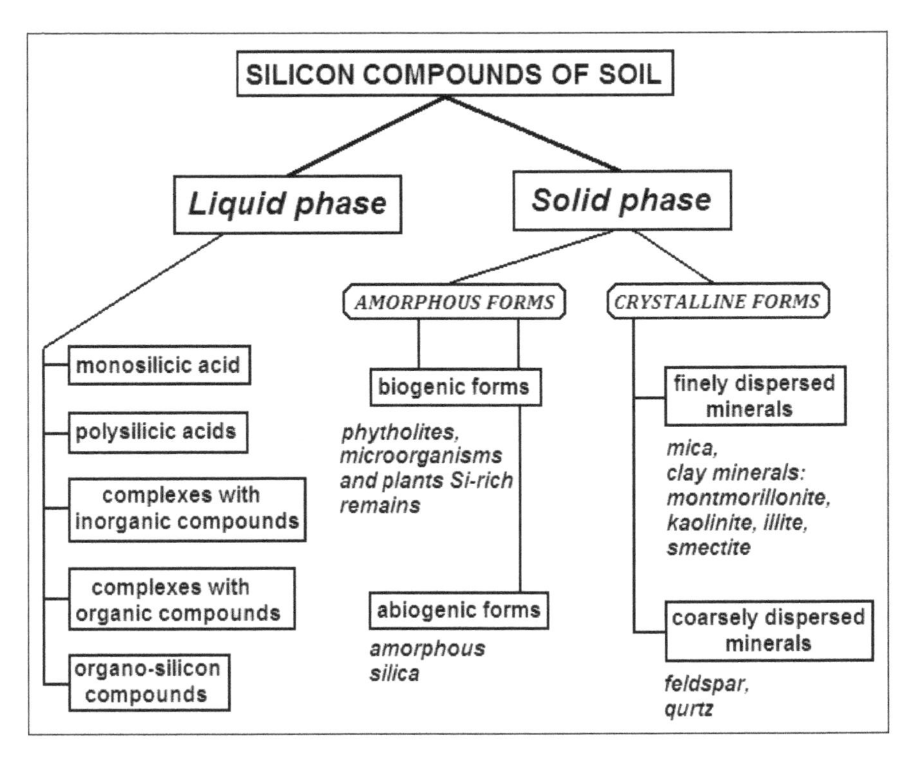

Figure 1: Forms of silicon compounds in soil (Matichenkov *et al.*, 1999*a*).

Polysilicic acids are compounds which consist of two or more silicon atoms linked by siloxane bonds in a line [-Si-O-Si-], with hydroxyl groups bonded with silicon atoms (Iler, 1979).

There are two groups of polysilicic acids in nature: 1) oligomers of silicic acid are contained from two till several hundred of silicon atoms per one molecule. They also were named as low-molecular weight polysilicic acids. It was believed that in diluted solutions (in most of natural waters, soil waters, lakes, rivers, and ocean waters), these compounds are unstable, and are found as intermediates between monosilicic acid and high-molecular weight polysilicic acids (Iler, 1979; Dietzel, 2000; 2002); 2) high-molecular weight polysilicic acids are containing till several thousands of silicon atoms per one molecule (Davies, 1964; Varshal *et al.*, 1980; Stumm and Morgan, 1996). These acids are chemically inert, but nonetheless can influence the physical properties of soils (Yatsynin, 1989; Matichenkov *et al.*, 1994).

Chemical inertness of polysilicic acids is specified by the ability of their molecules "to twist" and by this way to compensate the negative charges, which are formed during dissociation of hydroxyl groups (Iler, 1979). Polysilicic acids are also able to form silicon bridges between soil particles (Yatsynin, 1994). These bridges, due to permanent changing of soil humidity, undergo-dehydration with formation of silicon dioxide.

The processes of periodical drying-wetting of soils are formatting the siliceous films on a surface of practically all soil minerals. And exactly this film is the main source of monosilicic acid, when moisture comes upon the soil.

Formation of polysilicic acids occurs as the monosilicic acid's molecules condensation (Iler, 1979). Though this process is poorly investigated, however, it is known, that in open water-bases the process can go, if the concentration of monosilicic acid is above 100 mg Si/liter (Iler, 1979). But usually, in the soil solutions the concentration of monosilicic acid is not higher than 1–50 mg Si/liter (Matichenkov, 1990). Theoretically, in such diluted solutions polysilicic acids should be depolymerized. Nevertheless, that does not happen (Varshal *et al.*, 1980; Alvarez and Sparks, 1985; Matichenkov, 1990; Matichenkov and Shnaider, 1996; Matichenkov *et al.*, 1997).

We suppose that the formation of polysilicic acids and their stable existence are due to the location of anti-ions in the diffusion layer surrounding the mineral's surface. And only in this layer is it possible to permanently keep the high concentration of monosilicic acid, which is formed during the dissolvment of siliceous films following condensation to polysilicic acids (Frolov, 2004).

High molecular-weight polysilicic acids are also able to form the gels with stable structure, and the colloids containing nucleus with silica microcrystals inside it (Iler, 1979).

The ratio between mono- and polysilicic acids is usually stable in various soils or water-mineral systems (Matichenkov and Shnaider, 1996; Matichenkov *et al.*, 1997; Matichenkov and Bocharnikova, 2001). However, this ratio could be varied by coming in or out of soluble silicon compounds, as well as by changing the mineralogical composition or the conditions for-secondary minerals formation. Similar processes take place on the surface of soil microorganisms (Matichenkov, Calvert *et al.*, 2001).

The discussion about presence of organo-silicon compounds and soluble complexes with non-organic and organic ligands in soil solutions has been expounded in some publications as well (Fotiev, 1971; Matichenkov and Snyder, 1996).

SILICON CYCLE IN SOIL–PLANT SYSTEM

Biological cycle of silicon on our planet is the most intensive in terrestrial ecosystems where plants uptake silicon in the range from 20 to 7000 kg Si/he/year (Matichenkov and Bocharnikova, 1994). Silicon is the 4th most abundant element in plant biomass, after oxygen, carbon, and hydrogen (Kovda, 1956; Perelman, 1975; Bazilevich, 1993).

As noted above, the plants and soil microorganisms are able to uptake only monomers of silicic acid and its anions (Yoshida, 1975; Ma, 2003). In larger plants, this process takes place through the roots and leaves. However, the silicon is distributed within the plant irregularly, according to the needs of the organism (see below).

Absorbed molecules of monosilicic acid may be polymerized and by this way participate in the formation of organo-silicon compounds (Figure 2). In their turn, polysilicic acids in plants are able to be dehydrated with the formation of *phytholites* – amorphous silicon dioxide with complicate configuration. These phytholites are located within plant cells and in the intercellular space (Dobrovolsky *et al.*, 1988; Gol'eva, 2004).

It was demonstrated that the size, structure, and quantity of phytholites depend not only on the presence of silicon-forms available in soil, but also on the soil humidity, temperature, and the availability of other nutritional elements (Gol'eva, 2001; Hodson *et al.*, 2005). It should be noted that the formation and rate of growth of phytholites in plants are controlled by specific proteins (Harrison, 1996; Perry and Keeling-Tucker, 2000).

After the plants decay or their pulling-down, the biogenic silicon-forms (polysilicic acids, phytholites, organo-silicon compounds, etc.) are transported into the soil, where it is subjected to processes of dilution and/or decomposition. The final product of the silicon cycle process is the monosilicic acid, which once again involves the biological cycle and the influences of a number of physical-chemical and biological soil properties (Matichenkov and Bocharnikova, 1994; Bashkin, 2005).

It was distinguished that *eluvial* and *accumulative* types of silicon cycles, in which biologically active silicon forms are either accumulated in soil or are permanently moving out of the soil (Matichenkov and Bocharnikova, 1994). Both of these processes are initiated by the anthropogenic and natural factors (Matichenkov and Ammosova, 1994, Matichenkov *et al.*, 1994, Matichenkov and Bocharnikova, 1994).

The phenomenon of *silicatization* in nature should also be noted. Under this process, the replacement of carbon atoms takes place in the soil or ground (from the remains of plants and animals) by silicon atoms (Milnes and Twidale, 1983; Cid *et al.*, 1988). The opportunity of such replacement is conditioned by the high concentration of monosilicic acid in the soil solution.

Silicon and Flora

The intensity and type of silicon cycles in native *soil-plant* systems are connected not only with the presence of active silicon forms in the parent material (Olier, 1990; Bashkin, 2005), landscapes, and secondary minerals (Kovda, 1956), but also with the type of plant association (Bazilevich, 1993; Matichenkov and Bocharnikova, 1994). For example, the plant associations with the predomination of scanty silicon species such as pine forest and meadow non-crop grasses are formed on the active-silicon depleted soils (Lanning and Eleuterius, 1981; Bocharnikova and Matichenkov, 2006). On the other hand, the intensive silicon cycle embraces a lot of active silicon forms in

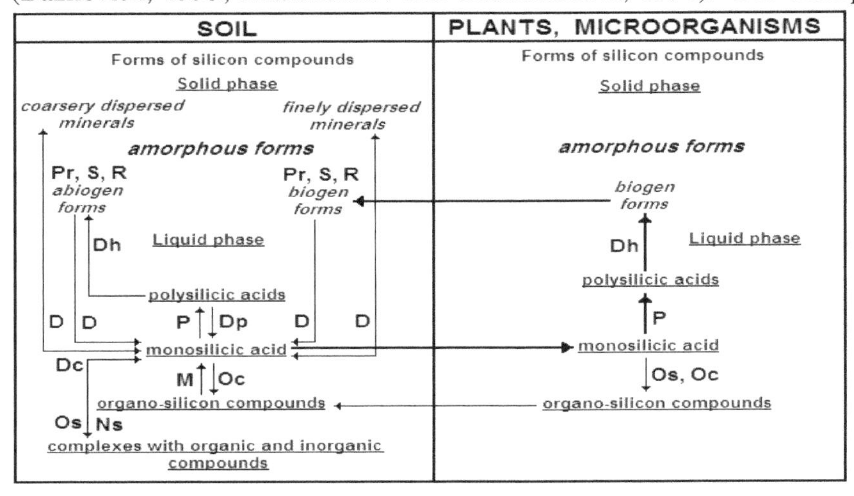

Figure 2. Silicon cycle in system soil–plant–microorganisms.

Note: D – dissolving, Pr – precipitation, P –polymerization, Dp – depolymerization, Dh – dehydratation, S – salt formation, R – replacement of inorganic anions, Oc – formation of organo-silicon compounds, Ns – formation complexes with inorganic compounds, Os – formation complexes with organic compounds, Dc – decomposition of complexes, M – mineralization of organo-silicon compounds.

formation of the crop associations or mixed wood (Bazilevich, 1993; Matichenkov and Bocharnikova, 1994). The accumulation and moving out of the soil silicon compounds determine the succession of plant communities (Bocharnikova and Matichenkov, 2006).

This determination causes increasing stress sensitivity of silicon-accumulative plants at silicon-deficient conditions. On the other hand, a sufficient amount of active silicon forms grants for *silicophilic* plants more chances in competitive activity with silicon-non-accumulative species.

The majority of agricultural lands of the world have a deficiency in active silicon forms (Matichenkov and Bocharnikova 1994; Matichenkov *et al.*, 1999*a*; 1999*b*; 2001; Matichenkov and Bocharnikova, 2004*a*; 2004*b*; Snyder *et al.*, 2006) due to the principle of agricultural activity – removing nutrient compounds from the soil. Usually, deficiency of macroelements such as P, N, K, and Ca is partially restored by mineral and organic fertilizers, but the lack of silicon is not compensated. It's estimated that the movement of silicon from the ground is ranged between 30–7000 kg Si/ha/year. As a result, the upper soil horizon has a deficiency of active silicon; and, consequently, the trophic chain: *agricultural plants–farm animals–people* feels a serious deficiency of silicon.

Localization of Silicon in Plants

Distribution of silicon between plant organs is not equal and may vary from 0.001% in the pulp of fruit to 10–15% in the epidermal tissues (Voronkov *et al.*, 1978). According to the total silicon content, the plants are subdivided into two big groups: *silicophiles* [silicon-accumulative plants] and *non-silicophiles* [silicon-non-accumulative plants]. *Silicophiles* are the species which has 1% or more of silicon per total dry mass, while *non-silicophiles* has less then 1% of silicon per dry mass (Epstein, 1999; Ma and Takahashi, 2002). Usually, the main amount of silicon is located in needles, husks, bark, and some others organs (Ma and Takahashi, 2002; Snyder *et al.*, 2006). The high content of silicon was also shown in root caps (Hodson, 1986; Ma *et al.*, 2001), where monosilicic acid is polymerized into polysilicic acids, which further are crystallized into microcrystals of amorphous silica. It is possible that microcrystals of silica play a role of "mechanical drill" in the root system. Indirectly, it was confirmed by the fact that under the deficiency of silicon nutrition, the formation of the secondary and tertiary plant's roots is sharply decreased (Matichenkov *et al.*, 1999*b*; Matichenkov *et al.*, 2001*b*; Matichenkov andBocharnikova, 2004*b*).

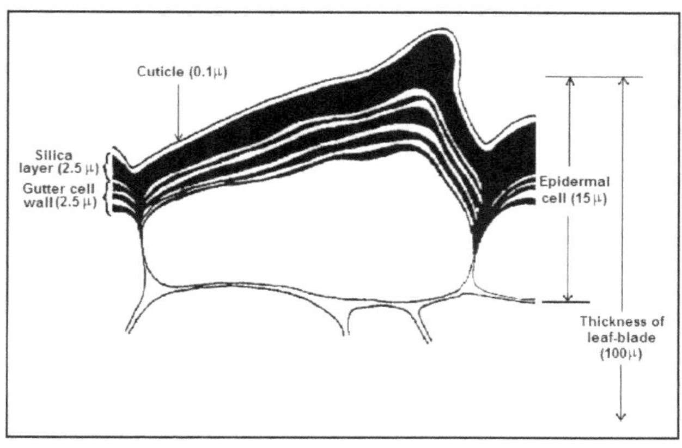

Figure 3. Scheme of epidermal cell of the leaf of *Oriza sativa* (Yoshida, 1975).

Absorbed molecules of monosilicic acid are accumulated in the epidermal tissues (Yoshida, 1975; Hodson and Sangster, 1989) and form the silicon-cellulose envelope where silicon is bonded with pectin and calcium (Waterkeyn *et al.*, 1982). As a result, the double cuticular layer (Figure 3) protecting and mechanically strengthening the plants is formed (Yoshida, 1975; Ma, 2003). This layer also prevents dehydration of the leaves (Emadian and Newton, 1989).

We speculated that the higher plants have a special mechanism for selective uptake of monosilicic acid from the soil solution. Indirectly, this idea was confirmed in various experiments. For example, Ma and Takahashi (2002) showed that *Oriza sativa* L. and *Lycopersicum esculentum* Mill. [grown hydroponically with presence of silicon compounds] may decrease concentration of monosilicic acid in nutrient solution from 50 to 5 mg Si/liter. In our experiment, *Hordeum vulgare* L. absorbed monosilicic acid from soil solution (2-20 mg Si/liter) and collected this compound in roots up to 500-520 mg Si/liter plant sap (Matichenkov *et al.*, 2005). Therefore, plants are really able to absorb monosilicic acid from the diluted solutions.

Because roots are containing not only monosilicic acids, but also polysilicic acids (Table 1), and, at the same time, it is known that plants are not able to absorb polysilicic acids from solutions (Epstein, 1999; Ma and Takahashi, 2002). Thus, it may be logically assumed that polymerization of the monosilicic acid molecules occurs inside of the root cells; and this process is started immediately after the uptake of monosilicic acid by the roots (Figure 4).

Forms of Silicon in Plants

Until recently, it was considered that about 90% of total silicon is present in plants as amorphous silicon dioxide – phytholites and epidermal films (Savant *et al.*, 1997; Epstein, 1999; Ma and Takahashi, 2002). This point of view was a subject of much controversy and was based on the "chemical rules" and chemical properties of silica (Iler, 1979). According to silicon chemistry, the occurrence of monosilicic and polysilicic acids of high concentrations is impossible in water or other solutions. In particular, it was known that at pH < 8 the amount of monosilicic acid in natural solutions is usually ranged between 0.1–100 mg Si/liter. Further increase of the monosilicic acid concentration initiates its polymerization or its precipitation in a form of amorphous silica (Iler, 1979).

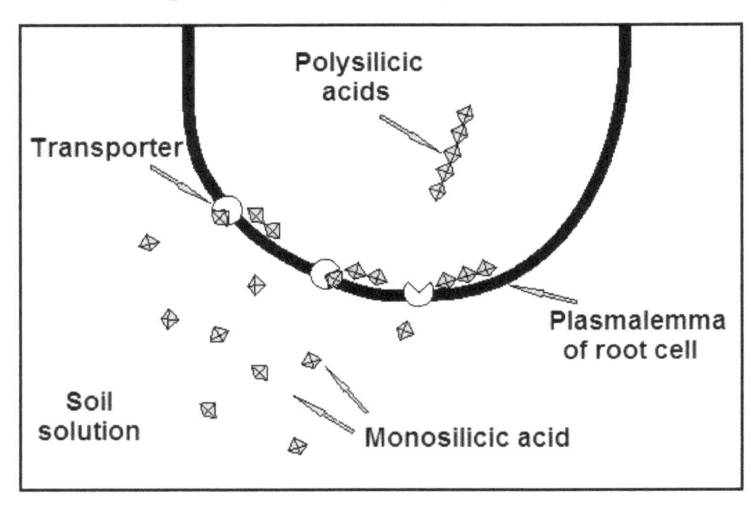

Figure 4. Hypothetical scheme of silicon uptake by plants: molecules of monosilicic acid penetrate into the cells of root filaments and are immediately polymerized into polysilicic acids.

However, it was supposed that such conditions are not characteristics for the intracellular and intercellular spaces of terrestrial plants (Medvedev, 2004), and therefore practically all polysilicic acids in plants should coagulate with the formation of microcrystalline lattice of silicon dioxide (Iler, 1979; Yatsinin, 1994). As a result, despite the absence of actual data about the real concentration of soluble silicon in plants, physiology has "created" the point of view that plants tissues have a very low concentration of mono- and polysilicic acids.

And relatively recently, Japanese scientists demonstrated in the sap of *Oriza sativa* the existents of mono- and polysilicic acids at concentrations of 100–600 and 2000–3000 mg Si/liter, respectively (Ma, 1990; Ma and Takahashi, 2002). Our measurements (Table 1) are in agreement with these data. Table 1 shows that 30-70% of the total silicon compounds from different parts of *H. vulgare* are presented as polysilicic acids; content of monosilicic acid also is high and can be around 1-5% of the total plant's silicon.

Table 1. Content of monosilicic and polysilicic acids in different parts of *Hordeum vulgare* L.

Part of the plant	Monosilicic acid		Polysilicic acids		Total silicon content, mg Si/kg dry mass
	mg Si/liter plant sap	% from total silicon content	mg Si/liter plant sap	% from total silicon content	
Leaf	100-400	1.0-5.0	2700-5000	30-55	9000-12000
Node of stem	130-200	1.6-2.5	3800-5000	40-62	8000-9000
Interstitial space	350-400	4.3-5.0	3500-5700	40-70	8000-9500
Root	500-520	3.0-4.0	5000-8000	30-53	15000-19000

Note: Plant material [fresh mass of 0.1-0.2 g] was ground with a mortar and pestle in distilled water at room temperature and centrifuged for 15 min at 6000 g [Centrifuge T-30, Janetzky, Germany] to precipitate colloids and solid particles. Concentration of monosilicic acid was immediately determined in the supernatant samples by the method developed by Mallen and Raily (see, Iler, 1979). This method allows eliminating the affect of phosphate anions on the monosilicic acid measuring. Our modification allows to measure monosilicic and polysilicic acids separately. For this remained supernatant was treated as described earlier (Matichenkov *et al.*, 1997) to depolymerize polysilicic acids, and concentration of monosilicic acid was determined again as noted above. The difference between the second and the first measurements shows amount of polysilicic acids in tested solution. Concentrations of silicon compounds are expressed per liter plant sap; sap volume per kg fresh mass of the leaves was measured especially. Total silicon content was analyzed in the dried plant material according to Elliot and Snyder (1991). Each measurement was made in 8-12 independent replications.

Thus, we can establish a fact that up to 75% of silicon in plant occurs in the form of water-soluble mono- and polysilicic acids. Undoubtedly, the high concentration of these compounds is stipulated for the important role of silicon in vital plant functions.

Silicon and Water Storage in Plants

An important property of silicon is the ability to accumulate and store the water within the organism throughout, forming polysilicic acids and their gels. The model experiment showed that one atom of silicon is able to hold up to 146 molecules of water (Table 2). We estimated that about 20–30% of the total silicon in organism may be involved in the process of maintaining internal water reserve. For example, under optimal silicon nutrition, the medium-latitude culture *H. vulgare*, usually containing around 1.2–1.4% of silicon per dry mass, is able to store only with silicon assistance from 6-37 g water per 100 g fresh mass of the individual plant.

The capability of silicon to hold water is very important for tropical and arid plants, which are usually *silicophiles* (Ahmad *et al.*, 1992; pers. obs., 1986-2007). From this stand point, it makes sense to assume, that *silicophiles* have a high tolerance to drought, in part because during the drought season, they can use "extra" water, which was collected in plants with silicon "assistance" during humid season. At least, some of our data is in agreement with this.

Table 2. Content of water molecules and silicon atoms in different gels, formed from silicic acid.

Gel number	Silicon, % from total gel mass	H_2O, % from total gel mass	H_2O/Silicon, molecules/atom
1	1.28	97.9	146
2	1.60	97.7	119
3	2.40	97.1	82
4	3.20	96.6	63
5	4.80	95.3	40

Note: Standard silicic acid [16% silicon, product of TerraTech LLC] at 24 °C and atmospheric pressure of 765 mm mercury was mixed with water and mineral acids [HCI, H_2SO_4, and others] in different proportions until steady-state gels were formed. Final amounts of water molecules and silicon atoms in gels were calculated on the basis of initial amounts of the compounds, which participated in the gel(s) formation, n=10.

In particular, the special experiment with woody tropical species *Hura crepitans*, *Hibiscus elatus*, *Ceiba pentandra*, and *Clitoria racemosa*, preserved during the long-term drought and maintained in the tropical mesocosms of Biosphere 2 Center [Columbia University, Arizona, USA], has argued for such believing. The analyses showed that water potential (Rascher *et al.*, 2004) and total water content in leaves of these trees, even in the middle of the day(s) (Figure 5) were not decreased of the essence, even when the rainy season was changed by the long-term season of the air and soil drought.

SILICON AND PLANTS RESISTANCE TO EXTREME ENVIRONMENTS

The number of publications, related to how active silicon compounds influence on plant growth and development, suggest the existence of the total mechanism, by which silicon improves plant stress-tolerance (Cherif *et al.*, 1994; Savant, *et al.*, 1997; Ma and Takahashi, 2002; Belanger, 2005). This mechanism seems to be universal, that even, for example, optimization of silicon nutrition, it will induce plant resistance against biogenic [infections, mycosis, insect injurious, etc.] and abiogenous [low or high temperature, salt stress, heavy metals, drought, etc.] factors (Bocharnikova *et al.*, 1999; Savant *et al.*, 1997; Matichenkov and Bocharnikova, 2004*a*; 2004*b*; Belanger, 2005; Snyder *et al.*, 2006).

There are several suppositions about the protective role of silicon in plants. In particular, the thickening of the epidermis, increasing chemical stability of DNA, RNA, and molecules of chlorophyll(s), functional activation of organelles, optimization of transport and redistribution of compounds within the plant, and others (Voronkov *et al.*, 1978; Aleshin, 1982; Aleshin *et al.*,

1987; Matichenkov, 1990; Savant *et al.*, 1997; Ma and Takahashi, 2002; Matichenkov and Kosobryukhov, 2004) can support the involvement of silicon in numerous adaptive responses of the plants.

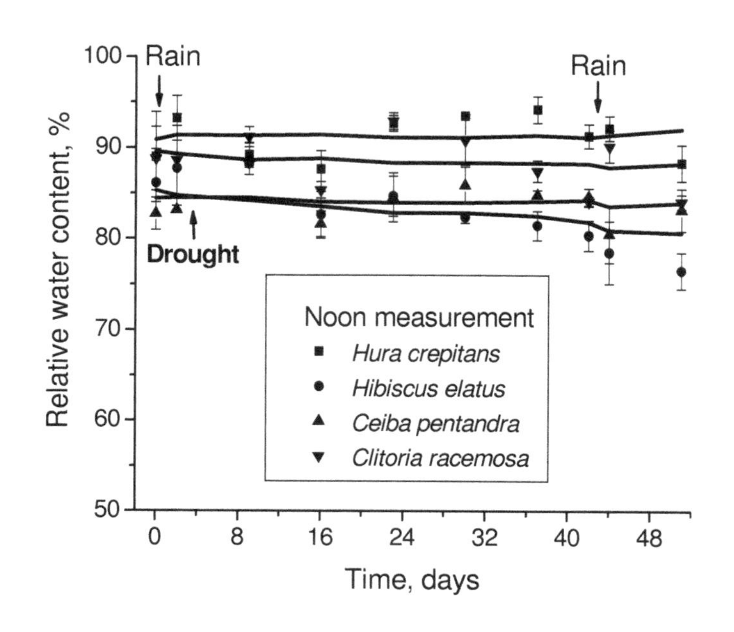

Figure 5. Relative water content in the middle of the days in the leaves of tropical woody cultures *Hura crepitans*, *Hibiscus elatus*, *Ceiba pentandra*, and *Clitoria racemosa* during the long-term drought, maintained in tropical mesocosms of Biosphere 2 Center [Columbia University, Arizona, USA]. Data points represent the mean ± SE, *n*=3-4.

But we speculate that all of these "adaptations" are the consequence of functioning of uniform protective mechanism, part of which is connected with the active silicon forms. Our opinion is indirectly supported by the fact that stress-tolerance of plants is increasing with rising silicon content within the tissues (Matichenkov *et al.*, 1999*b*; 2000). Moreover, as we found in the experiment with *Distichlis spicata* growing under high salinity, the highest amount of silicon was collected in the organs, which were more "suffering" over salty stress (Matichenkov, Bocharnikova *et al.*, 2007). We also hypothesize that living organisms have a special ability to coordinate absorption and rapid redistribution of silicon into the parts of the body, exposed to stress.

Silicon as Matrix for Organic Compounds Synthesis

In the mid 20th century, Russian scientists (Strelko *et al.*, 1963; Vysotsky *et al.*, 1967) had discovered and determined the possibility of catalytic synthesis of the organic compounds on silicon gel matrix. According to authors' data, the challenge of synthesized molecules from silicon gel matrix and conservation of matrix allowed to realize non-limiting low-temperature synthesis of initial molecules from more primitive organic "row material". On the basis of that V.V. Strelko and co-workers (1963) even advanced an idea that "*...the reactions of amino acids polycondensation, concentrated by adsorption on the silica surface, could lead to formation of pre-proteins and might play an important role in origin of life on the Earth*".

In modern organic chemistry, the polycondensation technology is developed in greater detail. According to general opinion, if during formation of silicic-gel certain organic molecule are present in the solution, the the silicic-gel surface has a possibility to hold the "print" matrix from this initial "molecule-former". Afterwards, if the "molecule-former" would be removed, such "print" on the silicic-gel surface becomes to be a "matrix or the center of replication" for synthesis of initial molecules. It is important to note that the "silicic acid-gel-matrixes" are playing the role of catalyzer in a number of reactions which are necessary for organic synthesis.

Some of them are: cycle formation, cyclization, rearrangements, reduction and/or oxidation, condensation, hydratation and/or dehydratation, formylation, protection of functional groups, and others (Banerjee *et al.*, 2001). At present time, many chemical and pharmaceutical companies are using this approach for synthesis of complex molecules.

HYPOTHESIS ON SILICON PARTICIPATION IN PROTECTION OF LIVING ORGANISMS UNDER STRESS CONDITIONS

Premises of Hypothesis

The possibility of biochemical compounds synthesis on the gels of polysilicic acids within living cell was not considered until now. The general reason for that, as we mentioned above, was conception, that in living organisms the polysilicic acids are existing in very low concentrations, or even do not exist totally.

However, the new data allow being in doubt about such conclusion. In particularly , the results, presented in Table 3, showed the high concentration of mono- and polysilicic acids in the leaves and stems of *D. spicata*, even when the plants were grown in a sand which had a very low concentration of water-soluble forms of silicon; it was ranged around 1.0 – 2.5 mg Si/kg sand.

In this experiment, the content of silicic acids [sum of mono- and polysilicic acids] of leaves and stems was 514 mg Si/kg fresh mass and 893 mg Si/kg fresh mass, respectively; inside the cells [in symplast] the content of silicic acids was higher than outside the cells [in apoplast] (Table 3).

During next experiment we examined the same parameters, but after 12 days of silicon fertilization throughout *D. spicata'* root system. In this case, the content of silicic acids of leaves and stems increased more than 40-67% (Table 3).

The Figures 6 and 7 illustrate the dynamics of mono- and polysilicic acids release from apoplast and symplast of *D. spicata* leaves and stems. It is clear that silicon fertilization induces accumulation of silicic acids in stem apoplast, and in symplast of the leaves and stems. Peculiarities of silicon accumulation and redistribution between plant organs we will describe in detail in a special publication (Matichenkov *et al.*, 2007). But here it is important to note, that shoots of plants are able to accumulate the water-soluble silicon forms in amount, sufficiently enough for silicon creative job in organism.

The additional pre-suppositions for organic compounds synthesis on gels of polysilicic acids in living cell are following:

- Active silicon forms moving into the living systems increase their stress-tolerance (Cherif *et al.*, 1994; Savant, 1997; Epstein, 1999; Ma and Takahashi, 2002; Matichenkov and Bocharnikova, 2004*a*; 2004*b*; Belanger, 2005);
- It has been found a specific protein (transporter), which regulates absorption of silicon from environment, and its redistribution within the plants (Ma *et al.*, 2006);
- Plants absorb silicon from the environment only in the form of monosilicic acid (Yoshida, 1975; Ma and Takahashi, 2002);
- Depending on plant species and type of tissue, the total silicon content in plants varies, on average, from 0.2 to 1.5% of dry mass (Epstein, 1999; Ma, Takahashi, 2002);
- In plants the silicon exists in the forms of mono- and polysilicic acids, amorphous silica [phytholites], and organo-silicon compounds. Percentage of mono- and polysilicic acids may rich 1-2% and 10-70% from the total silicon content, respectively (see Table 1);

- Water-soluble silicon compounds in plants purposefully redistributed directly into the parts and/or organs, which are exposed to stress (Matichenkov *et al.*, 2000; Matichenkov and Bocharnikova 2004*a*; 2004*b*; Matichenkov, Bocharnikova *et al.*, 2007);
- Polysilicic acids gels are able to form matrixes from organic "molecule-formers". These "matrixes" could carry out low-temperature synthesis of new molecules which are equal to "molecule-formers". The precursors for this synthesis may be chemically more simple organic components (Strelko *et al.*, 1963; Vysotsky *et al.*, 1967; Banerjee *et al.*, 2001).

Table 3. Content of mono- and polysilicic acids in leaves and stems of *Distichlis spicata.*

Part of the plant	Monosilicic acid		Polysilicic acids	
	Apoplast	Symplast	Apoplast	Symplast
	Silicon, mg Si /kg fresh mass			
No silicon fertilization				
Leaf	34	165	326	368
Stem	19	117	58	320
After silicon fertilizers				
Leaf	42	223	227	487
Stem	20	197	104	395
*LSD_{05}	5	15	15	20

Note: *Distichlis spicata* plants were grown in plastic pots (20×50×20 cm) with sand at temperatures of 22-24 °C /17-19 °C [day/night], and natural illumination. Soil salinity was simulated by irrigation of water with 10 g/liter of Na^+ as NaCI. Plants were watered by regular water containing no more than 0.10-0.15 mg Si/liter. Fertilizers (MiracleGro: www.miracle-gro.com) were added into ground 1-2 times per month. For the experiments, leaves and stems were cut into 2.0-2.5 cm length segments and placed to the flasks with distilled water; the plant material/water ratio was ranged from 1/100 to 1/200 (w/w). The flasks were shaken up by mixer for 25 h at room temperature. Content of mono- and polysilicic acids in apoplasts of leaves an stems was determined after sequestrating sampling of water aliquots (1-5 ml) during 0, 1, 5, 15, 30, 60, 180, 360, and 1440 (min) shaking. Then, plant material was grinded with mortar and pestle till homogenous substances and mixed with initial solutions; the flasks were shaken additionally for 60 min [total time for extraction of water-soluble fractions of silicon was 1500 min]. Content of mono- and polysilicic acids was measured again. The difference of the values between after grinding of solid material and before grinding of solid material was considered as the concentration of these compounds in the symplast. * LSD_{05}, the least standard deviation, *n*=4.

Hypothesis

During endeavor (ontogenesis) of total realization of genetic information, the organisms (Biel *et al.*, 1990; Biel and Fomina, 2005) usually feel deficiency of nutrient elements or energetic resources, or are affected by negative climatic factors, diseases, and other stressful pressures. At such conditions the living systems are spending substantial part of energy for maintenance, adaptation and defense against internal and external problems. As a result of that, the plants [and others organisms as well] are usually slowed in development, and do not realize interior potential. In principle, the situation may be improved by supplying organism with essential nutrition and by creating optimal environment and internal conditions. However, in real life this way is not practicable, especially in a large-scale agriculture.

Improvement could be achieved, if the responsibility for the part of repairing, metabolic and protective functions would be shifted from genetic apparatus to the other system(s).

a

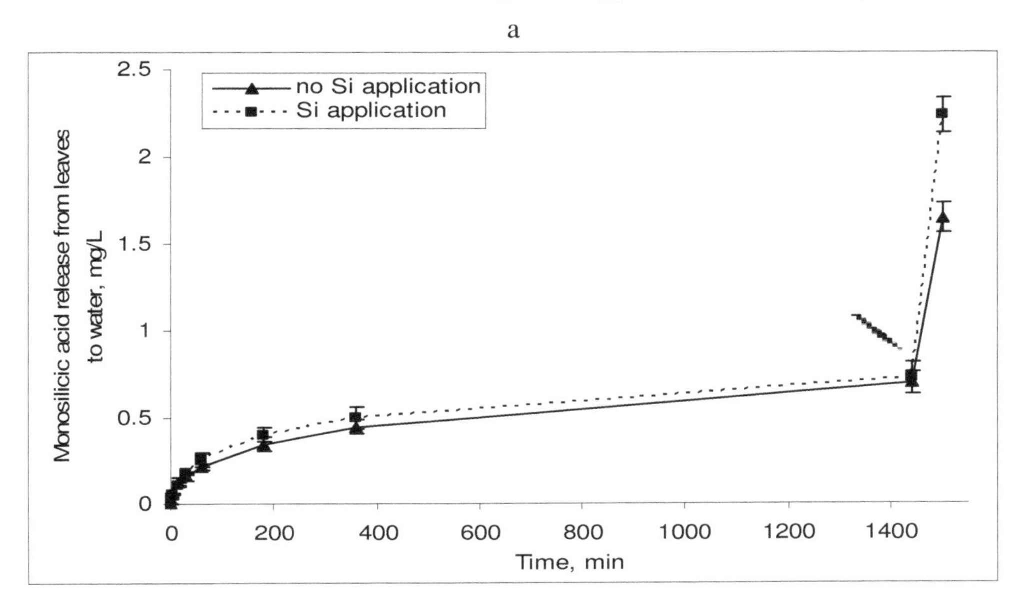

b

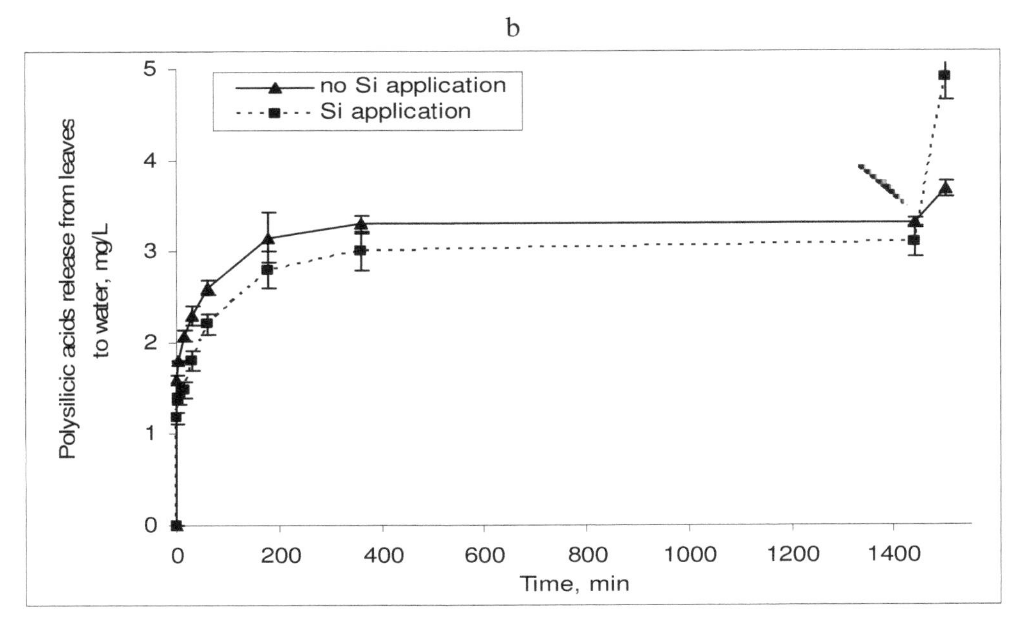

Figure 6. Dynamics of washing out of mono- (a) and polysilicic (b) acids from the leaves of *Distichlis spicata* grown without silicon fertilizers (▲) and with silicon fertilization (■), added to substrate 12 d before measurements.

Note: Arrow shows the moment of grinding of the leaves and further returning leaf homogenate to initial water-soluble fraction. The total silicon content has been measured after 1 h in the same volume. Description of details is presented in the note of Table 3. Data points represent the mean ± SE, n=4.

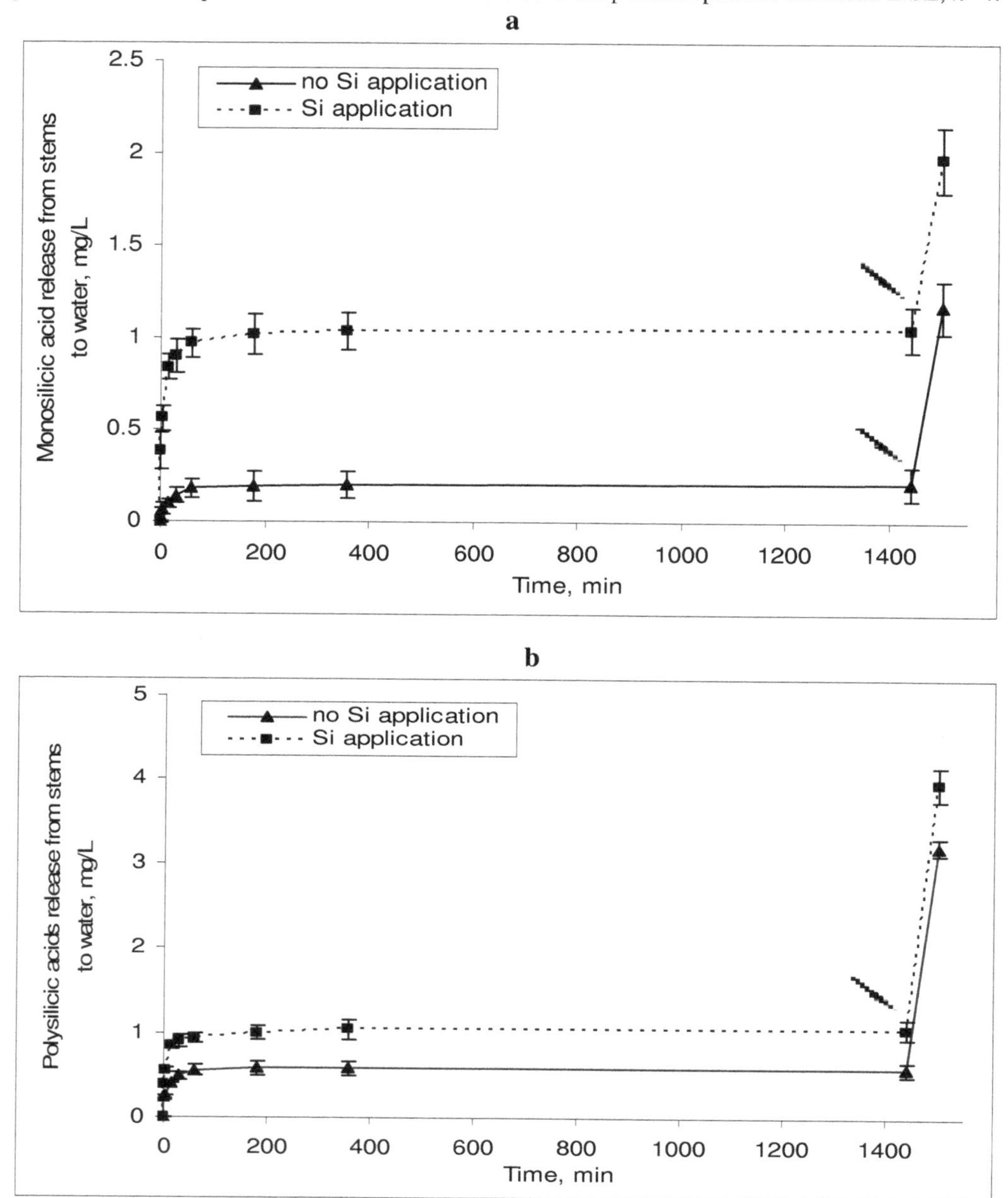

Figure 7. Dynamics of washing out of mono- (a) and polysilicic (b) acids from the stems of *Distichlis spicata* grown without silicon fertilizers (▲) and with silicon fertilization (■), added to substrate 12 d before measurements.

Note: Arrow shows the moment of grinding of the stems and further returning stem homogenate to initial water-soluble fraction. The total silicon content has been measured after 1 h in the same volume. Description of details is presented in the note of Table 3. Data points represent the mean ± SE, n=4.

Using the plants like example, we'll consider one of such system, although it is most likely that this mechanism is universal and functioning in all organisms, which are living now on our Planet.

Silicon uptake – Molecules of monosilicic acid, located in soil, penetrate across the root's plasmalemma [cell "sluice"] inside of the cells and are immediately condensed into active chains of polysilicic acids (Figure 4). Then the silicon compounds are distributed within all plant.

Silicon distribution – Some part of silicon compounds goes to epidermal layer, root caps, cell walls, and other tissues and organs for synthesis of phytholites and other silicon-containing structures. The second part of silicon is used to form polysilicic acids gels – the basis of further low-temperature synthesis of organic compounds. The third part of absorbed silicon is preserved and stored at the normal conditions as polysilicic acids or their gels with non-active surface within the cells and in the intercellular space.

Synthesis of organic compounds on the polysilicon matrixes at optimal conditions – Silicon compounds, located inside of the cells, are shaped into polysilicic acid gels. At certain moment, into activated polysilicon-gel is delivered the appropriate molecule-former [protein, simple or complicated protective compounds, and so on], which creates in polysilicic-gel the replica-matrix of itself (Figure 8). After printing and moving out replicating substance, modified polysilicic acid as a "gel-plate" becomes to be able to do catalytic synthesis of molecule-formers' copies from simpler structural blokes and components, located in cytoplasm.

Silicon-dependent synthesis of protective compounds during stress – After external irritation, the signal system is switching on the mechanism of stress identification. Simultaneously, organism additionally uptakes the silicon from the environment and transports it to stressed area. After receiving the information about type of stress, the nuclear founds adequate response for providing synthesis of correct protective compounds; it could be stress-proteins, antioxidant enzymes, antioxidants and others products (Figure 8). Then, the synthesized stress protective molecules are transported to the stressed aria(s) where they play protective roles, and are also used as molecule-formers for additional synthesis of their copies on the polysilicon matrixes.

However if an organism has a strong stress or many different stresses simultaneously, the rate and quantity of synthesizing stress-response-material(s) may not be sufficient enough to solve the problem because some other vitally-important functions should be done in organism at the same time. As a result of escalating energy deficiency and informational resources, the synthesis of "routine" compound, essential for cells is slowed down or even ceased.

We suppose that in living cell some of the protective compounds can move to the activated polysilicic acids gels, and are printed to form "matrixes" as molecule-former(s). Then, these "former(s)", after creation of itself replica in polysilicic acid gel, are transported to stressed-zones. At the end, the silicon-gel-matrixes begin to clone the same molecules from the "at hand material" of cytoplasm. At such cooperation the silicon-gel-matrixes give to organism the possibility [on a background of the stress] to release informational-commanding resources and part of energy for cell functioning in former "before-stress" regime. Thus, additional synthesis of protective compounds can carry out on polysilicic acids-gel-matrixes without direct participation of the genetic apparatus.

New Technologies

High pollution level of the environment and other negative consequences of human civilization bring up a question about the manner and conditions of humanity's survival. The application of synthetic drugs and pesticides initiates a number of negative consequences, which are now sensed by everybody. For example, antibiotics are not only killing pathogenic microorganisms, but also destroying the patients's immune system. Fungicides kill in soil and on the plants surface, together with pathogenic fungus, the useful organisms. Absolutely unwarrantable to apply in agriculture insecticides: while trying to protect the plants against insects, people provide penetration of the toxins inside of animals and, as a consequence – to human body. And, it is absolutely unallowable [in modern stage of genetic engineering development] to introduce genetically modified organisms into agriculture, food industry, pharmacology, and other fields because such organisms are the harbor for a number of unstudied consequences and frequently negative for humanity.

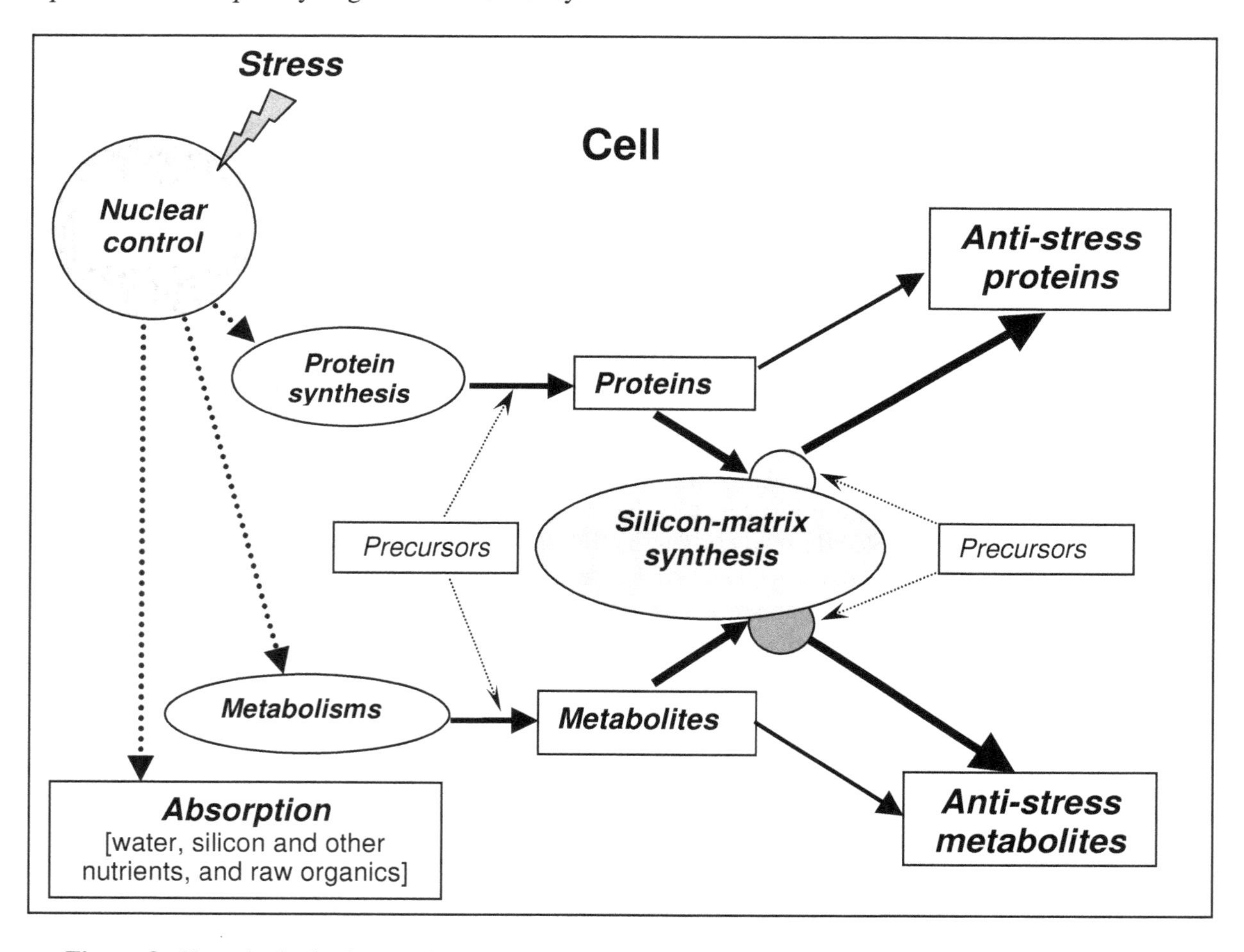

Figure 8. Hypothetical scheme of the universal auxiliary stress-protective mechanism of living systems with participation of the movable silicon compounds.

The way out from this situation, in some cases, could be using of silicon compounds, although we are far a way from conclusion that silicon is "panacea from all troubles". We are sure that competent application of silicon compounds may help to solve some large-scale problems, and to modify whole series of economically-important technologies. Below we have presented some of really solvable actual problems:

- Planetary problem of fresh water deficiency [the most of which is used for irrigation] can, in part, be solved by corrective application of silicon fertilizers, because the "activated" silicon facilitates a decrease of irrigation water amount on 30–50% without diminution of quantitative and qualitative crops characteristics (Matichenkov and Kosobryukhov, 2004; Snyder *et al.*, 2006)
- The application of silicon compounds, which facilitates plants tolerance to infection and are often more affective than fungicides and insecticides, will allow a decrease in the use the pesticides in agriculture (Datnoff *et al.,* 1991; Savant *et al.,* 1997);
- The application of silicon fertilizers promotes plant tolerance to high/low temperatures, saline intoxication, and etc., and as a result, permits guaranteed yields enriched with silicon compounds under extreme environment (Matichenkov and Kosobryukhov, 2004; Snyder *et al.*, 2006);
- Consumption of the foodstuffs with high level of silicon compounds makes people and animals stronger against diseases and unfavorable environment. Furthermore, an overdose of silicon compounds is not possible in principle, because the extra dose is transforming into an inert form – silica, which quickly moves out from the organism (Voronkov *et al.*, 1978; Iler, 1979; D'yakov *et al.*, 1990).

CONCLUSION

Silicon increases the level of resistance of living systems to stress(es), and does not toxically influence the organism. Active silicon forms are possible purposefully with low cost and without interference of the structure of genetic apparatus, to create conditions which are necessary for optimal use of the Program of Realization of Genetic Information (Biel et al., 1990; Biel and Fomina, 2005). Our proposition can be one of the directing criteria for detection of the way(s) "to switch on" the silicon-protective function in cell with using natural materials and internal reserves of organism.

Acknowledgements

We thank Dr. E. A. Bocharnikova (Institute of Physical and Chemical Soil-Science Problems of Russian Academy of Science, Pushchino, Russia) for participation in some experiments.

We dedicate this paper to memory of our friend and colleague Dr. Nicholas Patrick Yensen succumbed to battle with pancreatic cancer on August 24, 2006.

REFERENCES

1. Ahmad R., Zaheer S., Ismail S. 1992. Role of silicon in salt tolerance of wheat (*Triticum aestivum L.*) // Plant Sci., 85: 43–50.
2. Al Wajeeh K. 1999. Studies on the Microbiology of Silicon (*Ph.D. Thesis*). – University of Sheffield, England.
3. Aleshin N.E. 1982. Silicon content in rice RNA // Dokl. VASKhNIL (Proceed. Soviet Agr. Acad.) [in Russian], №. 6: 6–7.

4. Aleshin N.E., Avakyan E.P., Dyukunchak S.A., Aleshin E.P., Barushok V.P., Voronkov M.G. 1987. The role of silicon in protection of rice against diseases // Proceed. Acad. Sci. USSR. 291(2): 217–219.
5. Alvarez R., Sparks D.L. 1985. Polymerization of silicate anion in solution at low concentrations // Nature, Vol. 318: 649–651.
6. Anokhin S.M., Panichev A.M., Bgatov V.I. 1997. Basics of adaptation of animal organisms to alternative geochemical environment // Natural Materials for Man's Benefit. – Novosibirsk [in Russian]: 141–142.
7. Banerjee A.K., Laya Mimo M.S., Vera Vegas W.J. 2001. Silica gel in organic synthesis // Russian Chemical Reviews, 70 (11): 971–990.
8. Bashkin V.N. 2005. Biogeochemistry – "Nauchnii Mir" Press [in Russian], Moscow, 584 p.
9. Bazilevich N.I. 1993. Biological Productivity of Ecosystems in Northern Eurasia – "Nauka" Press [in Russian], Moscow: 293 p.
10. Belanger R.R. 2005. The role silicon in plant-pathogen interaction: toward universal model // Proceed. 3rd Silicon Agricultural Conference, G.H. Korndorfer, ed., October 22-26, 2005. Umberlandia, Universodade Federal de Uberlandia: 34–40.
11. Bgatov V.I., Motovilov K.Y., Speshilova M.A. 1987. Functions of Natural Minerals in metabolic processes of the farming poultries // Selkhoz. Biol. (Agricultural Biol.) [in Russian], 7: 98–102.
12. Biel K.Y., Fomina I.R. 2005. Self-regulation in plant living systems: methanol as example of volative mediators providing stress-tolerance, hypothesis // The Complex Systems under Extreme Conditions, R.G. Khlebopros and V.G. Soukhovolsky, eds. Krasnoyarsk, Russia: 116–133.
13. Biel K.Y., Kolmakov P.V., Lastivka A.N., Titlyanov E.A. 1990. Photosynthetic carbon metabolism in ontogenesis of green algae *Ulva fenestrata* // Biologiya Morya (Marine Biology) [in Russian], № 5: 52-58.
14. Birchall J.D., Chappel J.S. 1988. The chemistry of aluminium and silicon in relation to Alzheimer's disease // Clin. Chem. 23: 154–156.
15. Bocharnikova E.A., Matichenkov V.V. 2007. Influence of plant association on the Si cycle mobile Si in forest soil under various plant associations // Applied Soil Ecology, in press.
16. Bocharnikova E.A., Matichenkov V.V., Pinsky D.L. 1995. The influence of soluble silica acids on behavior of heavy metals in soil and natural waters // Pollution in Large Cities / Proceed. World-Wide Symp. – Venice-Padova, Italy: 43–50.
17. Bocharnikova E.A., Matichenkov V.V., Snyder G.H. 1999. The management of heavy metal behavior and mobility in the soil-plant system // Proceed. 31st Mid-Atlantic Industrial and Hazardous Waste Conference, June 1999: 614–622.
18. Brown C.A. 2002. The use of silicon gel for treating children's burn scars in Saudi Arabia: a case study // Occup. Ther. Int. 9(2): 121–30.
19. Brummer F. 2003. Living inside a glass box – silica in *Diatoms* // Silicon Biomineralization, W.E.G. Muller, ed. – Springer, Berlin: 3–10.
20. Camilleri C., Markich S.J., Noller B.N., Turley C.J., Parker G., van Dam R.A. 2003. Silica reduces the toxicity of aluminium to a tropical freshwater fish (*Mogurnda mogurnda*) // Chemosphere 50 (3): 355–364.
21. Carlisle E.M. 1972. Silicon. Essential element for the chick // Science (Wash. D.C.) 178: 154–156.
22. Carlisle E.M. 1984. Silicon // Biochem. Elem. (Biochem. Essent. Ultratranc. Elem.) 3: 257–291.

23. Carlisle E.M. 1997. Silicon // Handbook of Nutritionally Essential Mineral Elements, O'Dell B.L. and Sundle R.A. Marcel Dekker, eds. New York: 603– 618.
24. Carlisle E.M., McKeague J.A., Siever R., van Soest P.J. 1977. Silicon // Trace Elements Related to Health and Disease, H.C. Hopps, ed. – Geochem. and Environ., National Academy of Science, Washington: 54–72.
25. Cha J.N., Shimizu K., Zhou Y., Christianssen S.C., Chmelka B.F., Stucky G.D., Morse D.E. 1999. Silicatein filaments and subunits from marine sponge direct the polymerization of silica and silicones *in vitro* // Proc. Natl. Acad. Sci. USA 96: 361–365.
26. Cherif M., Asselin A., Belanger R.R. 1994. Defense responses induced by soluble silicon in cucumber roots infected by *Pythium* spp. // Phytopathology 84: 236–242.
27. Cid M.S., Detling J.K., Brizuela M.A., Whicker A.D. 1988. Patterns in grass silicification: response to grazing history and defoliation // Oecologia 80: 268–271.
28. Datnoff L.E., Raid R.N., Snyder G.H., Jones D.B. 1991. Effect of calcium silicate slag on blast and brown spot intensities and yields of rice // Plant Dis. 75: 729–732.
29. Davies S. 1964. Silica in streams and ground water // Am. J. Sci. 262: 870–876.
30. Dietzel M. 2000. Dissolution of silicates and the stability of polysilicic acid // Geochim. Cosmochim. Acta 64(19): 3275–3281.
31. Dietzel M. 2002. Interaction of polysilicic and monosilicic acid with mineral surfaces // Water-Rock Interaction, Stober and K. Bucher, eds.: 207–235.
32. Dobrovolsky G.V., Bobrov A.A., Gol'eva A.A., Shoba S.A. 1988. Opalescent phytholites in taiga biogeocenosis of middle taiga // Biol. Nauki (Biol. Sci.) [in Russian] 2: 96–101.
33. D'yakov V.M., Matichenkov V.V., Chernyshev E.A., Ammosova Y.M. 1990. Using of Silicon Compounds in Agriculture. Actual Problems of Chemical Science, Technology, and Preservation of the Environment (issue 7), Publ. NIITEKhIM [in Russian], Moscow, 32 p.
34. Elliot C.L., Snyder G.S. 1991. Autoclave-induced digestion for the colorimetric determination of silicon in rice straw // J. Agric. Food. Chem. 39: 1118–1119.
35. Emadian S.F., Newton R.J. 1989. Growth enhancement of loblolly pine (*Pinus taeda* L.) seedlings by silicon // J. Plant Physiol. 134(1): 98 –103.
36. Epstein E. 1999. Silicon // Annu. Rev. Plant Physiol. Plant Mol. Biol. 50: 641– 664.
37. Exley C., Pinnegar J.K., Taylor H. 1997. Hydroxyaluminosilicates and acute aluminium toxicity in fish // J. Theor. Biol. 198: 133–139.
38. Fotiev V.A. 1971. About nature of aqueous humus // Dokl. AN SSSR (Proceed. Acad. Sci. USSR) [in Russian] 1199(1): 198–201.
39. Frolov Y.G. 2004. Course of Colloid Chemistry. Surface Phenomena and Dispersion Systems. – "Al'yans" Press [in Russian], Moscow, 464 p.
40. Gol'eva A.A. 2001. Phytholites and Their Informational Role in Investigation of Natural and Archeological Objects. – Moscow-Syktyvkar-Elista [in Russian], 120 p.
41. Gol'eva A.A. 2004. Biogeochemistry of amorphous silica in plants and soils // Soils, Biogeochemical Cycles, and Biosphere, N.F. Glazovsky, ed. "Tov. Nauch. Izd. KMK" Press [in Russian], Moscow: 137–159.
42. Harrison C.C. 1996. Evidence for intramineral macromolecules containing protein from plant silicas // Phytochemistry 41: P. 37–42.
43. Hodson M.J. 1986. Silicon deposition in the roots, culm and leaf of *Phalaris canariensis* L. // Annals of Bot. 58(2): 167–177.
44. Hodson M.J., Sangster A.G. 1989. X-ray microanalysis of the seminal root of sorghum bicolor with particular reference to silicon // Annals of Bot. 64(6): 659–675.

45. Hodson M.J., White P.J., Mead A., Broadley M.R. 2005. Phylogenetic variation in the silicon (Si) composition of plants // Annals of Bot. 96: 1027–1046.
46. Iler R.K. 1979. The Chemistry of Silica. – Wiley, New York
47. Jugdaohsingh R., Anderson S.H., Tucker K.L., Elliot H., Kiel D.P., Thompson R.P., Powell J.J. 2002. Dietary silicon intake and absorption // Am. J. Clin. Nutr. 75(5): 887–893.
48. Kaufmann K. 1995*a*. Silica: the Amazing Gel. – Alive Books, 208 p.
49. Kaufmann K. 1995*b* Silica: the Forgotten Nutrient – Alive Books, 106 p.
50. Kovda V.A. 1956. Mineral composition of plants and forming of soil // Pochvoved. (Soil Sci.) [in Russian] 1: 6–38.
51. Kovda V.A. 1985. Soil Biogeochemistry. – “Nauka” Press [in Russian], Moscow, 263 p.
52. Kuzin N.M., Davani S.A., Kashevarov S.B., Leont’eva M.S., Koroleva I.M., Gorbunov A.S., Guznov I.G., Zainiddinov F.A. 2000. Using of adjustable silicon bandage in horizontal gastroplasty for patients with morbid adiposity // Khirurgiya (Surgery) [in Russian] 10: 16–19.
53. Lanning F.C., Eleuterius L.N. 1981. Silica and ash in several marsh plants // Gulf Research Reports **7**: 47–52.
54. Leibigh J. 1840. Chemistry in its application to agriculture and plant physiology (from the manuscript of the author by Lyon Playfair). Taylor and Walton, London, 345 p.
55. Ma J.F. 1990. Studies on beneficial effects of silicon on rice plants (*Ph. D. Thesis*). Kyoto University
56. Ma J.F. 2003. Function of silicon in higher plants // Progress in Molecular and Subcellular Biology, W.E.G. Muller, ed. Springer-Verlag, Berlin, Heidelberg 33: 127–147.
57. Ma J.F., Goto S., Tamai K., Ichii M. 2001. Role of root hairs and lateral roots in silicon uptake by rice // Plant Physiol. 127: 1773–1780.
58. Ma J.F., Takahashi E. 2002. Soil, Fertilizer, and Plant Silicon Research in Japan. Elsevier, the Netherlands, 281 p.
59. Ma J.F., Tamai K., Yamaji N., Mitani M., Konishi S., Katsuhara M., Ishiguro M., Murata Y., Yano M. 2006. Silicon transporter in rice // Nature 440: 688–691.
60. Martin-Jezequel V., Lopez P.J. 2003. Silicon – a general metabolite for diatom growth and morphogenesis // Silicon Biomineralization, W.E.G. Muller, ed. Springer, Berlin, Germany: 99–126.
61. Matichenkov V.V. 1990. Amorphous Silicic Oxide in Sody-Podzolic Soil and Its Influence on Plants (*Ph. D. Thesis*). Moscow State University [in Russian], Moscow, 26 p.
62. Matichenkov V.V., Ammosova Y.M. 1994. Effects of amorphous silica on some properties of Sody-Podzolic soils // Pochvovedenie (Soil Science) [in Russian] 7: 52–61.
63. Matichenkov V.V., Ammosova Y.M., Bocharnikova E.A. 1997. A method for measuring of the available for plants silicon forms in soil // Agrokhimiya (Agrochem.) [in Russian] 1: 76–84.
64. Matichenkov V.V., Ammosova Y.M., Bocharnikova E.A. 2001*a*. Influence of silicon fertilizers on plants and soil // Agrokhimiya (Agrochemistry) [in Russian] 12: 30–38.
65. Matichenkov V.V., Bocharnikova E.A. 1994. Total and partial biogeochemistry cycle of Si in various ecosystems // Sustainable Development: the View from the Less Industrialized Countries, J. Monge-Najera and San Jose, eds. UNED, Costa Rica: 467–481.
66. Matichenkov V.V., Bocharnikova E.A. 2001. The relationship between silicon and soil physical and chemical properties // Silicon in Agriculture, L.E. Datnoff, G.H. Snyder, and G.H. Korndorfer, eds. Elsevier, Amsterdam, The Netherlands: 209–219.

67. Matichenkov V.V., Bocharnikova E.A. (Eds.) 2004*a*. Silicon in Food, Agriculture and Environment / Proc. Int. Conf., 2–5 August 2004, Pushchino, Russia, 64 p.
68. Matichenkov V.V., Bocharnikova E.A. 2004*b*. Si in horticultural industry // Plant Mineral Nutrition and Pesticide Management, R. Dris and S.M. Jain, eds. 2: 217–228.
69. Matichenkov V.V., Bocharnikova E.A., Biel K.Y. 2007. Location of mono- and polysilicic acids in the different compartments of *Distichlis spicata* (Poaceae) shoots // Russian Plant Physiol.: in preparation.
70. Matichenkov V.V., Bocharnikova E.A., Yensen N.P., Biel K.Y. 2007. About location of the silicon within different organs of *Distichlis spicata* (Poaceae) and *Hordeum vulgare* (Poaceae), and its redistribution under high salinities // Plant Cell and Environ: in preparation.
71. Matichenkov V.V., Calvert D.V., Snyder G.H. 1999*a*. Comparison study of soil silicon status in sandy soils of south Florida // Proc. Soil Crop Sci. Florida 59: 132–137.
72. Matichenkov V.V., Calvert D.V., Snyder G.H. 1999*b*. Silicon fertilizers for citrus in Florida // Proc. Fl. St. Hort. Soc. 112: 5–8.
73. Matichenkov, V.V., Calvert D.V., Snyder G.H. 2000. Prospective silicon fertilization for citrus in Florida // Proc. Soil Crop Sci Florida 59: 137–141.
74. Matichenkov V.V., Calvert D.V., Snyder G.H. 2005. Minimizing nutrient and pollutants leaching from sandy agricultural soils and optimization of plant nutrition. Final report // DEP Direct Contract. Fort Pierce, IRREC, University of Florida, FL, USA, 494 p.
75. Matichenkov V.V., Calvert D., Snyder G.H., Bocharnikova E.A. 2001. Effect of Si fertilization on growth and P nutrition of Bahiagrass // Proc. Soil Crop Sci. Florida 60: 30–36.
76. Matichenkov V.V., Korenevsky A.A., Beveridge T.J. 2001*b*. Adsorption of soluble silicon compounds by *Pseudomonas aeruginosa* // Proceed. V.M. Goldschmidt Conf., May 20-24, 2001. Hot Springs, Virginia, USA, P. 3831.
77. Matichenkov V.V., Kosobryukhov A.A. 2004. Si effect on the plant resistance to salt toxicity // Proceed. 13th Int. Soil Conservation Organization Conference (ISCO), 4-9 July 2004. Brisbane, Australia: 287–295.
78. Matichenkov V.V., Pinsky D.L., Bocharnikova E.A. 1994. Influence of soil mechanical solidity on the state and forms of available silicon // Pochvovedenie (Soil Science) [in Russian] 11: 71–76.
79. Matichenkov V.V., Snyder G.H. 1996. Mobile silicon compounds in some soils of south Florida // Pochvovedenie (Soil Science) [in Russian] 12: 1448–1453.
80. Mayras Y., Riberi P., Cartier F., Allan P. 1980. Increase in blood silicon concentration in patients with renal failure // Biomedicine 33(7): 228–230.
81. McNaughton S.A., Bolton-Smith C., Mishra G.D., Jugdaohsingh R., Powell J.J. 2005. Dietary silicon intake in post-menopausal women // Br. J. Nutr. 94 (5): 813–817.
82. Medvedev S.S. 2004. Plant Physiology. Publ. St-Petersburg State University [in Russian], St-Petersburg, 336 p.
83. Milnes A.R., Twidale C.R. 1983. An overview of silicification in Cainozoic landscapes of arid central and southern Australia // Australian J. Soil Res. 21(4): 387–410.
84. Mohanty B.K., Gosh S., Mishara A.K. 1990. The role of silicon in *Bacillus licheniformis* // J. Appl. Bacteriol. 68: 55–60.
85. Nazarov A.G. 1976. Biogeochemical cycle of silica // Biogeochemical Cycles in Biosphere, V.A. Kovda, ed. "Nauka" Press [in Russian], Moscow: 199–257.
86. Olier K. 1990. Erosion. "Nedra" Press [in Russian], Moscow, 347 p.
87. Orlov D.S. 1985. Soil Chemistry. Publ. Moscow State University [in Russian], Moscow, 376 p.

88. Pellenard P. 1969. Use of silica gel and microorganisms for growing Niphargus virei (*hypogeal Crustacea Amphipoda*) in artificial media // Ann. Nutl. Aliment. 23(4): 283–288.
89. Perelman A.I. 1975. Geochemistry of Landscape. – "Visshaya Shkola" Press [in Russian], Moscow, 341 p
90. Perez-Grandos A.M., Vaquero M.P. 2002. Silicon, aluminium, arsenic and lithium: essentiality and human health implications // J. Nutr. Health Aging. 6(2): 154–162.
91. Perry C.C., Keeling-Rucker T. 2000. Biosilicification: the role of the organic matrix in structure control // J. Biol. Inorg. Chem. 5: 537–550.
92. Polynov B.B. 1952. Geochemical Landscapes. "Geographgiz" Press [in Russian], Moscow, 400 p.
93. Rascher U., Bobich E.G., Lin G.H., Walter A., Morris T., Naumann M., Nichol C.J., Pierce D., Bil' K., Kudeyarov V., Berry J.A. 2004. Functional diversity of photosynthesis during drought in a model tropical rainforest – the contributions of leaf area, photosynthetic electron transport and stomatal conductance to reduction in net ecosystem carbon exchange // Plant Cell and Environ. 27: 1239–1256.
94. Reimers N.F. 1990. Nature Management. "Mysl" Press [in Russian], Moscow, 640 p.
95. Savant, N.K., Snyder, G.H., Korndorfer G.H. 1997. Silicon management and sustainable rice production // Advance Agronomy 58: 151–199.
96. Siegel B.Z., Siegel S.M. 1973. The chemical composition of algal cell walls // CRC Crit Rev Microbiol. 3(1): 1–26.
97. Snyder G.H., Matichenkov V.V. Datnoff L.E. 2006. Silicon // Handbook of Plant Nutrition. Massachusetts University: 551–568.
98. Sokolova T.A. 1985. Clay Minerals in Soils of Humid Zones of the USSR. "Nauka" Press [in Russian], Novosibirsk, 252 p.
99. Soomro F.M. 2000.Effect of silicon compounds on microbial transformation in soil // Ph.D. thesis, University of Sheffield, England.
100. Sripanyakorn S., Jugdaohsingh R., Elliot H., Walker C., Mehta P., Shoukru S., Thompson R.P., Powell J.J. 2004. The silicon content of beer and its bioavailability in healthy volunteers // Br. J. Nutr. 91(3): 403–409.
101. Strelko V.V., Gushchin P.P., Vysotsky Z.Z. 1963. About interaction of some amino compounds with dehydrated silica gel // Dokl. AN SSSR (Proceed. Acad. Sci. USSR) [in Russian] 153(3): 619–621.
102. Stumm W., Morgan J.J. 1996. Aquatic Chemistry – Chemical Equilibria and Rates in Natural Waters (3rd Ed). – Wiley-Interscience, New York, 1022 p.
103. Trevors J.T. 1997. Bacterial evolution and silicon // Antonie Van Leeuwenhoek 71: 271–276.
104. Tuner R.E., Rabalais N.N., Justic D. 2006. Predicting summer hypoxia in the northern Gulf of Mexico: riverine N, P, and Si loading // Mar. Pollut. Bull. 52(2): 139–148.
105. Van Dyck K., Robbercht H, Van Cauwenbergh R., Van Vlasaer V., Deelstra H. 2000. Indication of silicon essentiality in humans: serum concentrations in Belgian children and adults, including pregnant women // Biol. Trace Elem. Res. 77(1): 25–32.
106. Varshal G.V., Dracheva I.N., Zamkina N.S. 1980. About forms of silicic acids and the methods of their determination in natural waters // Chemical Analysis of Marine Sediments. – "Nauka" Press [in Russian]: 156–186.
107. Velikova V., Cociasu A., Popa L., Boicenco L., Petrova D. 2005. Phytoplankton community and hydrochemical characteristics of the Western Black Sea // Water Sci. Technol. 51(11): 9–18.

108. Vladimirov V.L., Kirilov M.P., Fantin V.M. 1998. Metabolism and productive properties of bullheads at feeding by mixed fodder with zeolite // Dokl. RASKhN (Proceed. Russian Agricultural Academy) [in Russian] 4: 38–40.
109. Voronkov M.G., Zelchan G.I., Lukevits A.Y. 1978. Silicon and Life – "Zinatne" Press [in Russian], Riga, 578 p.
110. Vysotsky Z.Z., Danilov V.I., Strelko V.V. 1967. Properties of the polysilicic acid gels and conditions of the beginnings and development of life // Uspekhi Sovr. Biol. (Progr. Modern Biol.) [in Russian] 63(3): 362–379.
111. Wainwright M. 2005. Silicon and microbiology aspects – overview // Proceed. 3rd Silicon Agricultural Conference, G.H. Korndorfer, eds., October 22-26, 2005. Universodade Federal de Uberlandia, Umberlandia: 19–22.
112. Wainwright M., Al-Wajeh K., Grayston S.I. 1997. Effect of silicic acid and other silicon compounds on fungal growth in oligotrophic and nutrient-rich media // Mycol. Res. 101: 933–938.
113. Waterkeyn L., Bientait A., Peeters A. 1982. Callose et silice epidermiques rapports avec la transpiration culticulaire // La Cellule. 73: 263–287.
114. Yatsynin N.L. 1989. Genesis of the Residual Solonetzs. "Nauka" Press [in Russian], Alma-Ata, 125 p.
115. Yatsynin N.L. 1994. Colloidal-High-Polymeric Systems of Solonetzs in the Northern Kazakhstan (*Abstr. Dr. Sci. Dissertation*). Tashkent [in Russian], 37 p.
116. Yoshida S. 1975. The physiology of silicon in rice // Tech. Bull. n. 25, Food Fert. Tech. Centr., Taipei, Taiwan.
117. Yoshon T. 1990. Growth accelerating effect of silicon on *Pseudomonas aeruginosa* // Saitama Med. Sch. 17: 189–198.

PROBIOTIC: AS A FUNCTIONAL FOODS AND HUMAN HEALTH BENEFITS

Seema Kafley and Keiichi Shimazaki

Laboratory of Dairy Science, Faculty of Agriculture
Hokkaido University, Sapporo-0608589, Japan

Corresponding author: Seema Kafley. Email: itssseema4@hotmail.com

Keywords: probiotics, benefits of probiotic, functional foods

ABSTRACT

Probiotics for gastrointestinal implication can be defined as the ingestion of living microorganisms in certain numbers to confer health benefits that exceed those associated with basic nutrition and emerging as new popular functional foods. There is increasing commercial interest in probiotic products, especially in developed countries where products containing probiotics are produced in a variety of formats like probiotic-containing fermented milk products; conventional food formulations that serve primarily as delivery vehicles for probiotic bacteria, such as dietary Bifidobacterium strains. Properly formulated probiotic-containing foodstuffs offer consumers a low risk, and essential dietary component that has the potential to improve health in a variety of ways.

Key words: Probiotics; Microorganisms; Functional foods; Dairy products.

INTRODUCTION

Probiotics are currently defined as live microorganisms which, when consumed in sufficient quantities, confer health benefits to the host. Probiotics can also be defined as live microorganisms, such as *Lactobacillus* and *Bifidobacterium* bacterial species, as well as yeasts that have beneficial effects upon ingestion by restoring the balance of the host's intestinal microflora. It is important to maintain a healthy community of bacteria in the gastrointestinal (GI) tract. The microfloral colony in the digestive tract is ordinarily composed of both benevolent and potentially harmful bacteria. When the delicate balance between these bacteria becomes adversely affected, the body becomes exposed to excessive toxin levels, which then manifests as various types of diseases. Probiotic bacteria are used to alter this intestinal microfloral balance in such a way that, by being more abundant than the harmful bacteria, they inhibit the growth of harmful bacteria, promote good digestion, boost immunity, and increase resistance to infection. At present, probiotic bacteria are added to yogurts, various fermented milk products and dietary supplements during production. The majority probiotics currently marketed as nutritional supplements and used to supplement functional foods are strains of *Bifidobacterium* and *Lactobacillus* (IDF Bulletin, 2003). Most of the probiotics available at present are bacteria whereas *Saccharomyces boulardii* is one example of probiotic yeast. Organisms should meet the following basic requirements for consideration as a probiotic food product:

- Exhibit biosafety characteristics
- Ability to survive in the human stomach
- Beneficial for host health
- Easy to process and cost effective (stability/long shelf life, commercial scale production).

Role of probiotics for health benefits

Probiotics may exhibit antimicrobial, immunomodulatory, anticarcinogenic, antidiarrheal, antiallergenic and antioxidant activities. Considerable ongoing research is currently being undertaken focusing on the development of target-specific probiotics containing well-characterized bacteria that are selected for their health-enhancing characteristics. These probiotic products are entering the marketplace in the form of nutritional supplements and functional foods, such as yoghurt and other functional food products (Mattila-Shadolam et al., 1999). Probiotic bacteria also produce substances called bacteriocins, which act as natural antibiotics that kill undesirable microorganisms. Relief of lactose maldigestion symptoms and decreasing the duration of rotavirus diarrhea are health benefits conferred by many probiotic strains (IDF Bulletin, 2003). The potential benefits of probiotic foodstuffs are as follows:

Acidification of the colon:

Benevolent microflora produces lactic and acetic acids, which are essential in combating the growth of potentially harmful bacteria such as *Salmonella, Shigella* and *coliform* species in the gastro intestinal tract.

Vitamin production:

The intestinal flora enhances the production of several vitamins that are necessary for human health, including folic acid, B2, biotin, pantothenic acid, B6, B12 and vitamin K.

Elimination of intestinal gas and bloating:

The benevolent bacteria prevent harmful bacteria from becoming dominant in the colon where they often produce foul-smelling waste products and create painful intestinal gas and bloating. Excess gas resulting in chronic flatulence is often a direct result of having too few benevolent flora in the colon, which in turn allows harmful bacteria and yeasts to proliferate and multiply very quickly

Normalizes bowel movements:

Constipation is often alleviated by ingesting large quantities of benevolent bacteria because they promote the processing of bodily wastes considerably; decreasing the amount of time it takes for waste products to travel through the digestive system.

Improves immune system functioning:

The benevolent bacteria act to provide a powerful natural "internal vaccination" by stimulating the gut associated lymphoid tissue (GALT), which in turn combats harmful substances and organisms in the body that might otherwise cause infections and disease.

Deactivation of various carcinogenic compounds:

Numerous carcinogenic compounds derived from ingested foods and byproducts of other organisms are scavenged by the activities of benevolent bacteria.

Regulation of cholesterol levels:

Through a complex process, the benevolent flora helps to reduce the accumulation of cholesterol in the bloodstream and promote its excretion through the bowels.

Additional benefits associated with probiotic foodstuffs:

The benevolent microflora of probiotic foods work in conjunction with other species of beneficial intestinal microorganisms "native to the human body" to produce the following additional benefits:

- Creating a protective barrier against invasion by harmful microorganisms by colonizing and coating the intestinal mucosa
- Aiding in the production of natural beneficial antibiotic substances such as acidophillin, bifidin, hydrogen peroxide, etc.
- Breaking down carbohydrates, fats and proteins while rendering their toxic byproducts inert

Probiotics are important for recolonizing the intestine during and after antibiotic use. Probiotics also promote healthy digestion as the enzymes produced by probiotic bacteria aid digestion. *Lactobacillus acidophilus* is a source of lactase, the enzyme needed to digest milk sugar, which is not present in lactose-intolerant people. Since lactic acid bacteria produce acids, peroxides and bacteriocins, it is expected that they could influence the incidence of cancer.

Market status and current trends in probiotic food products

Probiotic products have entered the market in the form of nutritional supplements and functional foods. The probiotic market, especially that of dairy products such as yoghurts and other fermented dairy products, has experienced rapid growth in Japan, Europe and the USA. Fermented milk drinks have been available in Japan for several decades and the market for probiotic products is now mature. Probiotic beverages and foodstuffs have begun to diversify into juices in the European market. In addition, while probiotic products may already have become a vital component in the diets of consumers' in the developed world, they have yet to enter the markets of the developing world.

In general, probiotic bacteria are sold in two different formats: food and dietary supplements. Food products containing probiotic bacteria are almost exclusively dairy products, capitalizing on the traditional association of lactic acid bacteria with fermented milk. The supplement market contains many different product formats and contents, including capsules, liquids, tablets and even food-like formats. Importantly, dietary supplements are designed to supplement the diet and not replace any of the components of a diet.

In developed countries, market competition among producers of dairy products supplemented with probiotics is intense. Functional foods, dietary fiber, prebiotic and probiotics are the principal functional ingredients in many of these dairy food products. Common dairy probiotic products in Japan are listed in Table 1. It is expected that prebiotic and probiotics will continue to be among the major functional food ingredients for the foreseeable future in the world.

Table 1. Japanese companies, products, used probiotic bacteria and expected health promotional effect.

Company	Poducts	Probiotics bacteria	Expected effects
Calpis Ajinomoto Danone Co. Ltd.	Danone BIO BE80	*Bifidobacterium* DN-173 010	Control of intestinal function
Chichiyasu Co. Ltd.	Reuteri yogurt	*Lactobacillus reuteri* SD2112	
Glico Dairy Products Co. Ltd.	Glico choushoku yogurt	*Bifidobacterium* GCL2502	
Kagome Co. Ltd.	Labre	*Latobacillus brevis* subsp. *coagulans* KB290 (plant origin)	
Koiwai Dairy Products Co., Ltd.	KW lactic acid bacteria yogurt	*Lactobacillus paracasei* KW3110	vs. allergy
Kyodo Milk Industry Co. Ltd.	Meito onakani-oishii yogurt	*Bifidobacterium lactis* LKM512	
Meiji Dairies Corporation	Bulgaria yogurt LB81	*L. Bulgaricus* 2038 *L. thermophilus* 1131	
	Meiji probio yogurt LG21	*Lactobacillus gasseri* OLL2716 (LG21)	vs. *H. pylori*
Morinaga Milk Industry Co., Ltd.	Morinaga bifidus plain yogurt, Lactoferrin 200	*Bifidobacterium longum* BB536	Immune-stimulation
Nestle Group	Nestle LC1 yogurt	*Lactobacillus johnsonii* LC1	Control of intestinal function
Nippon Milk Community Co. Ltd.	Megmilk, Nature megumi, yoplait	*Lactobacillus* gasseri *Bifidobacterium* SP	Microcapsule
Ohayo Dairy Products Co. Ltd.	L-55 yogurt	*Lactobacillus acidophilus* 55	
Takanashi Milk Products Co. Ltd.	Takanashi yogurt onakae GG!	*Lactobacillus rhamnosus* GG	vs. atopy
Yakult Honsha Co. Ltd.	Bifiene V	*Bifidobcterium breve* yakult	
	Yakult, Sofuhl, Purela, Joie	*Lactobacillus casei* shirota	Immune-stimulation
Yotsuba Inc.	Yotsuba yogurt	*Bifidobacterium lactis* Bb-12	

CONCLUSIONS

In conclusion, the development of successful probiotic products will be contingent on both the proof of a probiotic effect and the development of foods that harbor high numbers of viable organisms at the time of consumption. At the start of the 21st century, global probiotic research efforts are focusing on the development of target-specific probiotic products containing well-characterized bacteria selected for their specific health-enhancing characteristics. Reports have already demonstrated number of beneficial effects associated with the consumption of probiotics and sales are increasing worldwide as consumers become more aware of their benefits.

In Japan, consumer demand for probiotic products is high, which is well suited to the traditional Japanese concept of gaining good health through the consumption of natural products. American and European consumers also perceive the health benefits of probiotics and have increased their consumption of fermented dairy products considerably in recent years. Due to the increased awareness among consumers regarding health, probiotic products are likely to enjoy very good prospects in both developed and developing nations.

Acknowledgement

The authors would like to thanks Dr. Megh Raj Bhandari, Research Scientist, Hokkaido University, Japan, for his valuable comments and suggestion on this manuscript.

REFERENCES:

1. Arunachalam, K. D. (1999). Role of bifidobacteria in nutrition, medicine and technology. Nutr. Res. 19: 1559-1597.
2. Gittlement, A. L. (2005). Probiotics for the 21st century. Naturedoc. http://www.naturodoc.com/library/detox/probiotics.htm
3. IDF Bulletin No. 380 (2003). Health effects of probiotics and culture-containing dairy products in humans. International Dairy Federation.
4. Matilla-Sandholm, T., Blum, S., and Collins, J. K. (1999). Probiotics: towards demonstrating efficacy. Trends Food Sci. Technol., 10: 393-399.
5. Scheinbach, S. (1998). Probiotics: Functionality and commercial status. Biotech. Advances, 16 (3): 581-608.
6. Shortt, C. (1999). The probiotic century: historical and current perspectives. Trends Food Sci. Technol., 10: 411-417.
7. Stanton, C., Gardiner, G., Meehan, H., Collins, K., Fitzgerald, G., Brendan, L., and Ross, R. P. (2001). Market potential for probiotics. Am. J. Clin. Nutr., 73: 476S-83S.
8. Symposium (2000). Probiotic bacteria: implications for human health. J. Nutr., 130: 382S-409S.

METABOLIC SYNDROME AND ITS MANAGEMENT THROUGH FUNCTIONAL FOODS

M. Jahangir Alam Chowdhury[1], Ashkhen D. Martirosyan[2] and M. Rafiqul Islam[3]

[1]Bangladesh Agricultural Research Institute, Gazipur, Bangladesh.
[2] Functional Foods Center, Richardson, TX, 75080 USA
[3] ENT Department, Dhaka Medical College, Bangladesh.

Corresponding author: M. Jahangir Alam Chowdhury, Email: jahangirbari@yahoo.com

Keywords: metabolic syndrome, functional foods, diabetes, insulin resistance, high blood pressure, high triglycerides, obesity

Metabolic syndrome is described by a cluster of symptoms including high blood sugar, insulin resistance, high blood pressure, high triglycerides, lower HDL and obesity, which tend to appear together in some individuals and increase their catastrophic health problem such as diabetes, hypertension, heart disease, Alzheimer's, cancer, and other age-related diseases. It is estimated that about 25% people over 35 years old are suffering from this syndrome (1). The term was coined by Gerry Reaven of Stanford University in the late 1980s. Syndrome X is a hidden but life-threatening metabolic disorder waiting for anyone.

Biochemistry of Syndrome X

Syndrome X develops slowly over time, often over a course of 20 years or more. It is the end result from years of consuming first foods, breads, soft drinks, refined or polished rice, refined wheat flour, which is high in simple carbohydrates. Once taken, these foods trigger a rapid increase of sugar levels in blood, and the body responds by raising levels of insulin secretion that move the sugar into the cells. Inside the cell, glucose is either used for energy or stored for future use in the form of glycogen in liver or muscle cells. The more carbohydrates are eaten, the more pancreas releases insulin to lower the excessive blood sugar. While insulin levels rise and fall with each meal and is part of the normal metabolic process, chronic carbohydrate overload causes chronic insulin overload. The cells recognize that excessive sugar as toxin. Cells try to shut down the influx of sugar into the cells and this state is called insulin resistance (2). Usually a healthy body cell contains 20,000 insulin receptors but hyperinsulinemia decreases that number. Some people produce two; three or four times the normal amount of insulin. Hyperinsulinemia results the hypertension by stimulating sympathetic nervous system activity and sodium and water reabsorption by the kidneys (3,4). Cells lost their sensitivity due to reduced insulin receptors or binding inability to insulin, they require even more insulin to maintain normal glucose levels. Excessive insulin stimulates the liver to produce triglyceride rich VLDL thereby increases triglyceride level in blood. The excess triglycerides make muscle cells insulin-resistance. In advanced stages of insulin resistance, the pancreas becomes exhausted and can no longer maintain the insulin production resulting in adult onset diabetes mellitus (also called type 2 diabetes)(2). In insulin resistance, both the blood sugar and blood insulin levels are high. Too high Ca/Mg ratio in cytoplasm is the another cause of Syndrome X. Mg suppresses the release of adrenal and Ca increases it which cause stress and elevates blood pressure. Ca also constricts the artery but mg relaxes (5,6,7). Over time, high blood sugar and high insulin cause a myriad of destructive damages to almost every tissue they touch. These symptoms collectively are called Syndrome X and it may be the cause of up to fifty percent of all heart attacks.

Cause of Insulin Resistance

- Eating excessive simple carbohydrates and sweets
- Highly processed foods as polish rice and refined flours
- Foods grown in mineral depleted soils which is common in developing countries
- Fast foods
- Preservatives
- Abdominal obesity
- Genetic Predisposition
- Sedentary Lifestyles
- High levels of physiological stress

Consequence of Syndrome X

The individual with Syndrome X is insulin resistance or hyperinsulinemia. The term hyperinsulinemia refers to higher than normal levels of insulin in the blood. Excessive insulin causes damage to the whole body including-

- **Glucose intolerance:** Blood glucose level is usually higher those individuals with Syndrome X. People with insulin resistant produce large quantities of insulin to maintain near normal blood glucose levels. In Syndrome X, VLDL, chylomicrons and their metabolic remnants (chylomicron and VLDL remnants) are removed more slowly from the plasma by virtue of their increased concentrations, resulting increased postprandial lipidemia. Unfortunately, the increased VLDL also reduces the ability to remove postprandial newly absorbed chylomicrons. More often those individual have impaired glucose tolerance (IGT)(8).
- **Dyslipidemia:** With high blood insulin level (insulin resistant), the liver produces more triglyceride rich VLDL. The amount of triglycerides therefore increases. Cholesterol ester transfer protein (CETP) transfers cholesterol from HDL to VLDL. As a result, the HDL cholesterol falls and triglycerides increases (8).
- **Hypertension:** Excessive insulin (insulin resistant) leads to sodium retention by kidney resulting excessive fluid in the body and ultimately hypertension. This condition is present in 50% of those with Syndrome X. Sympathetic nervous system activity is increased in insulin resistant individuals. Systolic pressure is often greater than 130 mmHg, and diastolic pressure higher than 85 mmHg. This contributes hypertension (8).
- **Endothelium:** The inner lining of the arterial walls comes under attack by excess insulin which increase the risk of arterial blood clots eventually cause heart attacks or strokes. The damages include reduced nitrous oxide activities (which could lead to hypertension), increased platelet and monocyte adhesion, increased pro-coagulant activity, impaired fibrinoloytic activity, and impaired degradation of glycosylated fibrin. The net result is increased blood pressure, increased formation of atherosclerotic plaques, increased

thrombus formation, angina and heart attack. The risk of cardiovascular disease is significantly increased when the endothelium is damaged.

- **Beta cell damage:** Long time excessive insulin production burns out beta cells. Ultimately needed insulin injection for sugar control (8).
- **Coronary heart disease:** Syndrome X changes fibrynolytic activity there by increase Plasminogen activator inhibitor 1 (PAI-1). When PAI-1 is high, dissolution of blood clot is reduced, and fibrinogen and thrombus formation increases. The increase in fibrinogen tends to increase coagulation. This plays a role in the development of coronary heart disease (8).
- **Polycystic ovary syndrome:** Ovaries being exposed to consistently higher levels of insulin increases its testosterone secretion accordingly, the ovary being insulin sensitive. It is a major factor in the development of polycystic ovary syndrome.
- **Cancer:** Some research indicates it might also increase the risk of prostate, colon and breast cancer.
- **Obesity:** Obesity is a common feature. The body mass index (BMI) is often greater than 25 kg/sq.m.
- **Premature Aging:** Syndrome X also generates high levels of cell-damaging free radicals and causes premature aging. Some researchers believe it can also increase the risk of Alzheimer's disease. In the mid-1970s, biologist Anthony Cerami discovered that chronically high blood glucose levels is the main trigger in a chemical process of Advanced Glycosylation End Products (AGES), which were implicated in normal and advanced aging and age-enhanced diseases. AGEs form at accelerated rates whenever blood-sugar levels are high with age. AGEs involve a chemical reaction between sugar and protein molecules. Serious damage occurs to cell membranes and collagen fibers. This cross-linking leads to the stiffening of connective tissue and hardening of arteries, leading to pre-mature aging and hypertension. As cross-links increasingly reduce the flexibility and permeability of tissues and cells, cellular communications and repair processes also begin to break down. A compensatory inflammatory response may be launched by the body, especially in the endothelium. This leads to a cascade of damaging events resulting in fatty streaks and atherosclerosis. Eventually, the tissues of the body become irreversibly transformed, and the expected result is aging, disease and finally death. It is well known that bathing cells in high sugar (in diabetics) causes premature aging. This is because this sugar-driven damage acquires quick speed, raising their levels of AGE-infused collagen to those of elderly people. Diabetics suffer a very high incidence of nerve, artery and kidney damage because high blood sugar levels in their bodies markedly accelerate the chemical reactions that form advanced glycation products. The endothelium of diabetic patients also secretes unwanted growth factors that lead to blood vessel hypertrophy and reduced lumen size. This reduces the blood flow, exacerbating the already compromised insulin delivery and further increases the chances of insulin resistance. The reduced blood flow leads to reduced oxygen delivery to needy tissues, resulting in increased peripheral neuropathy commonly seen in diabetes (9).
- **Antioxidant Depletion:** Low levels of antioxidant vitamins, DHEA (dehydroepiandrosterone) and high cortisol levels are commonly found in people with Syndrome X. It is due to the increased free radical activity, and concurrent reduction in the endogenous antioxidant level as the body tries to neutralize the free radical activities.

It is observed that atherosclerotic plaques not only contain cholesterol but also oxidized ascorbate (vitamin C). The body deposits the antioxidant ascorbate there in an attempt to overcome the free radical damage.

Diagnosis of Syndrome- X

According to the Adult Treatment Panel III criteria, the metabolic syndrome is identified by the presence of these components

- Insulin resistance - Triglyceride to HDL cholesterol ratio of more than 2 is a warning sign. If the ratio is over 4, it is a good indicator of insulin resistance.
- Central obesity- excessive fat tissue in the abdominal region
- Body Mass Index (BMI) is 25kg/m^2 or higher
- Fasting blood triglycerides is 150 mg/dl or higher
- Total cholesterol above 240 mg/dl
- Blood HDL cholesterol:
  Men — Less than 40 mg/dl
  Women — Less than 50 mg/dl
- Blood pressure- 130/85 mmHg or higher
- Fasting blood glucose- 110 mg/dL or higher

If any one has 3 or more of the above, he/she should consider either having or at high risk of Syndrome X (10,11,12,13).

Conventional Treatment of Syndrome X

Physicians have been concentrating on treating the symptoms of Syndrome X such as hyperglycemia, hypertension and dyslipidemia rather than concentrating on the underlying problem, which is insulin resistance. Syndrome X is usually reversible without drugs. This can be done by low glycemic index diet together with a nutritional supplementation program designed to slow carbohydrate absorption, increase insulin sensitivity, and normalize blood sugar levels. However, the majority of patients with syndrome X cannot accomplish these goals. In these cases, each metabolic disorder associated with syndrome X needs to be treated individually and aggressively. A short-term treatment with drugs may be needed and useful but long-term treatment is not effective.

Natural Treatment of Syndrome X

The nutritional diseases develop primarily from eating the wrong foods for a long time. Specifically, mineral deflected refined carbohydrates are the main dietary culprit behind the condition. That means, it can be prevented and reversed with a change in diet and a good supplements regimen. The proper diet combats insulin resistance and Syndrome X. There are no drug treatments that can directly reverse the insulin resistance as well as syndrome X. In fact a way to reverse the insulin resistance and Syndrome X are as follows (14-18)

❖ **Diet**- A model diet should have the following characteristics for Syndrome X sufferers
  - Rich in dietary fiber both soluble and insoluble
  - Less simple carbohydrates and high amount of complex carbohydrates
  - Should have sufficient amount of minerals and vitamins

- Low glycemic Index

- **Nutritional Supplementation**- Some minerals as K, Ca, Mg, Zn, Se, Cr, Mn & Cu and vitamins as Thiamin, Riboflavin, Niacin, Pantothenic acid, Pyridoxin ,Biotin & Folic acid with optimum amount improve the condition. In addition optimum doses of Vitamin A, C, and E helps to stabilize plaques, improve vascular tone and reduce thrombus (3).
- **Exercise**- Epidemiological studies have shown that modest exercise is beneficial. However, unequivocal metabolic benefits from exercise will not be achieved from a casual walk a couple of nights a week. Significant, regular, chronic exercise is required to see improvements in insulin action, triglycerides, and HDL cholesterol. The majority of patients with syndrome X cannot accomplish these goals
- **Weight Management**-Every attempt should be made to reduce total body weight calculated for age and height. This way Syndrome X will improve significantly.
- **Other lifestyle Factors**- Smoking is unequivocally bad, associated with high triglycerides, low HDL cholesterol and insulin resistance.

Reversing Syndrome X through Functional foods

Metabolic Syndrome X can be prevented or reversed through functional foods which are fortified with essential vitamins and minerals as well as low glycemic index. But we do not have any proper foods for its management. Many researchers showed that Syndrome X suffering patients are deficient in some minerals and vitamins, which play key role in metabolic, enzymatic as well as hormonal activity. Relative or absolute lack of minerals and vitamins leads to insulin resistance or uncontrolled condition. Diet for Syndrome X should have high fiber, rich complex carbohydrate, less simple carbohydrate, optimum amount of protein, fat, vitamin and minerals as well as low Glycemic Index (14-18). But traditional cereal products like wheat flour, refine wheat flour and refined rice do not have mentioned elements at optimum level even high Glycemic Index. Considering the nutritional requirement of such patients, a low glycemic index cereal flour has been developed through intensive research which is fortified with K, Ca, Mg, Zn, Se, Cr, Mn & Cu and vitamins as Thiamin, Riboflavin, Niacin, Pantothenic acid, Pyridoxin, Biotin & Folic acid. This functional cereal flour delay digestion and absorption of sugar, increase binding capacity of insulin receptors as well as increase the activity of insulin that produces naturally. The nobility of this flour is glycemic index 44 compare to conventional wheat flour had glycemic index 91. Finally it can address all the problems, associate with Syndrome-X. A trial showed that this functional food improves sugar level 32-56% after consumption. Blood total cholesterol decreased 35-56%, Triglyceride decreased 35-51% and LDL cholesterol decreased 45-68% by 60 days. Having potent efficacy in preventing and reversing Syndrome X it may the first choice for future attempt (19).

SUMMARY

Metabolic syndrome includes diabetes, insulin resistance, high blood pressure, high triglycerides, decreased HDL and obesity which tend to appear together in some individuals at early age. The main cause of this health condition is eating wrong food for long time. Especially excessive simple carbohydrates, sweets, processed foods, polish rice, refined flour, grain from mineral depleted soils, fast foods and preservatives. In insulin resistance condition cells lost their sensitivity to insulin so body require more insulin to maintain normal glucose levels. Excessive insulin stimulates the liver to produce triglyceride rich VLDL thereby increases triglyceride level in blood which makes cells insulin-resistance. Over time, high blood sugar and high insulin damage almost every tissue they touch. Excessive insulin or insulin resistance causes damage to the whole body including glucose intolerance, dyslipidemia, hypertension, increased formation of atherosclerotic plaques, heart attack, beta cell damage, Type 2 diabetes, polycystic ovary syndrome, obesity and low levels of antioxidant activity. Conventional treatments for Syndrome X have been concentrating on drugs but it is not effective for long time. Recent study showed that Syndrome X could be prevented or reversed through functional foods without drugs. The key approach is diet with high fiber, low glycemic index, sufficient minerals and vitamins. A fortified flour already been developed that contains high fiber, high complex carbohydrate, less simple carbohydrate, optimum amount of protein, fat, vitamin and minerals as well as low Glycemic Index (GI=44) that flour delay digestion and absorption of sugar, increase binding capacity of insulin receptors as well as increase the activity of insulin that produces naturally.

REFERENCES

1. Kendall P, Syndrome X and Insulin Resistance. Food Science and Human Nutrition. Colorado State University. 1997
2. Lam, M. Metabolic Syndrome. Preventive and Anti-Aging Medicine. Academy of Anti-Aging Research, U.S.A
3. Reaven GM, and Hoffman BB. A role for insulin in the aetiology and course of hypertension? Lancet. 1987; 2(8556): 435-437.
4. Reaven GM, Banting lecture 1988. Role of insulin resistance in human disease. Diabetes 1988; 37:1595- 607.
5. Boullin DJ, The action of extracellular cations on the release of the sympathetic transmitter from peripheral nerves. J Physiol. 1967; 189:85-99.
6. Johansson BW, Juul-Moller S, Ruter G, Soes-Pedersen U. Effect of i.v. adrenaline infusion on serum electrolytes, including S-Mg. Magnesium Bull 1986; 8:259-260.
7. Seelig MS, Consequences of magnesium deficiency enhancement of stress reactions; preventive and therapeutic implications. J Am Coll Nutr 1994; 13:429-446.
8. Reaven G, Strom TK, & Fox B, (2000). Syndrome X: Overcoming the silent killer that can give you a heart attack. New York: Simon & Schuster.
9. Lam M, The Five Proven Secrets to Longevity. Academy of Anti-Aging Research. U.S.A
10. Grundy SM, Brewer HB Jr, Cleeman JI, Smith SC Jr, Lenfant C. Definition of metabolic syndrome: Report of the National Heart, Lung, and Blood Institute/American Heart Association conference on scientific issues related to definition. Circulation. 2004;109:433-438.
11. Expert Panel (ATP III). Executive Summary of The Third Report of The National Cholesterol Education Program (NCEP) Expert Panel on Detection, Evaluation, and Treatment of High Blood Cholesterol in Adults (Adult Treatment Panel III). JAMA. 2001;285:2486-2497.
12. Alberti KG, Zimmet PZ. Definition, diagnosis and classification of diabetes mellitus and its complications. Part 1: Diagnosis and classification of diabetes mellitus provisional report of a WHO consultation. Diabet Med. 1998;15:539-553.
13. Paolisso G, Barbagallo M: Diabetes mellitus, and insulin resistance: the role of intracellular magnesium. Am J Hypertens 1997; 10:346-355.
14. Holt, S. Combat Syndrome X, Y and Z Newark, NJ: Wellness Publishing, 2002.

15. Braaten JT, Wood PJ, Scott FW, et al. Oat beta-glucan reduces blood cholesterol concentration in hypercholesterolemic subjects. Eur J Clin Nutr. 1994; 48:465-474.
16. Inglett GE. Nutrient patent for beta glucan from cereals. U.S. Patent No. 6,060,519, 2000.
17. Glore SR, Van Treeck D, Knehans AW. Guild M: Soluble fiber and serum lipids. A literature review. J Am Diet Assoc 1994; 94:425-436.
18. Hallfrisch J, Schofield DJ and Behall KM. Diets containing soluble oat extracts improves glucose and insulin responses of moderately hypercholesterolemic men and women. Am J Clin Nutr 1995; 61:379-82.
19. Chowdhury MJA, Rokeya B, et. al. Low glycemic flour for diabetes management. Diabetes and Endocrine Journal.2007; (in press).

OBESITY-ENHANCED COLON CANCER: A FUNCTIONAL FOOD APPROACH FOR PREVENTION

J. Vanamala, C.C. Tarver, P.S. Murano

Center for Obesity Research and Program Evaluation, Department of Nutrition and Food Science, Texas A&M University, College Station, TX, USA.

Corresponding author: Jairam Vanamala, E-mail: jairam@tamu.edu

Keywords: obesity, colon cancer, functional foods

INTRODUCTION

Over the past few decades, the prevalence of obesity has reached epidemic proportions in the U.S. and in many other countries around the world [1]. In the U.S. alone, obesity has increased from 15% to nearly one-third of the population since 1980 [2]. It is common knowledge that obesity-related diseases such as type II diabetes and certain cancers are increasingly becoming important public health problems.

A common feature of many obesity-related diseases is insulin resistance, which is defined as impaired biological response to the action of insulin [3]. Insulin resistance is characterized by i) tissue resistance to insulin action, ii) compensatory hyperinsulinemia to maintain normal blood glucose levels, and iii) excessive circulation of free fatty acids. These features lead to the development of several abnormal conditions associated with insulin resistance, which include: i) dyslipidemia, ii) elevated blood pressure, iii) glucose intolerance, iv) increased levels of inflammatory markers, v) prothrombotic changes in hemostatic factors, and vi) abdominal obesity [4]. The progression of insulin resistance to type II diabetes in obese individuals further enhances the risk of colorectal cancer [5-7].

Colorectal cancer is the second leading cause of cancer death (~52,180/yr) in the U.S., and effects men and women equally [8]. Although the exact causes of colorectal cancer remain unknown, this cancer is strongly correlated with obesity, particularly in men [9-15].

Despite the grim reality of colorectal cancer, diets rich in fruits and vegetables are protective against colorectal cancer [16-18]. However, it is important to understand how fruits and vegetables influence the molecular mechanisms involved in colorectal cancer to achieve optimal and safe chemoprevention.

Colorectal cancer is a multi-step process reflecting genetic alterations that drive the progressive transformation of normal cells into highly malignant ones [19]. The clonally expanding cell population is generally termed a preneoplastic lesion. Genetic alterations within the cell are initiated by the activation of a mutation in oncogenes or inactivation of tumor suppressor genes [20]. These alterations often result in dysregulation of cell proliferation and/or apoptosis [21-23], which can lead to the formation of aberrant crypt foci (ACF), a purported preneoplastic lesion of colorectal cancer. Rodent models of ACF have been used extensively to screen chemoprotective agents [24, 25].

Aberrant crypt formation is one of the best-characterized biomarkers for colorectal cancer in humans and carcinogen-induced animal models, and it involves expansion of the proliferative zone and inhibition of apoptosis [26]. Areas with four or more aberrant crypts are termed as high multiplicity aberrant crypt foci (HMACF) and correlate particularly well with the incidence of colorectal adenomas and carcinomas in rat models utilizing azoxymethane (AOM), a colon-specific carcinogen [27]. Both elevated insulin levels – a hallmark of insulin resistance and type II diabetes – and chronic colonic inflammation enhance the growth of aberrant crypt foci through various molecular pathways [28-32].

ELEVATED INSULIN LEVELS ENHANCE COLORECTAL CANCER

Accumulating evidence implicates insulin in the development of colorectal cancer [5-7, 28, 33-37]. Elevated levels of insulin are associated with increased risk of colorectal cancer and decreased apoptosis in normal rectal mucosa in humans [33, 34], and patients with type II diabetes are at an increased risk of colorectal cancer [5-7]. Furthermore, insulin injections enhance the multiplicity of ACF [35] and increase the number and size of intestinal tumors compared with saline injections [36] in rats. Direct measurements with hyperinsulinemic-euglycemic clamping strongly correlated insulin resistance with ACF multiplicity [28]. Tran et al. showed that intravenous infusion of insulin increased colorectal epithelial proliferation in rats in a dose-dependent manner [37]. In the same study, combining insulin with glucose and Intralipid (a commercially available emulsifier for i.v. use) infusion resulted in greater hyperinsulinemia and proliferation than infusion of insulin alone.

Current hypotheses hold that insulin may enhance colonocyte proliferation and decrease apoptosis via the insulin-like growth factor (IGF) system (Figure 1). The IGF system includes ligands (insulin, IGF-I and IGF-II), cell-surface receptors (IGF-IR, IGF-IIR, IR-A, IR-B, and hybrid IGF-IR/IR), and high-affinity-binding proteins (IGFBP-1 through 6) [38].

One common feature of many cancers is the overexpression of growth factor receptors [19], and IGF-IR is overexpressed in colorectal cancer [39]. IGF-IR mediates the anti-apoptotic activity of IGF ligands through the PI3K/Akt cascade and is abundantly expressed in colon cancer cells [40]. Experiments have shown that blocking IGF-IR inhibits tumor growth while enhancing chemotherapy-induced apoptosis [41]. Moreover, lower circulating IGF-I levels reduce the incidence of tumor growth and hepatic metastasis in a mouse model of colon cancer [42].

The “bioactive” theory of IGF-I suggests that freely circulating IGF-I is "bioactive" and able to activate IGF-IR. However, most IGF-I is complexed with IGF binding proteins, which prevent IGF-I from interacting with IGF-IR. Indeed, 70% to perhaps as much as 90% of total IGF-I is bound to IGFBP-3 [39, 43]. Although IGFBP-3 accounts for most of the bound IGF-I in the circulation, IGFBP-1 and -2 are apparently the key IGF binding proteins in terms of insulin regulation of the IGF system [reviewed in 43]. Insulin is the most important regulator of IGFBP-I synthesis (at the transcriptional level), and it also regulates IGFBP-2 (at the translational level) [reviewed in 43]. Hyperinsulinemia down-regulates IGFBP-1 and 2 levels [reviewed in 43] and may result in greater concentrations of free IGF-I that stimulate colonocyte proliferation and suppress apoptosis. Interestingly, mean concentrations of IGFBP-1 and 2 are lower and free IGF levels are higher in obese than non-obese individuals [44].

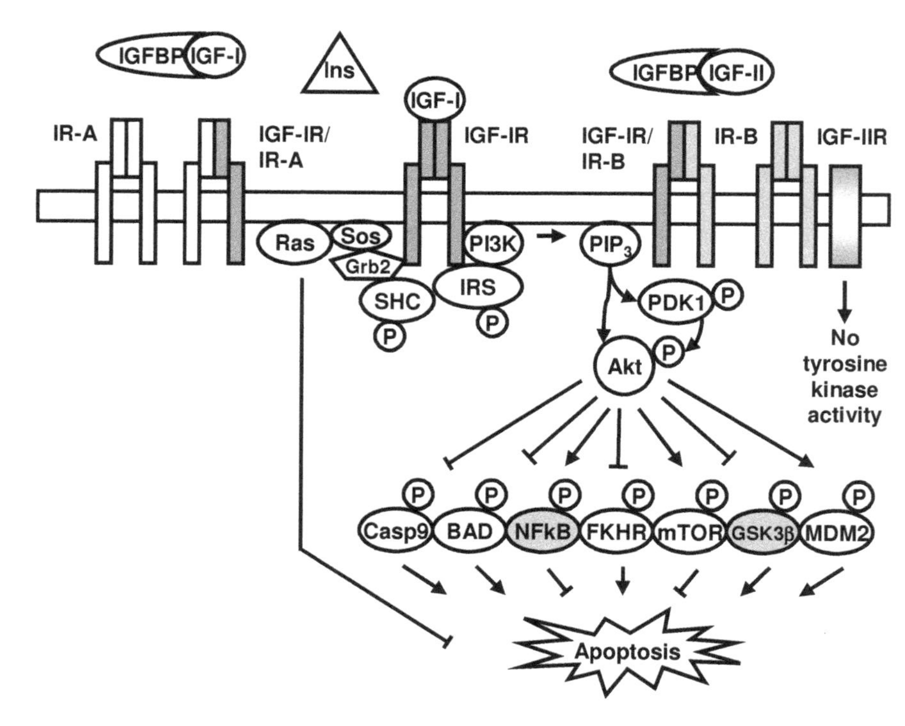

Figure 1. IGF system and pathways involved in apoptosis. Unless complexed to IGF binding proteins (IGFBPs), IGF ligands IGF-I, IGF-II, and insulin (Ins) can bind to the IGF system receptors with varying affinities. For simplicity, only signaling initiated by activated IGF-IR is shown. Activation of IGF-IR recruits and phosphorylates both insulin-receptor substrate (IRS) and SHC proteins, which then serve as anchors for other signaling molecules. The phosphorylated SHC binds growth factor receptor bound protein 2 (Grb2)-Son of sevenless (Sos) complex and activates the Ras-MAPK pathway. Phosphorylated IRS activates the PI3K/Akt pathway, which in turn activates several signaling pathways involved in apoptosis regulation. Similar pathways are activated by IR and the hybrid receptors. Binding of ligands to IGF-IIR results in no tyrosine kinase activity, and loss of IGF-IIR is thought to enhance interaction of IGF-II with IGF-IR.

The main signaling pathway for IGF-IR-mediated cell survival/anti-apoptosis is through the PI3K/Akt cascade [45-49]. This pathway leads to several anti-apoptotic endpoints as shown in Figure 1 [50]. One of these, glycogen synthase kinase 3β (GSK3β), links the IGF system to cell survival through stabilization of β-catenin, a down-stream effector of the Wnt pathway [32]. Insulin and IGF-I were shown to increase cytoplasmic β-catenin levels and subsequently stimulate TCF-dependent transcription through inhibition of GSK3β via the PI3K/Akt pathway (the β-catenin/TCF pathway is explained in greater detail below) [32]. The association of IGF-IR with the PI3K/Akt pathway also links the IGF system to the mitogenic and anti-apoptotic prostaglandin E2 (PGE2), an integral part of the prostaglandin pathway.

INFLAMMATION PROMOTES COLON CANCER

Inflammatory bowel disease (IBD), most commonly in the form of ulcerative colitis and to a lesser degree Crohn's disease, is a risk factor for colorectal cancer [51]. Ulcerative colitis is restricted to the colon, but Crohn's disease can affect any part of the digestive tract, though it most often affects the latter part of the small intestine (ileum) and the colon. Whereas ulcerative colitis affects only the mucosa of the colon in a continuous distribution, Crohn's disease can affect the entire thickness of the bowel wall and may be distributed in patches [51]. Interestingly, consumption of a high-fat diet is a risk factor for IBD [52-55].

Individuals suffering with ulcerative colitis are 20 times as likely to develop colorectal cancer compared to normal subjects [56]. In fact, chronic inflammatory conditions such as ulcerative colitis have been proposed as a possible pathway for the development of colorectal cancer [56, 57]. In this pathway, two pro-inflammatory enzymes, inducible nitric oxide synthase (iNOS) and cyclooxygenase-2 (COX-2) have both been implicated in colon carcinogenesis [31, 58-60]. Indeed, enhanced expression of iNOS and COX-2 have been shown in colorectal adenomas and adenocarcinomas in humans and chemically induced colonic tumors in rats [30, 61-63].

During chronic inflammation, iNOS activity is greatly induced and excessive nitric oxide (NO) production occurs [64]. NO has been suggested to induce DNA damage, act as a reactive nitrogen species, and modulate cell proliferation and apoptosis [64, 65]. In the ACF system, agents that inhibit iNOS also inhibit ACF formation, and thus may protect against colon tumorigenesis through this activity [66, 67]. Interestingly, NO is implicated in the regulation of COX-2 activity as well [68, 69]. Salvemini et al. [68] demonstrated in both E. coli lipopolysaccharide (LPS)-activated macrophages and in LPS-treated male Sprague-Dawley rats that NO inhibition resulted in attenuated prostaglandin (PG) synthesis, a major product of the COX enzyme. A later study put forward a possible link between NO and PG synthesis by both the COX-1 and -2 enzymes [69]. Peroxynitrite, a coupling product of NO and superoxide anion, increased COX activity and PG synthesis [69]. The authors suggested that peroxynitrite, an inorganic hydroperoxide, may act as a substrate for the peroxidase portion of the COX enzyme, thus inducing its activity [69].

Cyclooxygenases, also known as prostaglandin endoperoxidase (PGH) synthases, are key enzymes involved in the production of prostaglandins using arachidonic acid (AA) as a precursor [70]. PGH synthases have both a peroxidase activity and a cyclooxygenase activity at two distinct sites [70]. COX-2, the inducible isoform, is not usually detectable in normal cells, but its expression and activity are greatly induced in inflammation [70]. COX-2 also has been detected in 50% of colorectal adenomas and in up to 85% of colorectal cancer [30]. This and other accumulating evidence suggests that COX-2 may play an important role in colorectal carcinogenesis [71, 72]. COX-2 may facilitate colon cancer progression by stimulating cell proliferation and survival [73, 74], tumor cell invasiveness [75] and the production of angiogenic agents in colon cancer cells [76]. In rodent models of colon carcinogenesis, COX-2 inhibitors have been shown to reduce ACF and the incidence, multiplicity, and size of tumors [31, 70, 77]. Importantly, nuclear factor κB (NF-κB) – a transcription factor that regulates a variety of genes important in cellular responses, such as inflammation and colonocyte proliferation and apoptosis [78] – is a positive regulator of COX-2 expression in human colon adenocarcinoma cell lines exposed to LPS [79]. Furthermore, reduced prostaglandin biosynthesis through inhibition of COX-2 activity is the suggested molecular basis for the chemopreventive effects of nonsteroidal anti-inflammatory drugs (NSAIDs) on colorectal tumorigenesis in human and rodents [71, 72].

PGE2 is the major prostaglandin product of the cylooxygenase pathway found in many solid tumors, including colorectal tissue [80]. It is the product of COX-2 and prostaglandin E synthase action on AA, which is derived from diacylglycerol or phospholipids (Figure 2). PGE2 is implicated in abnormal cell proliferation, apoptosis inhibition, induction of angiogenesis, and also has been found at increased levels in colon cancer tissue [31, 70]. PGE2 protects colonocytes from programmed cell death, thus promoting tumor growth by activating peroxisome proliferator-activated receptor (PPAR) β/δ, a lipid-activated transcription factor that belongs to the nuclear receptor superfamily [80].

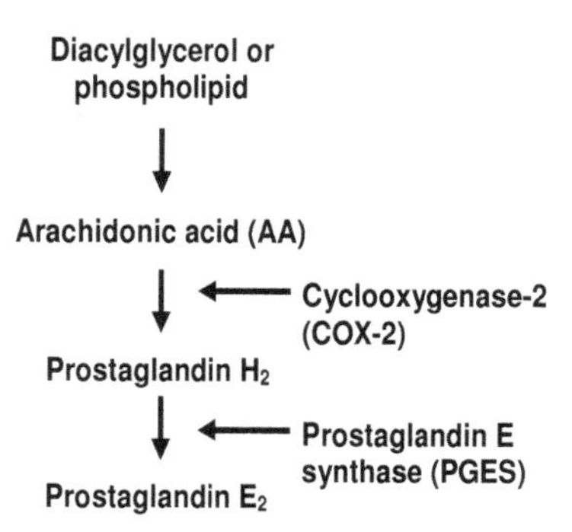

Figure 2. PGE2 production. Prostaglandin E2 (PGE2) is an important metabolite of prostaglandin pathway. The body converts diacylglycerol and phospholipids into fatty acids such as arachidonic acid (AA). Inflammation induces cyclooxygenase-2 to convert AA into Prostaglandin H2 (PGH2), which in turn converted to PGE2 by Prostaglandin E synthase (PGES) enzymes.

PPARβ/δ transcription is mediated by the Wnt/β-catenin pathway (Figure 3), which is activated in colon carcinogenesis in humans and experimentally-induced rodent models [29, 81]. Under homeostatic conditions, most β-catenin is bound to α-catenin and cadherin in the plasma

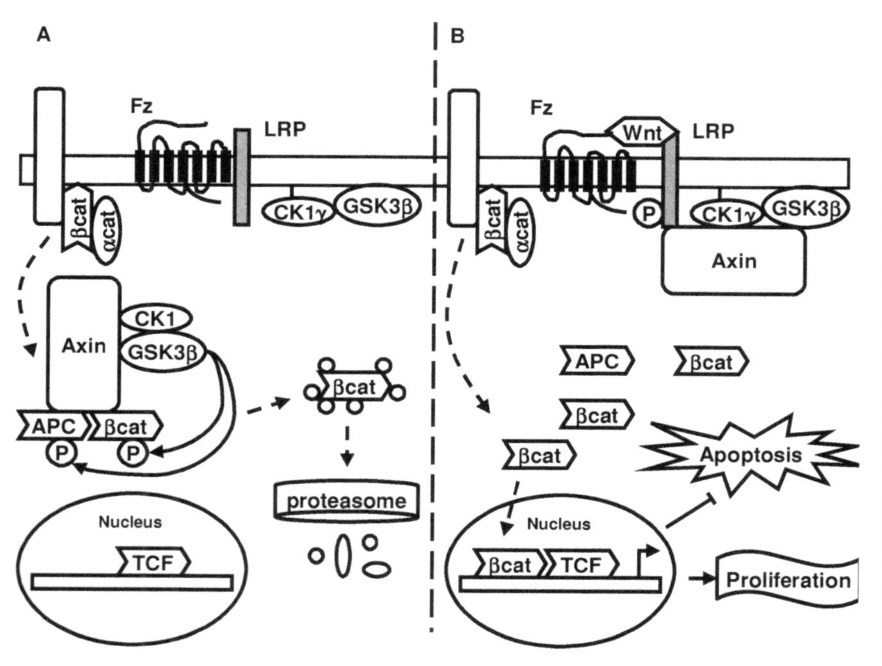

Figure 3. Canonical Wnt signaling. Most β-catenin (βcat) complexes with cadherin and α-catenin (αcat) in the plasma membrane, but low levels reside in the cytosol. A) In the non-stimulated system, Axin serves as a scaffold for adenomatous polyposis coli (APC), cytosolic β-catenin, casein kinase 1 (CK1) and glycogen sythnase kinase 3β (GSK3β). The latter two phosphorylate APC and β-catenin, preventing β-catenin from translocating into the nucleus. The phosphorylated β-catenin undergoes ubiquitination and subsequent degradation by the proteasome. B) A Wnt ligand binds to the Frizzled/LRP coreceptor complex. Casein kinase 1γ (CK1γ) and GSK3β are thought to phosphorylate LRP, which recruits Axin away from the destruction complex. Without the destruction complex to bind it, β-catenin stabilizes in the cytosol and translocates into the nucleus, where it binds TCF/Lef and promotes the transcription of mitogenic and anti-apoptotic Wnt genes. APC mutation or damage to APC or β-catenin by radiation or carcinogens such as AOM can circumvent the tightly controlled Wnt system. These factors prevent formation of the destruction complex and allow cytosolic β-catenin accumulation and nuclear translocation. In effect, this turns the Wnt system perpetually "on," leading to enhanced proliferation and decreased apoptosis which may result in colon carcinogenesis. "Solid" lines indicate positive or negative regulation, "dashed" lines indicate movement.

membrane. Steady-state levels of cytoplasmic or nuclear β-catenin are very low as it is strictly regulated [82]. When the Frizzled/LRP coreceptor complex is inactive, cytosolic β-catenin is bound to a destruction complex containing Axin, adenomatous polyposis coli (APC), and GSK3β [83]. GSK3β consitutively phosphorylates β-catenin, which cytoplasmic proteosomes then degrade. However, when a Wnt ligand binds to the Frizzled/LRP coreceptor, casein kinase 1γ (CK1γ) and GSK3β are thought to phosphorylate LRP and recruit Axin away from the destruction complex [83]. No longer phosphorylated by GSK3β, β-catenin now accumulates in the cytosol and translocates into the nucleus. Nuclear β-catenin binds to T-cell-factor 4 (TCF-4) and stimulates transcription of mitogenic genes including PPARβ/δ and cyclin D1 [83, 84].

APC or β-catenin mutation by radiation or carcinogens such as AOM can circumvent the tightly controlled Wnt system [85-87]. These factors prevent formation of the destruction complex and allow cytosolic β-catenin accumulation and nuclear translocation. In fact, the loss of functional APC or β-catenin is considered one of the earliest events in colon cancer [88].

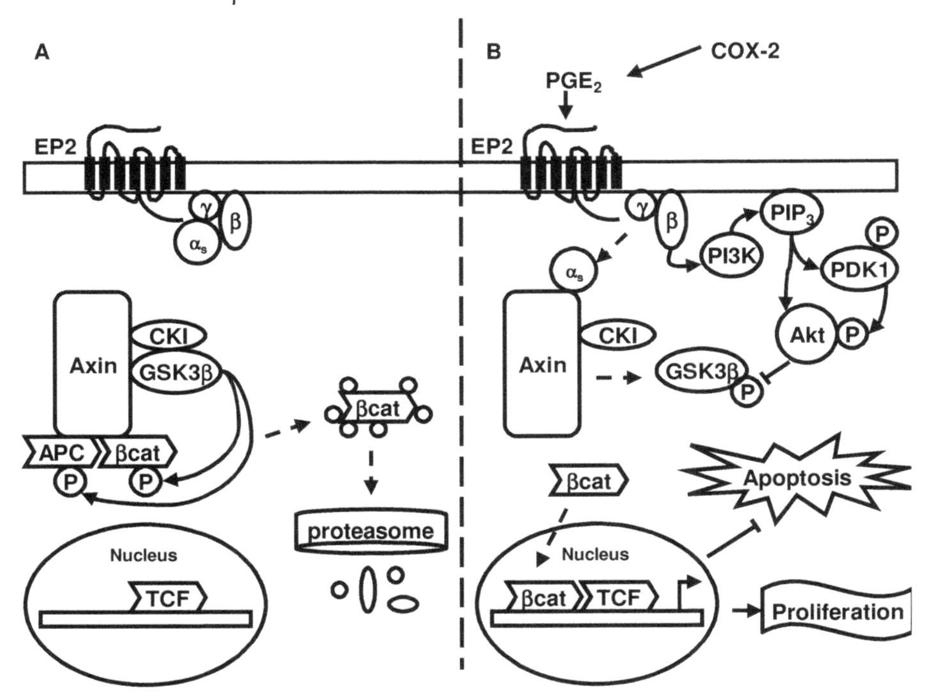

Figure 4. Interaction between PGE2 and the Wnt pathway. A) In the non-stimulated system, a complex containing Axin (which serves as a scaffolding protein), APC, CK1 and GSK3β prevents β-catenin (βcat) from translocating into the nucleus. β-catenin subsequently undergoes ubiquitination and degradation by the proteasome. B) Overexpression of COX-2 leads to the generation of PGE2, which can activate EP2 receptors. This activation causes the β and γ subunits of Gs protein to stimulate the PI3K/Akt pathway as the α subunit binds to Axin and promotes the release of GSK3β. Phosphorylation of GSK3β by the PI3K/Akt pathway protects β-catenin from phosphorylation, allowing it to stabilize in the cytosol and translocate into the nucleus. Once in the nucleus, β-catenin binds to TCF/Lef and promotes the transcription of mitogenic and anti-apoptotic genes. "Solid" lines indicate positive or negative regulation, "dashed" lines indicate movement.

As mentioned above, the association of IGF-IR with the PI3K/Akt pathway links the IGF system to the prostaglandin pathway. Di Popolo et al. showed that IGF-II up-regulates COX-2 mRNA expression and PGE2 synthesis in colon cancer cells, apparently through IGF-IR [89]. In that study, treatment of colon cancer cells with an IGF-IR blocking antibody suppressed COX-2 mRNA expression and PGE2 synthesis, by attenuating PI3K levels.

Elevated levels of PGE2 from IGF-II and/or colonic inflammatory stimuli are then able to activate genes such as PPARβ/δ and cyclin D1 indirectly through the PI3K/Akt cascade via autocrine action on the EP2 cell surface receptors, as shown in Figure 4 [29].

Therefore, colonic inflammation and elevated insulin levels resulting from insulin resistance or type II diabetes – both important risk factors for colorectal cancer – can operate through separate molecular pathways to promote colon tumor formation (Figure 5). These pathways converge on the Wnt/β-catenin system by promoting stabilization of cytosolic β-catenin in colonocytes. The β-catenin then translocates into the nucleus and initiates the transcription of mitotic and anti-apoptotic genes such as PPARβ/δ and cyclin D1.

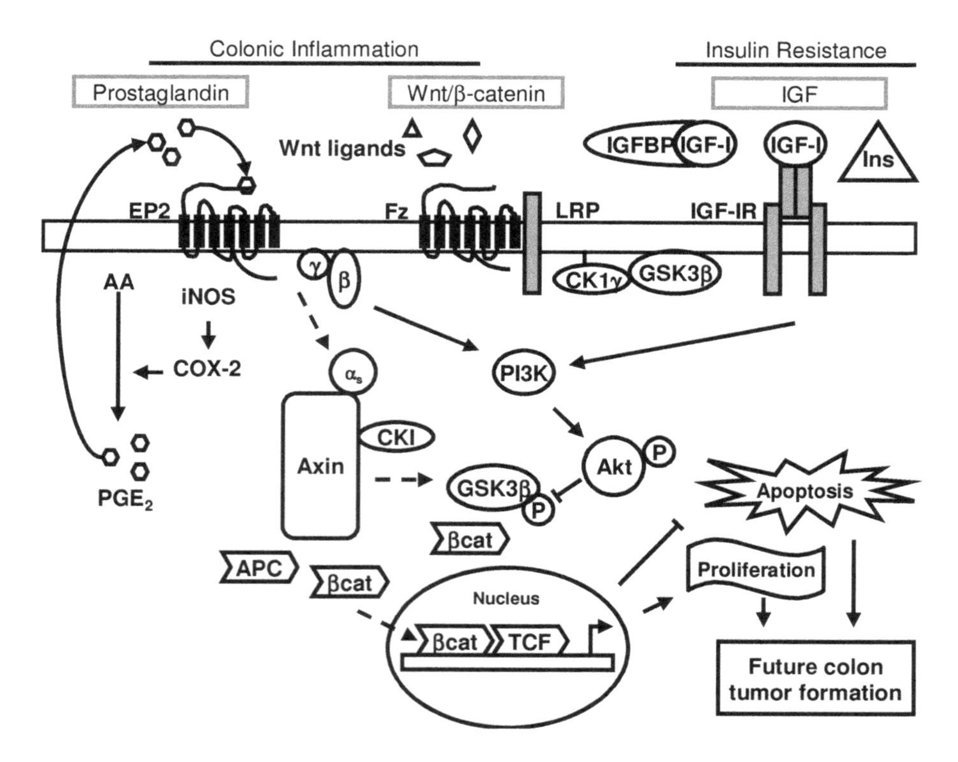

Figure 5. Molecular pathways linking colonic inflammation and insulin resistance to colon carcinogenesis. Colonic inflammation-induced stimulation of PGE2 production in the prostaglandin pathway and hyperinsulinemic stimulation of IGF receptors converge on the Wnt pathway to enhance β-catenin-mediated transcription of mitogenic and anti-apoptotic factors. Chronic activation of these pathways, which are implicated in obesity-enhanced colon carcinogenesis, can lead to future colon tumor formation. Fruits, vegetables, and their bioactive compounds may help suppress these pathways and protect

against colon cancer. Only key points highlighted in the text are shown here. "Solid" lines indicate positive or negative regulation, "dashed" lines indicate movement.

FRUITS AND VEGETABLES MAY HELP PREVENT COLORECTAL CANCER

Multiple experimental and epidemiological studies over the past four decades have shown that diets rich in fruits and vegetables are protective against various cancers, including colorectal cancer [16-18]. As Table 1 demonstrates, many of these fruits and vegetables act as antioxidants and/or affect the molecular pathways summarized above, suggesting potential mechanisms for their chemoprotective abilities. Others have been shown to affect biomarkers (e.g., ACF, tumors) or risk factors (e.g., colitis, hyperinsulinemia) for colorectal cancer, but their precise functions remain unknown.

Whole fruits and vegetables or their bioactive compounds may act synergistically. That is, their combined effect is greater than their individual effects. For example, oral administration of garlic and tomato together led to less colonocyte proliferation and greater apoptosis than garlic or tomato alone [90]. Franke et al. [91] showed that the combination of two isoflavones, genistein and daidzein, inhibited neoplastic transformation induced by 3-methylcholanthrene in C3H 10T1/2 murine fibroblasts better than each separately.

The synergistic/additive effects of bioactive compounds may be responsible for the greater chemoprotective ability of whole fruits than their isolated constituents. Seeram et al. found that whole pomegranate juice demonstrated greater antioxidant and antiproliferative activity in several human colon cancer cell lines than its constituents punicalagin, ellagic acid, and total tannins [92]. Similarly, pomegranate juice induced apoptosis in HT-29 cells while concentrations of the constituent compounds equalized to amounts found in pomegranate juice had no effect. We also found that whole grapefruit led to a greater colonocyte apoptotic index than isolated naringin, a major flavonoid, in an animal model of colon carcinogenesis [93].

CONCLUSION

In summary, fruits and vegetables exert chemoprotective effects by modulating molecular pathways that are important to the progression of colorectal cancer in normal and insulin resistant and/or obese individuals. Many of these compounds exhibit an additive and/or synergistic effect, which means that appropriate combinations and doses of these compounds prove more effective than individual compounds. Therefore, it is important to understand how fruits, vegetables, and their bioactive compounds influence the molecular mechanisms involved in obesity-enhanced colorectal cancer. This knowledge will enable to develop effective functional food based chemoprotective measures.

Table 1. Influence of functional foods on colon carcinogenesis

	Functional Food	Dose/Purity	Cell Line/Model	Measured Effect
	Apple			
[94]	Juice, "clear"*	*Ad libitum* (avg. 21.5 mL/day)	F344 rats, colon	↓ proliferation
[94]	Juice, "cloudy"	*Ad libitum* (avg. 22.9 mL/day)	F344 rats, colon	↓ proliferation, ↓HMACF
	Black cumin			
[95]	Seed oil	200 mg/kg body wt/day	F344 rats, colon	↓ ACF, ↓HMACF
	Blueberry (Briteblue)			
[96]	Crude extract	IC50: ~4000 μg/mL	HT-29, Caco-2	↓ proliferation
	Brussels sprouts			
[97]	Juice	5% (v/v) in water, daily	F344 rats, colon	↓ ACF
	Caraway			
[98]	Seed	30 – 90 mg/kg body wt/day	Wistar rats, colon	↓ ACF
	Carrot			
[99]	Whole carrot	10% diet (w/w)	BDIX/OrlIco rats	↓HMACF
	Fenugreek			
[100]	Seed powder	1% (w/w) diet	F344 rats, colon	↓ ACF, ↓ HMACF
	Garden Cress			
[101]	Juice	5% in water, daily	F344 rats	↓ ACF, ↓ HMACF
	Garlic			
[90, 102]	Aqueous suspension (2%)	20 mg/day	SD rats, colon	↓ ACF, ↓ proliferation, ↑ apoptosis, ↓ COX-2
	Grape			
[103, 104]	Red wine	50 mg total polyphenols/kg body wt/day	F344 rats, colon	↓ tumors, ↑ apoptotic index, ↓ COX-2 (mRNA), ↓ iNOS (mRNA)
	Grapefruit			
[105]	Juice	237 mL (8 oz.), 3 times/day	Obese humans	↑ wt loss
[105]	Whole fruit	1/2 grapefruit, 3 times/day	Obese humans	↑ wt loss, ↓ 2 h insulin
[93]	Whole fruit	13.5 g/kg diet	SD rats	↓ ACF, ↓ HMACF, ↓ COX-2, ↓ iNOS
	Olive			
[106]	Skin	IC50: ~100 μM; ~68 – 410 μM (apoptosis)	HT-29	↓ proliferation, ↑ apoptosis
	Orange			
[107]	Juice	*Ad libitum* (avg. 29.5 mL/day)	F344 rats, colon	↓ tumor incidence, ↓ proliferation
	Pomegranate			

	Functional Food	Dose/Purity	Cell Line/Model	Measured Effect
[92, 108]	Juice (concentrate)	6 – 100 μg/mL	HT-29	↓ proliferation, ↑ apoptosis, ↓ COX-2, ↓ Akt activity, ↓ NF-κB binding
	Sebestian plum			
[109]	Whole fruit	Homogenized and suspended in water; 12g/kg body wt/day	Wistar rats, colon, plasma	↓ inflammation, ↑ GSH-Px, ↑ SOD, ↑ selenium, ↑ plasma TAS
	Tea, black			
[103, 104]	Decaffeinated extract	50 mg total polyphenols/kg body wt/day	F344 rats, colon	↓ tumors, ↑ apoptotic index, ↓ COX-2 (mRNA), ↓ iNOS (mRNA)
	Tea, green			
[110]	Water extract	2 g tea/100 mL water, daily	Wistar rats	↓ ACF, ↓ p21 ras
[111]	Water extract	1.5% (w/v), daily	APCmin mice	↓ tumor multiplicity, ↓ β-catenin, ↓ cyclin D1, ↓ c-jun
[111]	Water extract	Containing 25 μM EGCG	HEK293	↓ β-catenin
	Tea, white			
[111]	Water extract	Containing 25 μM EGCG	HEK293	↓ β-catenin
	Tomato			
[90]	Aqueous suspension (2%)	20 mg/day	SD rats, colon	↓ ACF, ↓ proliferation, ↑ apoptosis, ↓ COX-2
	Turmeric			
[112]	Curcumin	0.5 – 4% diet; 77% curcumin, 17% demethoxycurcumin, 3% bisdemethoxycurcumin	CF-1 mice, colon	↓ adenomas, ↓ adenocarcinomas, ↓ adenoma volume, ↓ adenocarcinoma volume
[112]	Curcumin	2% diet; 97% pure	CF-1 mice, colon	↓ adenomas, ↓ adenocarcinomas, ↓ adenoma volume, ↓ adenocarcinoma volume
[113]	Curcumin	–	HT-29, SW480	↓ proliferation, ↑ apoptosis, ↓ COX-2, ↓ PGE2
[114]	Curcumin	1 – 50 μM; > 80% pure	HCA-7, IEC-18, SCC450, SKGT4	↓ COX-2 (protein, mRNA, and activity); ↓ PGE2

*Pectin removed from juice; **ACF**, aberrant crypt foci; **AMPK**, AMP-activated protein kinase; **COX-2**, cyclooxygenase 2; **GSH-Px**, glutathione peroxidase; **GSK3β**, glycogen synthase kinase 3β; **HMACF**, high multiplicity aberrant crypt foci (>4 ACF); **IC50**, concentration required for 50% inhibition; **iNOS**, inducible nitrogen oxide synthase; **NF-κB**, nuclear factor κB; **PARP**, poly(ADP-ribose) polymerase; **SD**, Sprague-Dawley; **SOD**, superoxide dismutase; **TAS**, total antioxidant status; **VEGF**, vascular endothelial growth factor

REFERENCES

1. Flier, J.S., *Obesity wars: molecular progress confronts an expanding epidemic.* Cell, 2004. 116(2): p. 337-50.
2. Flegal, K.M., M.D. Carroll, C.L. Ogden, and C.L. Johnson, *Prevalence and trends in obesity among US adults, 1999-2000.* JAMA, 2002. 288(14): p. 1723-7.
3. Morisco, C., G. Lembo, and B. Trimarco, *Insulin Resistance and Cardiovascular Risk: New Insights from Molecular and Cellular Biology.* Trends Cardiovasc Med, 2006. 16(6): p. 183-188.
4. Grundy, S.M., H.B. Brewer, Jr., J.I. Cleeman, S.C. Smith, Jr., and C. Lenfant, *Definition of metabolic syndrome: report of the National Heart, Lung, and Blood Institute/American Heart Association conference on scientific issues related to definition.* Arterioscler Thromb Vasc Biol, 2004. 24(2): p. e13-8.
5. Elwing, J.E., F. Gao, N.O. Davidson, and D.S. Early, *Type 2 diabetes mellitus: the impact on colorectal adenoma risk in women.* Am J Gastroenterol, 2006. 101(8): p. 1866-71.
6. Limburg, P.J., R.A. Vierkant, Z.S. Fredericksen, C.L. Leibson, R.A. Rizza, A.K. Gupta, D.A. Ahlquist, L.J. Melton, 3rd, T.A. Sellers, and J.R. Cerhan, *Clinically confirmed type 2 diabetes mellitus and colorectal cancer risk: a population-based, retrospective cohort study.* Am J Gastroenterol, 2006. 101(8): p. 1872-9.
7. Meyerhardt, J.A., P.J. Catalano, D.G. Haller, R.J. Mayer, J.S. Macdonald, A.B. Benson, 3rd, and C.S. Fuchs, *Impact of diabetes mellitus on outcomes in patients with colon cancer.* J Clin Oncol, 2003. 21(3): p. 433-40.
8. American Cancer S*ociety (ACS). How Many People Get Colorectal Cancer? 2006 [cited 2007 Jan. 23]; Available from: http://www.cancer.org/docroot/CRI/content/CRI_2_2_1X_How_Many_People_Get_Colorectal_Cancer.asp?sitearea=*
9. Vainio, H. and F. Bianchini, eds. *Weight Control and Physical Activity.* IARC handbooks of cancer prevention. Vol. 6, 2002, IARC Press: Lyon, France.
10. Shike, M., *Body weight and colon cancer.* Am J Clin Nutr, 1996. 63(3 Suppl): p. 442S-444S.
11. Giacosa, A., S. Franceschi, C. La Vecchia, A. Favero, and R. Andreatta, *Energy intake, overweight, physical exercise and colorectal cancer risk.* Eur J Cancer Prev, 1999. 8 Suppl 1: p. S53-60.
12. Murphy, T.K., E.E. Calle, C. Rodriguez, H.S. Kahn, and M.J. Thun, *Body mass index and colon cancer mortality in a large prospective study.* Am J Epidemiol, 2000. 152(9): p. 847-54.
13. Kono, S., K. Handa, H. Hayabuchi, C. Kiyohara, H. Inoue, T. Marugame, S. Shinomiya, H. Hamada, K. Onuma, and H. Koga, *Obesity, weight gain and risk of colon adenomas in Japanese men.* Jpn J Cancer Res, 1999. 90(8): p. 805-11.
14. Caan, B.J., A.O. Coates, M.L. Slattery, J.D. Potter, C.P. Quesenberry, Jr., and S.M. Edwards, *Body size and the risk of colon cancer in a large case-control study.* Int J Obes Relat Metab Disord, 1998. 22(2): p. 178-84.
15. Ford, E.S., *Body mass index and colon cancer in a national sample of adult US men and women.* Am J Epidemiol, 1999. 150(4): p. 390-8.
16. Block, G., B. Patterson, and A. Subar, *Fruit, vegetables, and cancer prevention: a review of the epidemiological evidence.* Nutr Cancer, 1992. 18(1): p. 1-29.
17. Donaldson, M.S., *Nutrition and cancer: a review of the evidence for an anti-cancer diet.* Nutr J, 2004. 3: p. 19.

18. Steinmetz, K.A. and J.D. Potter, *Vegetables, fruit, and cancer prevention: a review.* J Am Diet Assoc, 1996. 96(10): p. 1027-39.
19. Hanahan, D. and R.A. Weinberg, *The hallmarks of cancer.* Cell, 2000. 100(1): p. 57-70.
20. Lipkin, M., B. Reddy, H. Newmark, and S.A. Lamprecht, *Dietary factors in human colorectal cancer.* Annu Rev Nutr, 1999. 19: p. 545-86.
21. Bird, R.P., *Role of aberrant crypt foci in understanding the pathogenesis of colon cancer.* Cancer Lett, 1995. 93(1): p. 55-71.
22. Bird, R.P. and C.K. Good, *The significance of aberrant crypt foci in understanding the pathogenesis of colon cancer.* Toxicol Lett, 2000. 112-113: p. 395-402.
23. Bird, R.P., D. Salo, C. Lasko, and C. Good, *A novel methodological approach to study the level of specific protein and gene expression in aberrant crypt foci putative preneoplastic colonic lesions by Western blotting and RT-PCR.* Cancer Lett, 1997. 116(1): p. 15-9.
24. Cheng, L. and M.D. Lai, *Aberrant crypt foci as microscopic precursors of colorectal cancer.* World J Gastroenterol, 2003. 9(12): p. 2642-9.
25. Roncucci, L., M. Pedroni, F. Vaccina, P. Benatti, L. Marzona, and A. De Pol, *Aberrant crypt foci in colorectal carcinogenesis. Cell and crypt dynamics.* Cell Prolif, 2000. 33(1): p. 1-18.
26. Zheng, Y., P.M. Kramer, R.A. Lubet, V.E. Steele, G.J. Kelloff, and M.A. Pereira, *Effect of retinoids on AOM-induced colon cancer in rats: modulation of cell proliferation, apoptosis and aberrant crypt foci.* Carcinogenesis, 1999. 20(2): p. 255-60.
27. Magnuson, B.A., I. Carr, and R.P. Bird, *Ability of aberrant crypt foci characteristics to predict colonic tumor incidence in rats fed cholic acid.* Cancer Res, 1993. 53(19): p. 4499-504.
28. Tran, T.T., N. Gupta, T. Goh, D. Naigamwalla, M.C. Chia, N. Koohestani, S. Mehrotra, G. McKeown-Eyssen, A. Giacca, and W.R. Bruce, *Direct measure of insulin sensitivity with the hyperinsulinemic-euglycemic clamp and surrogate measures of insulin sensitivity with the oral glucose tolerance test: correlations with aberrant crypt foci promotion in rats.* Cancer Epidemiol Biomarkers Prev, 2003. 12(1): p. 47-56.
29. Castellone, M.D., H. Teramoto, and J.S. Gutkind, *Cyclooxygenase-2 and colorectal cancer chemoprevention: the beta-catenin connection.* Cancer Res, 2006. 66(23): p. 11085-8.
30. Eberhart, C.E., R.J. Coffey, A. Radhika, F.M. Giardiello, S. Ferrenbach, and R.N. DuBois, *Up-regulation of cyclooxygenase 2 gene expression in human colorectal adenomas and adenocarcinomas.* Gastroenterology, 1994. 107(4): p. 1183-8.
31. Watanabe, K., T. Kawamori, S. Nakatsugi, and K. Wakabayashi, *COX-2 and iNOS, good targets for chemoprevention of colon cancer.* Biofactors, 2000. 12(1-4): p. 129-33.
32. Desbois-Mouthon, C., A. Cadoret, M.J. Blivet-Van Eggelpoel, F. Bertrand, G. Cherqui, C. Perret, and J. Capeau, *Insulin and IGF-1 stimulate the beta-catenin pathway through two signalling cascades involving GSK-3beta inhibition and Ras activation.* Oncogene, 2001. 20(2): p. 252-9.
33. Palmqvist, R., G. Hallmans, S. Rinaldi, C. Biessy, R. Stenling, E. Riboli, and R. Kaaks, *Plasma insulin-like growth factor 1, insulin-like growth factor binding protein 3, and risk of colorectal cancer: a prospective study in northern Sweden.* Gut, 2002. 50(5): p. 642-6.
34. Keku, T.O., P.K. Lund, J. Galanko, J.G. Simmons, J.T. Woosley, and R.S. Sandler, *Insulin resistance, apoptosis, and colorectal adenoma risk.* Cancer Epidemiol Biomarkers Prev, 2005. 14(9): p. 2076-81.
35. Corpet, D.E., C. Jacquinet, G. Peiffer, and S. Tache, *Insulin injections promote the growth of aberrant crypt foci in the colon of rats.* Nutr Cancer, 1997. 27(3): p. 316-20.
36. Tran, T.T., A. Medline, and W.R. Bruce, *Insulin promotion of colon tumors in rats.* Cancer Epidemiol Biomarkers Prev, 1996. 5(12): p. 1013-5.

37. Tran, T.T., D. Naigamwalla, A.I. Oprescu, L. Lam, G. McKeown-Eyssen, W.R. Bruce, and A. Giacca, *Hyperinsulinemia, but not other factors associated with insulin resistance, acutely enhances colorectal epithelial proliferation in vivo.* Endocrinology, 2006. 147(4): p. 1830-7.
38. Sachdev, D. and D. Yee, *Disrupting insulin-like growth factor signaling as a potential cancer therapy.* Mol Cancer Ther, 2007. 6(1): p. 1-12.
39. Davies, M., S. Gupta, G. Goldspink, and M. Winslet, *The insulin-like growth factor system and colorectal cancer: clinical and experimental evidence.* Int J Colorectal Dis, 2006. 21(3): p. 201-8.
40. Weber, M.M., C. Fottner, S.B. Liu, M.C. Jung, D. Engelhardt, and G.B. Baretton, *Overexpression of the insulin-like growth factor I receptor in human colon carcinomas.* Cancer, 2002. 95(10): p. 2086-95.
41. Adachi, Y., C.T. Lee, K. Coffee, N. Yamagata, J.E. Ohm, K.H. Park, M.M. Dikov, S.R. Nadaf, C.L. Arteaga, and D.P. Carbone, *Effects of genetic blockade of the insulin-like growth factor receptor in human colon cancer cell lines.* Gastroenterology, 2002. 123(4): p. 1191-204.
42. Wu, Y., S. Yakar, L. Zhao, L. Hennighausen, and D. LeRoith, *Circulating insulin-like growth factor-I levels regulate colon cancer growth and metastasis.* Cancer Res, 2002. 62(4): p. 1030-5.
43. Frystyk, J., *Free insulin-like growth factors -- measurements and relationships to growth hormone secretion and glucose homeostasis.* Growth Horm IGF Res, 2004. 14(5): p. 337-75.
44. Nam, S.Y., E.J. Lee, K.R. Kim, B.S. Cha, Y.D. Song, S.K. Lim, H.C. Lee, and K.B. Huh, *Effect of obesity on total and free insulin-like growth factor (IGF)-1, and their relationship to IGF-binding protein (BP)-1, IGFBP-2, IGFBP-3, insulin, and growth hormone.* Int J Obes Relat Metab Disord, 1997. 21(5): p. 355-9.
45. Chakravarti, A., J.S. Loeffler, and N.J. Dyson, *Insulin-like growth factor receptor I mediates resistance to anti-epidermal growth factor receptor therapy in primary human glioblastoma cells through continued activation of phosphoinositide 3-kinase signaling.* Cancer Res, 2002. 62(1): p. 200-7.
46. Datta, S.R., H. Dudek, X. Tao, S. Masters, H. Fu, Y. Gotoh, and M.E. Greenberg, *Akt phosphorylation of BAD couples survival signals to the cell-intrinsic death machinery.* Cell, 1997. 91(2): p. 231-41.
47. Kulik, G., A. Klippel, and M.J. Weber, *Antiapoptotic signalling by the insulin-like growth factor I receptor, phosphatidylinositol 3-kinase, and Akt.* Mol Cell Biol, 1997. 17(3): p. 1595-606.
48. Remacle-Bonnet, M.M., F.L. Garrouste, S. Heller, F. Andre, J.L. Marvaldi, and G.J. Pommier, *Insulin-like growth factor-I protects colon cancer cells from death factor-induced apoptosis by potentiating tumor necrosis factor alpha-induced mitogen-activated protein kinase and nuclear factor kappaB signaling pathways.* Cancer Res, 2000. 60(7): p. 2007-17.
49. Zheng, W.H., S. Kar, and R. Quirion, *Insulin-like growth factor-1-induced phosphorylation of the forkhead family transcription factor FKHRL1 is mediated by Akt kinase in PC12 cells.* J Biol Chem, 2000. 275(50): p. 39152-8.
50. Hennessy, B.T., D.L. Smith, P.T. Ram, Y. Lu, and G.B. Mills, *Exploiting the PI3K/AKT pathway for cancer drug discovery.* Nat Rev Drug Discov, 2005. 4(12): p. 988-1004.
51. Crohn's and Colitis Foundation of America. *About Crohn's.* 2006 [cited 2007 Jan. 25]; Available from: http://www.ccfa.org/info/about/crohns.
52. Sakamoto, N., S. Kono, K. Wakai, Y. Fukuda, M. Satomi, T. Shimoyama, Y. Inaba, Y. Miyake, S. Sasaki, K. Okamoto, G. Kobashi, M. Washio, T. Yokoyama, C. Date, and H.

Tanaka, *Dietary risk factors for inflammatory bowel disease: a multicenter case-control study in Japan.* Inflamm Bowel Dis, 2005. 11(2): p. 154-63.

53. Geerling, B.J., P.C. Dagnelie, A. Badart-Smook, M.G. Russel, R.W. Stockbrugger, and R.J. Brummer, *Diet as a risk factor for the development of ulcerative colitis.* Am J Gastroenterol, 2000. 95(4): p. 1008-13.
54. Shoda, R., K. Matsueda, S. Yamato, and N. Umeda, *Epidemiologic analysis of Crohn disease in Japan: increased dietary intake of n-6 polyunsaturated fatty acids and animal protein relates to the increased incidence of Crohn disease in Japan.* Am J Clin Nutr, 1996. 63(5): p. 741-5.
55. Reif, S., I. Klein, F. Lubin, M. Farbstein, A. Hallak, and T. Gilat, *Pre-illness dietary factors in inflammatory bowel disease.* Gut, 1997. 40(6): p. 754-60.
56. Potter, J.D., *Colorectal cancer: molecules and populations.* J Natl Cancer Inst, 1999. 91(11): p. 916-32.
57. Philpott, M. and L.R. Ferguson, *Immunonutrition and cancer.* Mutat Res, 2004. 551(1-2): p. 29-42.
58. Crowell, J.A., V.E. Steele, C.C. Sigman, and J.R. Fay, *Is inducible nitric oxide synthase a target for chemoprevention?* Mol Cancer Ther, 2003. 2(8): p. 815-23.
59. Sporn, M.B. and N. Suh, *Chemoprevention of cancer.* Carcinogenesis, 2000. 21(3): p. 525-30.
60. Taketo, M.M., *Cyclooxygenase-2 inhibitors in tumorigenesis (Part II).* J Natl Cancer Inst, 1998. 90(21): p. 1609-20.
61. Ambs, S., W.G. Merriam, W.P. Bennett, E. Felley-Bosco, M.O. Ogunfusika, S.M. Oser, S. Klein, P.G. Shields, T.R. Billiar, and C.C. Harris, *Frequent nitric oxide synthase-2 expression in human colon adenomas: implication for tumor angiogenesis and colon cancer progression.* Cancer Res, 1998. 58(2): p. 334-41.
62. Sano, H., Y. Kawahito, R.L. Wilder, A. Hashiramoto, S. Mukai, K. Asai, S. Kimura, H. Kato, M. Kondo, and T. Hla, *Expression of cyclooxygenase-1 and -2 in human colorectal cancer.* Cancer Res, 1995. 55(17): p. 3785-9.
63. Takahashi, M., K. Fukuda, T. Ohata, T. Sugimura, and K. Wakabayashi, *Increased expression of inducible and endothelial constitutive nitric oxide synthases in rat colon tumors induced by azoxymethane.* Cancer Res, 1997. 57(7): p. 1233-7.
64. Ohshima, H. and H. Bartsch, *Chronic infections and inflammatory processes as cancer risk factors: possible role of nitric oxide in carcinogenesis.* Mutat Res, 1994. 305(2): p. 253-64.
65. Lopez-Farre, A., J.A. Rodriguez-Feo, L. Sanchez de Miguel, L. Rico, and S. Casado, *Role of nitric oxide in the control of apoptosis in the microvasculature.* Int J Biochem Cell Biol, 1998. 30(10): p. 1095-106.
66. Kawamori, T., M. Takahashi, K. Watanabe, T. Ohta, S. Nakatsugi, T. Sugimura, and K. Wakabayashi, *Suppression of azoxymethane-induced colonic aberrant crypt foci by a nitric oxide synthase inhibitor.* Cancer Lett, 2000. 148(1): p. 33-7.
67. Rao, C.V., T. Kawamori, R. Hamid, and B.S. Reddy, *Chemoprevention of colonic aberrant crypt foci by an inducible nitric oxide synthase-selective inhibitor.* Carcinogenesis, 1999. 20(4): p. 641-4.
68. Salvemini, D., S.L. Settle, J.L. Masferrer, K. Seibert, M.G. Currie, and P. Needleman, *Regulation of prostaglandin production by nitric oxide; an in vivo analysis.* Br J Pharmacol, 1995. 114(6): p. 1171-8.
69. Landino, L.M., B.C. Crews, M.D. Timmons, J.D. Morrow, and L.J. Marnett, *Peroxynitrite, the coupling product of nitric oxide and superoxide, activates prostaglandin biosynthesis.* Proc Natl Acad Sci U S A, 1996. 93(26): p. 15069-74.

70. Wendum, D., J. Masliah, G. Trugnan, and J.F. Flejou, *Cyclooxygenase-2 and its role in colorectal cancer development.* Virchows Arch, 2004. 445(4): p. 327-33.
71. Williams, J.A. and E. Shacter, *Regulation of macrophage cytokine production by prostaglandin E2. Distinct roles of cyclooxygenase-1 and -2.* J Biol Chem, 1997. 272(41): p. 25693-9.
72. Prescott, S.M. and F.A. Fitzpatrick, *Cyclooxygenase-2 and carcinogenesis.* Biochim Biophys Acta, 2000. 1470(2): p. M69-78.
73. Tsujii, M. and R.N. DuBois, *Alterations in cellular adhesion and apoptosis in epithelial cells overexpressing prostaglandin endoperoxide synthase 2.* Cell, 1995. 83(3): p. 493-501.
74. Sheng, H., J. Shao, J.D. Morrow, R.D. Beauchamp, and R.N. DuBois, *Modulation of apoptosis and Bcl-2 expression by prostaglandin E2 in human colon cancer cells.* Cancer Res, 1998. 58(2): p. 362-6.
75. Tsujii, M., S. Kawano, and R.N. DuBois, *Cyclooxygenase-2 expression in human colon cancer cells increases metastatic potential.* Proc Natl Acad Sci U S A, 1997. 94(7): p. 3336-40.
76. Tsujii, M., S. Kawano, S. Tsuji, H. Sawaoka, M. Hori, and R.N. DuBois, *Cyclooxygenase regulates angiogenesis induced by colon cancer cells.* Cell, 1998. 93(5): p. 705-16.
77. Kawamori, T., C.V. Rao, K. Seibert, and B.S. Reddy, *Chemopreventive activity of celecoxib, a specific cyclooxygenase-2 inhibitor, against colon carcinogenesis.* Cancer Res, 1998. 58(3): p. 409-12.
78. Dutta, J., Y. Fan, N. Gupta, G. Fan, and C. Gelinas, *Current insights into the regulation of programmed cell death by NF-kappaB.* Oncogene, 2006. 25(51): p. 6800-16.
79. Kojima, M., T. Morisaki, K. Izuhara, A. Uchiyama, Y. Matsunari, M. Katano, and M. Tanaka, *Lipopolysaccharide increases cyclo-oxygenase-2 expression in a colon carcinoma cell line through nuclear factor-kappa B activation.* Oncogene, 2000. 19(9): p. 1225-31.
80. Wang, D., H. Wang, Q. Shi, S. Katkuri, W. Walhi, B. Desvergne, S.K. Das, S.K. Dey, and R.N. DuBois, *Prostaglandin E(2) promotes colorectal adenoma growth via transactivation of the nuclear peroxisome proliferator-activated receptor delta.* Cancer Cell, 2004. 6(3): p. 285-95.
81. Wang, D., J.R. Mann, and R.N. DuBois, *WNT and cyclooxygenase-2 cross-talk accelerates adenoma growth.* Cell Cycle, 2004. 3(12): p. 1512-5.
82. Bienz, M. and H. Clevers, *Linking colorectal cancer to Wnt signaling.* Cell, 2000. 103(2): p. 311-20.
83. Clevers, H., *Wnt/beta-catenin signaling in development and disease.* Cell, 2006. 127(3): p. 469-80.
84. He, T.C., T.A. Chan, B. Vogelstein, and K.W. Kinzler, *PPARdelta is an APC-regulated target of nonsteroidal anti-inflammatory drugs.* Cell, 1999. 99(3): p. 335-45.
85. Segditsas, S. and I. Tomlinson, *Colorectal cancer and genetic alterations in the Wnt pathway.* Oncogene, 2006. 25(57): p. 7531-7.
86. Nakashima, M., S. Meirmanov, R. Matsufuji, M. Hayashida, E. Fukuda, S. Naito, M. Matsuu, K. Shichijo, H. Kondo, M. Ito, S. Yamashita, and I. Sekine, *Altered expression of beta-catenin during radiation-induced colonic carcinogenesis.* Pathol Res Pract, 2002. 198(11): p. 717-24.
87. Takahashi, M. and K. Wakabayashi, *Gene mutations and altered gene expression in azoxymethane-induced colon carcinogenesis in rodents.* Cancer Sci, 2004. 95(6): p. 475-80.
88. Sparks, A.B., P.J. Morin, B. Vogelstein, and K.W. Kinzler, *Mutational analysis of the APC/beta-catenin/Tcf pathway in colorectal cancer.* Cancer Res, 1998. 58(6): p. 1130-4.

89. Di Popolo, A., A. Memoli, A. Apicella, C. Tuccillo, A. di Palma, P. Ricchi, A.M. Acquaviva, and R. Zarrilli, *IGF-II/IGF-I receptor pathway up-regulates COX-2 mRNA expression and PGE2 synthesis in Caco-2 human colon carcinoma cells.* Oncogene, 2000. 19(48): p. 5517-24.
90. Sengupta, A., S. Ghosh, and S. Das, *Modulatory influence of garlic and tomato on cyclooxygenase-2 activity, cell proliferation and apoptosis during azoxymethane induced colon carcinogenesis in rat.* Cancer Lett, 2004. 208(2): p. 127-36.
91. Franke, A.A., R.V. Cooney, L.J. Custer, L.J. Mordan, and Y. Tanaka, *Inhibition of neoplastic transformation and bioavailability of dietary flavonoid agents.* Adv Exp Med Biol, 1998. 439: p. 237-48.
92. Seeram, N.P., L.S. Adams, S.M. Henning, Y. Niu, Y. Zhang, M.G. Nair, and D. Heber, *In vitro antiproliferative, apoptotic and antioxidant activities of punicalagin, ellagic acid and a total pomegranate tannin extract are enhanced in combination with other polyphenols as found in pomegranate juice.* J Nutr Biochem, 2005. 16(6): p. 360-7.
93. Vanamala, J., T. Leonardi, B.S. Patil, S.S. Taddeo, M.E. Murphy, L.M. Pike, R.S. Chapkin, J.R. Lupton, and N.D. Turner, *Suppression of colon carcinogenesis by bioactive compounds in grapefruit.* Carcinogenesis, 2006. 27(6): p. 1257-65.
94. Barth, S.W., C. Fahndrich, A. Bub, H. Dietrich, B. Watzl, F. Will, K. Briviba, and G. Rechkemmer, *Cloudy apple juice decreases DNA damage, hyperproliferation and aberrant crypt foci development in the distal colon of DMH-initiated rats.* Carcinogenesis, 2005. 26(8): p. 1414-21.
95. Salim, E.I. and S. Fukushima, *Chemopreventive potential of volatile oil from black cumin (Nigella sativa L.) seeds against rat colon carcinogenesis.* Nutr Cancer, 2003. 45(2): p. 195-202.
96. Yi, W., J. Fischer, G. Krewer, and C.C. Akoh, *Phenolic compounds from blueberries can inhibit colon cancer cell proliferation and induce apoptosis.* J Agric Food Chem, 2005. 53(18): p. 7320-9.
97. Uhl, M., F. Kassie, S. Rabot, B. Grasl-Kraupp, A. Chakraborty, B. Laky, M. Kundi, and S. Knasmuller, *Effect of common Brassica vegetables (Brussels sprouts and red cabbage) on the development of preneoplastic lesions induced by 2-amino-3-methylimidazo[4,5-f]quinoline (IQ) in liver and colon of Fischer 344 rats.* J Chromatogr B Analyt Technol Biomed Life Sci, 2004. 802(1): p. 225-30.
98. Kamaleeswari, M., K. Deeptha, M. Sengottuvelan, and N. Nalini, *Effect of dietary caraway (Carum carvi L.) on aberrant crypt foci development, fecal steroids, and intestinal alkaline phosphatase activities in 1,2-dimethylhydrazine-induced colon carcinogenesis.* Toxicol Appl Pharmacol, 2006. 214(3): p. 290-6.
99. Kobaek-Larsen, M., L.P. Christensen, W. Vach, J. Ritskes-Hoitinga, and K. Brandt, *Inhibitory effects of feeding with carrots or (-)-falcarinol on development of azoxymethane-induced preneoplastic lesions in the rat colon.* J Agric Food Chem, 2005. 53(5): p. 1823-7.
100. Raju, J., J.M. Patlolla, M.V. Swamy, and C.V. Rao, *Diosgenin, a steroid saponin of Trigonella foenum graecum (Fenugreek), inhibits azoxymethane-induced aberrant crypt foci formation in F344 rats and induces apoptosis in HT-29 human colon cancer cells.* Cancer Epidemiol Biomarkers Prev, 2004. 13(8): p. 1392-8.
101. Kassie, F., S. Rabot, M. Uhl, W. Huber, H.M. Qin, C. Helma, R. Schulte-Hermann, and S. Knasmuller, *Chemoprotective effects of garden cress (Lepidium sativum) and its constituents towards 2-amino-3-methyl-imidazo[4,5-f]quinoline (IQ)-induced genotoxic effects and colonic preneoplastic lesions.* Carcinogenesis, 2002. 23(7): p. 1155-61.

102. Sengupta, A., S. Ghosh, S. Bhattacharjee, and S. Das, *Indian food ingredients and cancer prevention - an experimental evaluation of anticarcinogenic effects of garlic in rat colon.* Asian Pac J Cancer Prev, 2004. 5(2): p. 126-32.
103. Luceri, C., G. Caderni, A. Sanna, and P. Dolara, *Red wine and black tea polyphenols modulate the expression of cycloxygenase-2, inducible nitric oxide synthase and glutathione-related enzymes in azoxymethane-induced f344 rat colon tumors.* J Nutr, 2002. 132(6): p. 1376-9.
104. Caderni, G., C. De Filippo, C. Luceri, M. Salvadori, A. Giannini, A. Biggeri, S. Remy, V. Cheynier, and P. Dolara, *Effects of black tea, green tea and wine extracts on intestinal carcinogenesis induced by azoxymethane in F344 rats.* Carcinogenesis, 2000. 21(11): p. 1965-9.
105. Fujioka, K., F. Greenway, J. Sheard, and Y. Ying, *The effects of grapefruit on weight and insulin resistance: relationship to the metabolic syndrome.* J Med Food, 2006. 9(1): p. 49-54.
106. Juan, M.E., U. Wenzel, V. Ruiz-Gutierrez, H. Daniel, and J.M. Planas, *Olive fruit extracts inhibit proliferation and induce apoptosis in HT-29 human colon cancer cells.* J Nutr, 2006. 136(10): p. 2553-7.
107. Miyagi, Y., A.S. Om, K.M. Chee, and M.R. Bennink, *Inhibition of azoxymethane-induced colon cancer by orange juice.* Nutr Cancer, 2000. 36(2): p. 224-9.
108. Adams, L.S., N.P. Seeram, B.B. Aggarwal, Y. Takada, D. Sand, and D. Heber, *Pomegranate juice, total pomegranate ellagitannins, and punicalagin suppress inflammatory cell signaling in colon cancer cells.* J Agric Food Chem, 2006. 54(3): p. 980-5.
109. Al-Awadi, F.M., T.S. Srikumar, J.T. Anim, and I. Khan, *Antiinflammatory effects of Cordia myxa fruit on experimentally induced colitis in rats.* Nutrition, 2001. 17(5): p. 391-6.
110. Jia, X. and C. Han, *Effects of green tea on colonic aberrant crypt foci and proliferative indexes in rats.* Nutr Cancer, 2001. 39(2): p. 239-43.
111. Orner, G.A., W.M. Dashwood, C.A. Blum, G.D. Diaz, Q. Li, M. Al-Fageeh, N. Tebbutt, J.K. Heath, M. Ernst, and R.H. Dashwood, *Response of Apc(min) and A33 (delta N beta-cat) mutant mice to treatment with tea, sulindac, and 2-amino-1-methyl-6-phenylimidazo[4,5-b]pyridine (PhIP).* Mutat Res, 2002. 506-507: p. 121-7.
112. Huang, M.T., Y.R. Lou, W. Ma, H.L. Newmark, K.R. Reuhl, and A.H. Conney, *Inhibitory effects of dietary curcumin on forestomach, duodenal, and colon carcinogenesis in mice.* Cancer Res, 1994. 54(22): p. 5841-7.
113. Lev-Ari, S., Y. Maimon, L. Strier, D. Kazanov, and N. Arber, *Down-regulation of prostaglandin E2 by curcumin is correlated with inhibition of cell growth and induction of apoptosis in human colon carcinoma cell lines.* J Soc Integr Oncol, 2006. 4(1): p. 21-6.
114. Zhang, F., N.K. Altorki, J.R. Mestre, K. Subbaramaiah, and A.J. Dannenberg, *Curcumin inhibits cyclooxygenase-2 transcription in bile acid- and phorbol ester-treated human gastrointestinal epithelial cells.* Carcinogenesis, 1999. 20(3): p. 445-51.

AUTHOR INDEX

A

B

C

D

E

F

G

H

I

SUBJECT INDEX

A

B

C

G

H

I

L

M

N

O

P

R

Functional Foods for Chronic Diseases

Advances in the Development of Functional Foods

Editor: Danik M. Martirosyan, Ph.D.

Editorial Assistants:

Megh Raj Bhandari, PhD

Undurti Das, PhD

Ashkhen D. Martirosyan

CPSIA information can be obtained
at www.ICGtesting.com
Printed in the USA
LVHW010816081122
732583LV00011B/309